D0989691

THE AMERICAN SYSTEM OF MANUFACTURES

The American System of Manufactures

THE REPORT OF THE COMMITTEE ON THE MACHINERY OF THE UNITED STATES 1855 AND THE SPECIAL REPORTS OF GEORGE WALLIS AND JOSEPH WHITWORTH 1854 EDITED WITH AN INTRODUCTION BY NATHAN ROSENBERG FOR THE UNIVERSITY PRESS

EDINBURGH

1969

22 George Square, Edinburgh 8
North America
Aldine Publishing Company
529 South Wabash Avenue, Chicago
Australia and New Zealand
Hodder & Stoughton Limited
Africa, Oxford University Press
India, P. C. Manaktala & Sons
Far East, M. Graham Brash & Son
85224 016 3
Library of Congress
Catalog Card Number 68-19879
Printed in Great Britain by
R. & R. Clark Ltd, Edinburgh

Contents

Contents

Special Report of Mr. George Wallis

Contents

Special Report of Mr. Joseph Whitworth

A note on the text of the reports

Each report has been reprinted here in its entirety. No substantive changes have been made in the original texts. The small changes which have been made are purely stylistic: subject headings have been added where they were thought to be helpful, some marginal guides have been dropped, the number of separate paragraphs has been compressed, etc. Numerous minor corrections have been made, involving such matters as misspelling, punctuation, and obvious typographical errors. A few small errors in addition have been allowed to remain (pp. 181-91) where it appeared probable that a total was wrong because some individual item had been incorrectly entered or where the total was consistent with a reference appearing elsewhere in the text. The temptation to improve upon syntax or to correct grammatical errors–strong upon several occasions–has been stoutly resisted.

Introduction, by Nathan Rosenberg

The reports which are reprinted in this volume are, for several reasons, of great historical interest. Considered together they provide one of the most detailed, and certainly one of the most reliable, descriptive accounts of American manufacturing methods in the middle of the nineteenth century. Their authors possessed impressive–one is tempted to say 'unsurpassed'–qualifications for examining and reporting upon the technical aspects of the American productive establishment. Since, furthermore, the authors were intimately familiar with the state of the industrial arts in Great Britain, their observations in these reports shed much light upon Anglo-American differences in technology. One *caveat* is justified: because the authors were members of committees which had crossed the Atlantic to seek instruction, they naturally concentrated attention on what was novel and distinctive and not necessarily on what was representative or average. Their primary interest, quite correctly from their point of view, was in what have come to be called 'best-practice' techniques.

The reports also symbolize, in an important sense, the emergence of the United States as a world economic power. Indeed, they were instrumental in initiating a sizeable reverse flow of technical knowledge which had for so long flowed from the Old World to the New; for these reports clearly and emphatically called attention to certain unique technical developments in the United States and urged their adoption in Great Britain. In this they were immediately successful, since the vast expansion of the Enfield Armoury, commencing in 1855, was undertaken with machinery which had been purchased by the ordnance committee during its tour of the United States. As this marked the beginning of the transatlantic adoption of what the English, recognizing its distinctive features, called the 'American System of Manufacturing,' it was an economic event of major importance. Thus the reports had an immediate effect in the shaping of events of great consequence. However, in spite of the fact that the reports themselves have long commanded the attention of scholars because of their authoritative treatment of American technology in the 1850s,[1] surprisingly little is generally known of the larger

[1] D. L. Burn forcefully called attention to the importance of the contents of these reports in his article 'The Genesis of American Engineering Competition, 1850–1870' *Economic History* II (January 1931) 292-311. Burn argued, in large measure on the basis of evidence provided by these reports.

episode of which they constituted a chapter. Even more surprising, practically nothing is known of the seminal contributions made by one of the authors, John Anderson, to the advance of British technology. These deficiencies will be repaired below.

I

Our proximate starting point is properly the Crystal Palace Exhibition of 1851. It is important, at the outset, that the landscape remain uncluttered with facile historical clichés. There is a natural tendency to impose order and unity upon the past so that it will be more readily comprehensible and manageable. Too often, however, the unity is only in the eye of the beholder–especially when he is a writer of history texts. Since the Great Exhibition occurred fully three-quarters of a century after Watt had transformed the steam engine into a commercially feasible proposition, it is easy to think of the steam engine as the sole prime mover of British industry. Yet in fact water wheels were placed prominently on display at the Exhibition and windmills had by no means passed entirely from the scene. If one wishes to regard Britain as the 'workshop of the world' one ought at least to keep an open mind upon the nature of the productive activities and the sort of technology employed inside the workshops. Similarly, if we accept Professor Rostow's use of 1851 as the benchmark date for the maturity of the British economy,[1] we should not allow such

---

that American engineering rivalry had begun in earnest earlier than was commonly recognized by English historians. More recently John Sawyer has drawn upon *The Report of the Committee on the Machinery of The United States of America* as well as the reports of other travellers and commissions, to emphasize the distinctive aspects of the American social and cultural environment and its role in producing 'The American System of Manufacturing'. John E. Sawyer, 'The Social Basis of the American System of Manufacturing' *Journal of Economic History* (December 1954) 361-79. H.J. Habakkuk, in his important book, *American and British Technology in the Nineteenth Century* (Cambridge 1962), draws extensively on these reports.

[1] 'After take-off there follows what might be called the drive to maturity. There are a variety of ways a stage of economic maturity might be defined: but for these purposes we define it as the period when a society has effectively applied the range of (then) modern technology to the bulk of its resources. . . . By, let us say, the Exhibition of 1851, Britain had mastered and extended over virtually the whole range of its resources all that the then modern science and technology had to offer an economy with the resources (and the population-resource balance) of mid-nineteenth century Britain.' W.W. Rostow *The Stages of Economic Growth* (Cambridge 1960) 59-61. Britain was not in fact mature by Rostow's definition in 1851. As the British were to discover at the Exhibition, there existed a 'modern technology' across the Atlantic which they had not yet seriously begun to exploit. Whether all of it was worth exploiting, in view of differences in factor prices, is a question which Rostow's last sentence suggests he was only parenthetically aware of. Presumably it would be possible to define 'maturity' in a way which would be meaningful from an engineering point of view. The economic meaning of the term is not clear.

metaphors to conjure up in our minds images only of–to use Clapham's phrase–'the wheels of iron and the shriek of escaping steam.'[1] For agriculture was still Britain's largest single industry, employing 26 per cent of the male population over 20 years of age. The next industry in terms of numbers was domestic service, employing over one million people, predominantly women–employing, in fact, 1 in 9 of all females in the population over 10 years of age. The building trade–after agriculture the largest single employer of men–was still virtually untouched by mechanization. The blacksmith was still much more numerous than the ironworker. The coal miner employed no machinery at the coal face and the vast shoe-making and clothing industries were as yet still unaffected by the sewing machine. In short, the proportion of the labour force which was directly engaged in productive operations involving the extensive use of power-driven machinery was still small.[2]

To say this, however, is to express a twentieth-century perspective. Contemporaries regarded the changes in the first half of the nineteenth century as momentous, and they were right. Perceptive observers of the nineteenth century were aware that the application of power-driven machinery was bringing about fundamental changes, not only in the form of economic activity, but in the whole fabric of social organization.

The Crystal Palace Exhibition, the first of the great international exhibitions of the second half of the nineteenth century, was conceived and carried out with royal support and in the face of widespread opposition. 'The opposition, vigorous and unrestrained in both houses of Parliament and among the public, held that England would be overrun with foreign rogues and revolutionaries and that the morals of the English and their loyalty to the crown would be endangered and coincidentally their trade secrets stolen.'[3]

---

[1] J.H.Clapham *An Economic History of Modern Britain* II (Cambridge 1938) 22. The year 1851 is frequently used as a convenient benchmark. Lewis Mumford (*Technics and Civilization* (New York) 1934, 154-5) used it to mark the highpoint of what he called (borrowing the term from Patrick Geddes) the 'Paleotechnic Phase' of history. G.H.Tupling ('The early metal trades and the beginnings of engineering in Lancashire' *Transactions of the Lancashire and Cheshire Antiquarian Society* 1949, 33-4) uses the date–and the event–to mark the arrival at puberty of the mechanical engineering profession.

[2] Clapham, op. cit. 22-5. See also Phyllis Deane and W.A.Cole *British Economic Growth 1688–1959* (Cambridge 1962) ch. 4.

[3] Guy Stanton Ford 'International expositions' *Encyclopedia of the Social Sciences* VI (New York 1931) 23. The more extreme form of opposition to the Exhibition came from rabid, die-hard protectionists, whose cause had so recently been dealt its mortal blow. None was more rabid than the egregious Colonel Sibthorp. '... the Protectionists might have found many better spokesmen, though none more picturesque, than Colonel Charles de Laet Waldo Sibthorp, the member for Lincoln.... Never yet had his country's impending doom seemed so demonstrable as when Colonel Sibthorp heard of the preparations for an exhibition of the industry of all nations. The country would be flooded with cheap foreign goods; hordes of foreigners

The entire exhibition was housed in a great glass building, resembling a greenhouse, in Hyde Park. The resemblance to a greenhouse was not, in fact, accidental. The building's architect, Sir Joseph Paxton, had himself been a gardener and designer of greenhouses. His most notable earlier edifice, to which the Crystal Palace bore a striking resemblance, was a greenhouse built on the enormous estate of the Duke of Devonshire at Chatsworth, where he had been the head gardener.[1] The new building, whose design was regarded with scepticism and apprehension on the part of the engineering profession, was a magnificent demonstration of the new architectural possibilities of glass and iron.[2]

> Some idea may be formed of the leading peculiarities of the building, by recalling the fact, that its main avenue, between the columns, is nearly double the width of the nave of St Paul's Cathedral, while its length is more than four times as great. The walls of St Paul's are 14 feet thick, those of the Hyde Park Building are 8 inches. St Paul's required thirty-five years to erect, the building will be finished in about half that number of weeks.[3]

Disraeli, employing a grandiloquence which characterized so many of the public utterances in connection with the Exhibition, referred to the Crystal Palace as 'that enchanted pile which the prescient philanthropy of an accomplished and enlightened Prince have raised for the glory of England and the delight and instruction of two Hemispheres'.[4]

---

would come over here to sell their wares: thieves and anarchists would flock to London from all over the world: secret societies were already being formed on the Continent to assassinate the Queen: worse still, there would be an influx of Papists, bringing with them idolatry, schism, bubonic plague and venereal disease. It was all too horrible to contemplate, and all the doing of that damned German Prince.' Christopher Hobhouse *1851 and the Crystal Palace* (New York 1937) 19-20.

[1] For background material on the genesis of the Crystal Palace, see Yvonne ffrench *The Great Exhibition: 1851* (London 1950) ch. 6; and Violet R. Markham *Paxton and the Bachelor Duke* (London 1935) esp. chs. 10, 11, and Appendix 3. Violet Markham was Paxton's granddaughter.

[2] See Matthew Digby Wyatt 'On the construction of the building for the exhibition of the works of industry of all nations in 1851' *Minutes of Proceedings of the Institution of Civil Engineers* x (1850–1) 127-65, and particularly the discussion of Wyatt's paper, 165-91. The paper was delivered on 14 January 1851. The Crystal Palace was moved from Hyde Park to Sydenham after the close of the Exhibition. It survived until its destruction by fire in 1936.

[3] Wyatt, op. cit. 133.

[4] As quoted in Asa Briggs *Victorian People* (Chicago 1955) 37. Several decades later the view was expressed that the instruction was solely of foreigners by Englishmen, and that Englishmen learned nothing from foreigners; and that, therefore, the subsequent decline of British eminence was due in part to this ill-conceived Exhibition. 'It is a moot point whether the growth of manufacturing industries abroad is a cause or an effect of the decay in British industries, but it is certain that the increase of this competition is an accelerating factor in the decline now prevailing. There is no doubt

Among the most interesting and least noticed aspects of the Crystal Palace were some of the techniques and principles employed in its construction.[1] There is a rich irony here. For it was at the Crystal Palace Exhibition that many Englishmen were first familiarized, through an examination of American products, with productive methods which seemed so novel and original that they were promptly dubbed 'The American System of Manufacturing.' And yet, little noticed, the very building within which this discovery was made had been constructed by methods directly incorporating this system. The building, which had been constructed with extraordinary speed, was a triumph of standardization and interchangeability of parts. Girders, columns, sash-bars, and gutters were produced to uniform specifications and were interchangeable throughout the building. The sash-bars, 200 miles of which were produced in a three-month period, were pre-cut to a uniform length so that they could be fitted easily into place on the roof without having to perform any carpentering operations; the gutters–Paxton gutters as they were called–although more complicated in form than the sash-bars, were completely shaped and pre-cut by machinery. The whole of the construction of the building was characterized by a highly imaginative use of specialized machinery. The sash-bars were cut by a machine which was a modification of one devised and used earlier by Paxton. The drilling of nail-holes was accomplished by a machine, attended by four or five boys, which placed the holes with unerring accuracy. Machinery was even devised for painting the sash-bars–a particularly tedious affair, as it involved the application of three coats of paint to 200 miles of bar. A special planing machine was employed which assured that floor planks and planks used elsewhere in the building were of uniform thickness. Even so trivial an item as handrails for staircases and galleries were produced in uniform lengths by machine.

For the enormous glazing operation specialized techniques were also introduced. The Crystal Palace consisted of an estimated 896,000 square feet of glass. The glazing of the nave roof was greatly facilitated by a mechanical contrivance specially conceived for the purpose:

> Seventy-six machines were constructed, each capable of accommodating two glaziers; these machines consisted of a stage of deal about 8 feet square, with an opening in its centre sufficiently

---

that from this point of view of the interests of the manufacturing classes the World's Exhibition of 1851 was a great error of judgment. The exhibits helped foreigners without benefiting English producers.' 'Artifex' and 'Opifex' *The Causes of Decay in a British Industry* (London 1907) 51.

[1] The following discussion draws upon the accounts of the construction of the Crystal Palace in *Official Description and Illustrated Catalogue of the Great Exhibition of the Works of Industry of All Nations* I (London 1851) esp. 52-6 and 71-7; and Society for Promoting Christian Knowledge *The Industry of Nations, as Exemplified in the Great Exhibition of 1851* I (London 1852) esp. 92-107. For historical details on the organization of the Exhibition see K.W. Luckhurst 'The Great Exhibition of 1851' *Journal of the Royal Society of Arts* 99 (20 April 1851) 413-56.

> large to admit of boxes of glass, and supplies of sash-bars, putty, etc., being hoisted through it. The stage rested on four small wheels, travelling in the Paxton gutters, and spanned a width consisting of one ridge and two sloping sides. In bad weather the workmen were covered by an awning of canvas, stretched over hoops for their protection.
>
> In working, the men sat at the end of the platform next to whatever work had been last done; from which they pushed the stage backward sufficiently far to allow them to insert a pane of glass, and as soon as that was completed they moved again far enough to allow of the insertion of another. In this manner each stage travelled uninterruptedly from the transept to the east and west ends of the building.[1]

The crowds attending the Exhibition, after its May 1st opening, were in no position to observe or appreciate the manner in which the building had been constructed. The building itself and its contents, on the other hand, were found to be inordinately interesting. Early reaction to the American exhibit was derisive and even contemptuous. The American commissioners had reserved an excessive amount of floor space and many of the consignments arrived late. A French writer, with endearing partisanship, made the following comments on the scene the day before the official opening:

> *Fervet Opus !* The work grows and advances with giant steps. You have seen in the warm rain of the month of March the bare tree clothe itself in one day with budding leaves; it seems as though we actually saw the verdure grow. Well, so in the Crystal Palace is seen to spring up on all the walls, at the corner of every avenue, a crop of splendid works swathed but a few moments back in the holland and brown paper of the bale-maker. It is the effect of a magician's wand, and the French portion is especially remarkable for the ease and rapidity with which the wand is obeyed. The space, which but now was empty, when you pass a second time you find filled by France. She waited till the last moment–her invariable custom–and then went ahead. At this very moment–and I have just left the spot–nothing is done on our side, nothing is ready, and yet nobody appears anxious, so certain is it that we shall be ready to-morrow. This is our strength. And it is our motto, '*Toujours prêts !*' The most complete branch of the Exhibition at the present moment is the American; it is complete; it is largely and solidly established. Order reigns in the American exhibition; but it is open to one objection–namely, the want of objects to exhibit.[2]

The shortcomings of Brother Jonathan's exhibit in the early days were apparently compounded by the gratuitous display of a great spread

---

[1] Society for Promoting Christian Knowledge, op. cit. I, 101-2.
[2] Ibid. 152.

eagle. The tastelessness of this display in the public eye seems to have been accentuated by the fact that the American exhibit was flanked by 'magnificent displays of Russian, Austrian and French art'.[1] The American exhibition area became known as 'prairie ground', a scathing reference to its apparent desolation and lack of interest. Certainly a visitor who came to the American exhibit to gratify his aesthetic sensibilities was wasting his time. It was severe and utilitarian in nature, and the visitor beheld a profusion of objects, most of which possessed no ornamental value whatever, but were contrived to cater to some specific human need. Characteristic of the American display were its ice-making machines, corn-husk mattresses, fireproofed safes, meat biscuits, india-rubber shoes and life-boats, railroad switches, nautical instruments, telegraph instruments, a special grease-removing soap which was usable either with salt or fresh water, and artificial eyes and legs. Gradually it became apparent that there was something different about the nature of the objects on display at the American exhibit.[2] Attention was focused more forcefully on its contents as a result of some spectacular performances. Perhaps the rudest shock to British pride was the remarkable victory of the yacht *America*, which completely outsailed its British competitors. 'Their yacht', said *The Times*, on 2 September 1851, 'takes a class to itself. Of all the victories ever won none has been so transcendent as that of the New York schooner. The account given of her performance suggests the inapproachable excellence attributed to JUPITER by the ancient poets, who describe the King of the Gods as being not only supreme, but having none other next to him. "What's first?"–"The America." "What's second?"–"Nothing."' More pregnant with economic consequences, however, was Cyrus McCormick's Virginia Reaper, which *The Times* described as a '. . . cross between a flying machine, a wheelbarrow and an Astley chariot'. The incredulity with which it was initially regarded was finally transformed to extravagant paeans. Although the McCormick Reaper had firmly established itself in the

---

[1] William T. Hutchinson *Cyrus Hall McCormick: Seed-time, 1809–1856* (New York 1930) 385.

[2] The official Catalogue shrewdly commented upon the distinctiveness of the American exhibit:

'The absence in the United States of those vast accumulations of wealth which favour the expenditure of large sums on articles of mere luxury, and the general distribution of the means of procuring the more substantial conveniences of life, impart to the productions of American industry a character distinct from that of many other countries. The expenditure of months or years of labour upon a single article, not to increase its intrinsic value, but solely to augment its cost or its estimation as an object of *virtu*, is not common in the United States. On the contrary, both manual and mechanical labour are applied with direct reference to increasing the number or the quantity of articles suited to the wants of a whole people, and adapted to promote the enjoyment of that moderate competency which prevails among them.' *Official Description and Illustrated Catalogue* III, 1431.

American market during the 1840s, the mechanical reapıng of grain was a novelty in English eyes in 1851.[1]

The Jury on Agricultural Implements decided during the summer to subject some of the agricultural machinery to a practical test. The tests were held on Mr Mechi's model farm at Tiptree Heath in Essex on 24 July 1851. The wheat was still green and the day was, as reported by the American Commissioner, B. P. Johnson, '. . . one of the favorite days of England–that is, *rain incessantly*'. Under these unfavourable circumstances, Hussey's Reaper, which was tested first, operated unsatisfactorily. The grain clogged the machine, which passed over the grain without cutting it. Although circumstances would have appeared to warrant postponement of the tests, the McCormick Reaper was placed in the field. In spite of the rain and soggy stalks it performed its work effectively and with dispatch, at a pace estimated to be the equivalent of twenty acres a day. At subsequent tests under more propitious conditions the McCormick Reaper continued to function well and to outclass the Hussey Reaper, which encountered mechanical difficulties.[2]

The effect of the practical demonstration of the American reaper was electric. A few days after the trials at Tiptree Hall, the American Commissioner B.P. Johnson wrote:

> You can hardly imagine how the tone is altered since we have had our implements tried. The 'Prairie Ground' is filled with inquirers, and some gentlemen have found out that there are some people who know what they are doing in some other parts of the Globe, as well as this little island, where it is most readily admitted there are many 'clever people.' The McCormick machine was put together in the Palace again, and yesterday it had more visitors, I believe, than the Ko-i-noor diamond itself.[3]

Even *The Times*, which Daniel Webster had once called 'the bitterest, the ablest, and the most Anti-American press in all Europe',[4] paid homage by predicting that 'the reaper will amply remunerate England for her outlay connected with the Great Exhibition'; it also expressed the belief that 'the reaping machine from the United States is the most valuable contribution from abroad, to the stock of our previous knowledge, that we have yet discovered'.[5]

Gradually evidence became more and more compelling that there was

---

[1] Hutchinson states that, in spite of the earlier work of Henry Ogle and the Reverend Patrick Bell, '. . . in 1848 there was probably not one reaper at work in all England'. Hutchinson, op. cit. 380.

[2] Charles T. Rodgers *American Superiority at the World's Fair* (Philadelphia 1852) 14-16. This volume is a compilation, mostly of newspaper and weekly materials, collected by the author who was himself a U.S. Commissioner. See also Hutchinson, op. cit. 388-93.

[3] Rodgers, op. cit. 15-16.

[4] Ibid. 61.

[5] *The Times*, 16 October and 27 September 1851, as quoted in Hutchinson, op. cit. 391.

much to learn at the American exhibit. Attention was directed to its locks by the remarkable performance of the American, Alfred C. Hobbs, lockpicker extraordinary, in picking the highly esteemed and supposedly 'unpickable' Bramah lock. Bramah's famous lock had been displayed in a Piccadilly shop window for about forty years with a notice offering 200 guineas to anyone who could pick it. Hobbs took up the challenge and created a sensation by picking the lock, but only after working at it for a total of fifty-one hours spread over sixteen separate days, between 24 July and 23 August.

'An unprejudiced person', wrote H. W. Dickinson,[1] 'must admit that a lock that can resist picking so long as this did, fulfills every practical purpose.' The official jurors made the same point, in an obvious reference to Hobbs' exploits. 'On the comparative security afforded by the various locks which have come before the Jury, they are not prepared to offer an opinion. They would merely express a doubt whether the circumstance that a lock has been picked under conditions which ordinarily could scarcely ever, if at all, be obtained, can be assumed as a test of its insecurity'.[2]

Hobbs, whose huckstering skill was second only to his dexterity with a lock, achieved great success by obscuring this point and generating a strong sense of insecurity among the mercantile community which must have been plagued, after Hobbs' appearance, by nightly visions of thieves absconding with their valuables. His favourite tactic in America appears to have been to secure agreement from bankers unfamiliar with his prowess that they would purchase Hobbs' lock if Hobbs could open their present locks.[3]

Hobbs reported that, after informing the American Consul of his lock-picking plans, the Consul replied: 'For goodness sake, do something to help us up at the exhibition. The Americans have about one-eighth of the building in London to exhibit in; there are about three barrels of shoe-pegs and a bundle of brooms. It is a total failure. Do something, if you can, to help us out.'[4]

Mr Hobbs' success with the Bramah lock occurred on the day (23 August 1851) on which the yacht *America* won her race at Cowes. The

---

[1] H. W. Dickinson 'Joseph Bramah and his inventions' *Transactions of the Newcomen Society* XXII 175.
[2] 'Exhibition of the works of industry of all nations' *Reports by the Juries* (London 1852) 500.
[3] An account of Hobbs' protracted encounter with the Bramah lock may be found in *Illustrated London News*, 6 September 1851, with illustrations of Hobbs' tools. For a thorough account of the lock controversy following Hobbs' success see George Price *A Treatise on Fire and Thief-Proof Depositories and Locks and Keys* (London 1856). Mr Hobbs revealed some of his trade secrets many years later in an article 'Locks and their failings' *Transactions of the American Society of Mechanical Engineers* VI (1884–5) 233-54. For a short, somewhat anecdotal account by Mr Hobbs of his opening of the Bramah lock, see *Transactions of the American Society of Mechanical Engineers* V, 123-6.
[4] Ibid. 124.

two events fluttered British coat-tails and considerably raised American stock at the Exhibition.

Some of Britain's most eminent lockpickers attempted to return the compliment by opening Hobbs' own lock, the Newell Parautoptic Lock, for which feat Hobbs offered a $1,000 reward. None succeeded. The lock was awarded the prize medal by the Exhibition Jurors.

What was really important about Hobbs' locks was the method of their production. For in spite of the fact that Bramah had several decades earlier devised (with the young Maudslay's help) a range of specialized machinery for producing the components of the famous Bramah lock, his machine methods seem to have had no effect on the rest of the lock industry, where handicraft techniques still predominated.[1]

> The eminent engineer, John Farey, stated that when young he '. . . had the good fortune to be intimate with Mr Joseph Bramah, and had acquired a knowledge of his locks, which were then in high repute. The secret workshops, wherein the locks were manufactured, contained several curious machines, for forming parts of the locks, with a systematic perfection of workmanship, which was at that time unknown in similar mechanical arts. These machines had been constructed by the late Mr Maudslay, with his own hands, whilst he was Mr Bramah's chief workman. . . . The machines, before mentioned, were adapted for cutting the grooves in the barrels, and the notches in the steel plates, with the utmost precision. The notches in the keys, and in the steel sliders, were cut by other machines, which had micrometer screws, so as to ensure that the notches in each key should tally with the unlocking notches of the sliders in the same lock. The setting of these micrometer screws was regulated by a system, which assured a constant permutation in the notches of succeeding keys, in order that no two should be made alike. Mr Bramah attributed the success of his locks to the use of those machines, the invention of which had cost him more study than that of the locks; without the machines, the locks could not have been made in any great number, with the requisite precision, as an article of trade. There was great originality in those machines, which were constructed before analogous cases (beyond the clock-maker's wheel-cutting machines) were in existence.'[2]

Despite this, Hobbs was so impressed with the backwardness of British productive methods in lockmaking that he remained in England after the exhibition and formed a partnership, Hobbs, Ashley and Company. He

---

[1] 'Exhibition of the works of industry of all nations' *Reports by the Juries* (London 1852) 500.

[2] John Chubb 'On the construction of locks and keys' *Proceedings of the Institution of Civil Engineers* 9 (1850) 310-28. The quotation is taken from 331-2.

H. W. Dickinson described a few of the tools which have survived from Bramah's 'secret workshop' in 'Joseph Bramah and his inventions,' op. cit. 173-4.

stayed in England until 1860. His company established a plant in Cheapside where it pioneered in the introduction of American machine methods to lock production.

The contrast between contemporary methods of lock production in Britain and those introduced by Hobbs was strikingly recorded in an article in *The Engineer* in 1859 by a writer who had just visited Hobbs' establishment. The tone, as well as the content of the article, suggest that he was overwhelmed by what he saw.

> There is, perhaps, no other branch of manufacturing industry in this country over which antiquated usages have held such sway, and in which the application of labour-saving machinery has had greater prejudices to contend with, than in the lock trade. In order, therefore, to understand the value of Mr Hobbs' improvements on the English system–or, rather, no system–of lockmaking, and to see the ground from which he started, a brief glance at the position of the lock trade in this country, even at the present day, may not be out of place.
>
> In Wolverhampton–the principal seat of the lock trade–the manufacture is still conducted for the most part by hand labour, or by the use of such primitive tools and appliances as have been in use from time immemorial. It is true, some of the largest firms have introduced steam-power into their workshops, but it is only applied to such simple operations as forging and stamping the raw material; but the shaping, finishing, and fitting of the whole of the special mechanism of the lock, even in the establishments of such large makers as the Messrs Chubb, is conducted by hand-labour, with the aid of the shears, the file, and the punch. To quote one who ought to be an authority on such subjects, Mr Price, in his book on locks and keys, says, speaking of a certain description of wooden-case lock: 'It is almost incredible that, even in making "stock" locks, at the present time, not a circular saw nor machinery of any kind is used. The cavity for the works of a double-handed lock (Steele's patent) is bored out with a hand-auger and finished with a chisel, the block of wood being tightly screwed in a vice. The wood used for these locks is oak (pipe staves); and when Quebec timber is employed, from its hardness and roughness, the labour in working it by hand is very great; yet not a single maker has had the temerity to introduce a circular saw, which in other manufactures has produced such beneficial results. In making the iron parts, not even a press is used, all the holes being punched out by hand. The plates are cut out by hand, and the bolts are all forged.'[1]

[1] 'Hobbs' Lock Manufactory' *The Engineer* (18 March 1859) 188. The 'Mr Price' who is quoted is George Price. The quotation is from his book, *A Treatise on Fire and Thief-Proof Depositories and Locks and Keys* (London 1856).

Hobbs himself, in referring to the description of lockmaking provided by

The author subsequently described the operations by which locks were produced in Hobbs' manufacturing establishment. He was deeply impressed by the specialized machinery, particularly the punching machines which enabled a finished component to be stamped instantly out of blank metal plates.

This was possible only by devoting meticulous attention to the preparation of the dies, a task to which only the most skilled workmen were assigned.[1] The use of 'automaton machinery . . . carefully adjusted to the standard gauges' eliminated reliance upon the skill of the labourer in determining the exact form and dimension of the final product. Moreover, it essentially eliminated the enormously tedious 'fitting' process by which the separate components were assembled into a finished product. The painstaking adjustments in fitting together the separate components, each produced by hand-labour, were one of the most time-consuming and expensive aspects of handicraft production.[2] Hobbs' methods, by contrast, were startling.

> There is a store-keeper standing behind a counter with a piece of paper before him with certain figures and queer terms upon it; he looks at the paper as a dispensing chemist does to a prescription, goes off to a particular drawer, and with a grocer's scoop takes out a certain quantity of lock elements, comes back to his prescription, and then goes off for another scoopful of screws, studs, bolts, etc. The prescription is an order for so many locks of some particular pattern, and he is literally compounding them in the manner we have described. When the proper quantity has been meted out, they are given out to the fitters, who in the course of a few minutes from these *disjecta membra*, have formed them into complete locks without the aid of any other tools than their fingers, and a small hammer to rivet them together.[3]

The author was also impressed with '. . . the care with which the raw materials are selected and prepared, so as to economise the labour required in their conversion to the special forms required in the various parts of the locks he manufactures'.[4] The rigid standardization of com-

---

M. Duhamel du Monceau in *Art du Serrurier*, published in 1767, makes the following observation: 'It is worthy of remark, that the tools described are the same as those which are used by the locksmith at the present day; showing how little improvement has been made in the means of producing locks.' *The Construction of Locks* compiled from the papers of A. C. Hobbs of New York, and edited by Charles Tomlinson (London 1868) 4.

[1] 'Mr Hobbs has carried this system of punching with compound dies to greater perfection than has ever, that we are aware of, been attempted before.' 'Hobbs' Lock Manufactory', op. cit. 189.

[2] 'This method of adapting one part to another by hand-labour was that which, till lately, was pursued in the Birmingham gun trade, and also at Wolverhampton in the manufacture of locks.' Ibid. 189.

[3] Ibid. [4] Ibid.

ponents furthermore provided maximum interchangeability among different locks, even of different sizes, and minimized the amount of special tools and machinery required to produce a particular range of locks.

'The factory is a compact brick building of three storeys . . . and may be readily distinguished at all times by the energetic snorting of a high-pressure steam engine, not a single whift of whose steam is allowed to escape without having completed its appointed tale of locks, bolts, bars, keys, screws, etc.'[1]

Hobbs' conception of standardization and interchangeability, made possible by the use of specialized machinery, was so sweeping and clearly articulated that a lengthy quotation, in his own words, seems to be justified.

> If we suppose that a lock of particular construction comprises twenty screws and small pieces of metal, and that there are required, for general disposal in the market, five sizes of such a lock; there would thus be a hundred pieces of metal required for the series, each one differing, either in shape or size, from every one of the others. Now, on the factory or manufacturing system, as compared with the handicraft system, forging, drawing, casting, stamping, and punching, would supersede much of the filing; the drilling machine would supersede the hand-worked tools. This would be done–not merely because the work could be accomplished more quickly or more cheaply–but because an accuracy of adjustment would be attained, such as no hand-work could equal, unless it be such special work as would command a high rate of payment. For any one size in the series, and any one piece of metal in each size of lock, a standard would be obtained which could be copied to any extent, and all the copies would be like each other. To pursue our illustration, the manufacturer might have a hundred boxes or drawers, and might supply each with a hundred copies of the particular piece of metal to which it is appropriated, all so exactly alike that any one copy might be taken as well as any other. Ten pieces, one from each of ten of these boxes, would together form a lock; ten, one from each of another ten boxes, would form a second lock, and so on; and there would be, in the whole of the boxes, materials for a thousand locks of one construction, a hundred of each size.
>
> Now the advantage of the machine or factory mode of producing such articles is this, that they can be made in large numbers at one time, whenever the steam-engine is at work; and that when so made, the pieces are shaped so exactly alike, the screws have threads so identical, and the holes are bored so equal in diameter, that any one of a hundred copies would act precisely like all the

---

[1] Ibid.

> others, thereby giving great advantages to the men employed in putting the lock together.
>
> These principles are being applied by Messrs. Hobbs and Co. in their London establishment. A number of machines, worked by steam-power, are employed in shaping the several pieces of metal contained in a lock; and all the several pieces are deposited in labelled compartments, one to each kind of piece. The machines are employed–in some cases to do coarse work, which they can accomplish more quickly than it can be done by men; and in other cases to do delicate work, which they can accomplish more accurately than men . . .[1]

Hobbs' London firm was successful, especially in the market for the cheaper locks, although he seems to have encountered difficulties in adapting his machinery. There were no ready imitators of his methods and his firm continued its existence in an industry where handicraft methods persisted. At the London Exhibition of 1862 it was stated that 'The patent locks, made by Messrs. Hobbs and Co. . . . are too well known to require description, and are now established articles in England. Their introduction has tended to reduce the price of the cheaper description of lever locks made by hand, but it does not appear that machinery has been introduced by any other maker. Much greater difficulties were met with by Mr Hobbs than were expected; a great deal of time was occupied in perfecting his machinery'.[2] The report went on to suggest that the proliferation of sorts and sizes of locks added significantly to his machinery costs.

Remarks which Hobbs made at a meeting of the Society of Arts suggest that he found British labour somewhat intractable for his purposes. Although the subject under discussion was the application of machinery to firearms, it seems fairly apparent that Hobbs was speaking, at least in part, of his personal experience.

> He (Mr Hobbs) would suggest, as a step preliminary to the introduction of machinery in any particular branch of manufacture, that they should endeavour to reform the minds of the workmen before they began. They had more to contend with in the workmen than in the want of capital. In America they might set to work to invent a machine, and all the workmen in the establishment would, if possible, lend a helping hand. If they saw any error they would mention it, and in every possible way they would aid in carrying out the idea. But in England it was quite the reverse. If the workmen could do anything to make a machine go wrong they would do it, and if the same amount of ingenuity on the part of the English workmen were exercised in producing labour-saving machinery as he (Mr Hobbs) had seen displayed in pro-

---

[1] The papers of A.C. Hobbs, op. cit. 162-3.

[2] *International Exhibition, 1862. Reports by the Juries* (London 1863). Class XXXI, 12.

> ducing the least amount of work in the longest possible time, England never need fear that other nations would go ahead of her in this respect. He thought the great obstacle in the way of the gunmakers of Birmingham in introducing machinery, was the opposition of the workpeople to such innovations.[1]

Several members of the Society rose, after Hobbs' comments, to confirm the evaluation of the British workmen presented by their 'transatlantic friend'.

Complaints of employers and their professional staffs about the quality of their labour force are hardly unusual or worthy of note. Comparative statements, however, based on experience in more than one country, deserve to be taken more seriously, and Hobbs' observations were echoed by others who had had similar opportunities on both sides of the Atlantic.

In his testimony before the Select Committee on Scientific Instruction in 1868 Mr A. Field, a hardware manufacturer who had lived in the United States for fourteen years, stated:

> ... the Englishman has not got the ductility of mind and the readiness of apprehension for a new thing which is required; he is unwilling to change the methods which he has been used to, and if he does change them, he makes demands of price by trade rules which actually oppose the change of the article, or certainly attach to it something in the shape of a fine or an extra demand beyond a fair price for the making of the article. An American readily produces a new article; he understands everything you say to him as well as a man from a college in England would; he helps the employer by his own acuteness and intelligence; and, in consequence, he readily attains to any new knowledge, greatly assisting his employer by thoroughly understanding what is the change that is needed, and helping him on the road towards it.[2]

Similar comparative observations are scattered through the reports reprinted in this volume.

There was one American exhibit which exceeded all others in capturing the fascinated attention of visitors: Colt's repeating pistols. Even the elderly Duke of Wellington, a frequent visitor, whose influence was not always felt on the side of innovation, was often observed at the exhibit forcefully asserting the advantages of repeating firearms to an audience of officers and friends. Colonel Colt himself was accorded[3] the

---

[1] *Journal of the Society of Arts* v, no. 219 (30 January 1857) 165. The comments were made after the delivery of John Anderson's paper 'On the application of machinery in the war department'.

[2] *Report from the Select Committee on Scientific Instruction* Parliamentary Papers, 15 (1867–8) Q. 6722.

[3] Colt's title of Colonel facilitated his entry into English circles where it was of undoubted assistance in the sale of his revolvers. The title–actually he was a Lieutenant-Colonel–was acquired only a short time before he sailed

singular honour of an invitation to address the Institute of Civil Engineers–apparently the first American to be so honoured.[1] In describing his American plant, Colt stated:

> Machinery is now employed by the Author, to the extent of about eight-tenths of the whole cost of construction of these firearms; he was induced gradually to use machinery to so great an extent, by finding that with hand labour it was not possible to obtain that amount of uniformity, or accuracy in the several parts, which is so desirable, and also because he could not otherwise get the number of arms made, at anything like the same cost, as by machinery. Thus he obtains uniformity as well as cheapness in the production of the various parts, and when a new piece is required, a duplicate can be supplied with greater accuracy and less expense, than could be done by the most skilful manual labour, or on active service a number of complete arms may be readily made up from portions of broken ones, picked up after an action.[2]

The meeting was something of a major event. It was attended by high-ranking members of the British military establishment, eminent mem-

---

for England to attend the Exhibition. It was an honorific title, bestowed upon him by his friend Thomas H. Seymour, as a token of gratitude for the assistance provided by Colt in electing Seymour governor of Connecticut in 1850 (William R. Edwards *The Story of Colt's Revolver*, Harrisburg, Pennsylvania, 1953, 268). Colt's chief function as Governor Seymour's aide-de-camp '. . . seems to have been making sure that Seymour got home from official parties not too drunk.' (Ibid.). Nevertheless, Colt was a clever salesman who was not oblivious to the effect of a title upon an audience of potential buyers. As a very young man Colt travelled around the country earning money by giving demonstrations on the practical effects of laughing-gas. During that period he had advertised himself in the newspapers and elsewhere as 'Dr Coult'.

One is irresistibly reminded of Charles Dickens' satirization of America's rather democratic attitudes respecting the use of titles. While dining at Mrs Pawkins's seedy New York boarding-house, Martin Chuzzlewit gradually surmised that his dinner companions were all men of rank. 'Pursuing his inquiries, Martin found that there were no fewer than four majors present, two colonels, one general, and a captain, so that he could not help thinking how strongly officered the American militia must be; and wondering very much whether the officers commanded each other; or if they did not, where on earth the privates came from. There seemed to be no men there without a title: for those who had not attained to military honours were either doctors, professors, or reverends.'

[1] Samuel Colt 'On the application of machinery to the manufacture of rotating chambered-breech fire-arms, and the peculiarities of those arms' *Minutes of Proceedings of the Institution of Civil Engineers* XI (Session 1851–1852) 30-50. The subsequent discussion of his paper (pp. 51-68) extended over two evenings.

[2] Ibid. 44-5.

bers of the engineering profession, and by such American dignitaries as Abbott Lawrence, the American ambassador, and Robert James Walker, former Secretary of the U.S. Treasury. Those familiar with Colt's shrewd salesmanship tactics may suspect that it was not entirely fortuitous that so many of his friends, anxious to deliver effusive testimonials to the murderous efficacy of his firearms, just happened to be present among that distinguished audience.[1]

Although the use of Colt's repeating arms was already widespread and had achieved considerable notoriety in campaigns against the Seminoles and Mexicans, they were a complete novelty in England at the time of the Exhibition. Among those who had at least heard of the repeating firearm there were complete misconceptions as to its structure and mode of performance. Some actually thought it was a revolver with six separate barrels which loaded at the muzzle.[2]

The repeating firearm clearly heralded something drastically new in military technology. Somewhat less spectacular than the Colt revolvers was a nearby display, at the American exhibit, of rifles produced by Robbins and Lawrence, of Windsor, Vermont. These rifles, 'the various parts made to interchange', as stated in the official catalogue, were the product of elaborate machine methods. The workmanship–or perhaps one should say 'machinemanship'–of the locks was particularly admired, and the interchangeability of the components was a source of considerable excitation. Robbins and Lawrence were awarded a medal for their rifles. The firearms exhibits suggested that American technology had progressed in directions which, at the very least, were worth examining.

Colt was so impressed with the potential market for his revolver in England and Europe that he determined to establish a factory in London. He returned to America after the Exhibition in February 1852 but sailed again for England in October of the same year. In establishing his London branch he found it necessary to bring with him both men and machinery from America, although he later used English labour. His plant, on the Thames bank at Pimlico near Vauxhall Bridge, was opened on 1 January

---

[1] Cf. Charles W. Sawyer *Firearms in American History* II (San Leandro, Calif. 1939) 29-30. Colt was awarded a Telford gold medal by the Institute.

[2] The *Maidstone Gazette*, published in the town where the chief cavalry depot was located, admitted to having held this quaint view before the Great Exhibition. 'In a recent article on the Kaffir war, we quoted a suggestion from a correspondent of *The Times*, that a Burgher corps should be formed, armed with Colt's revolver pistols. We had then the notion that the pistol was something like our ordinary six-barrel revolvers, which load at the muzzle, and are, consequently, very deficient in force. In examining the American department in the Great Exhibition, last week, we came upon a stall of arms exhibited by Mr Colt, the inventor. . . . The result of examination and inquiry convinced us if the Great Exhibition confers no other national benefit, that of having public attention called to this formidable arm, is one of no ordinary importance.' *Maidstone Gazette*, as quoted in Rodgers *American Superiority at the World's Fair* 31.

1853.[1] His factory quickly came to be looked upon as a model of the very latest and best in American machine technology and was universally and extravagantly admired by British visitors.[2] It will be referred to again below.

Before the Great Exhibition had ended the earlier disdain had been replaced by curiosity and a desire to learn more about Yankee wares and methods. The *Daily News* observed of the visitors at the Exhibition:

> A great change has taken place in the comparative attractiveness of the various departments. Formerly the crowds used to cluster most in the French and Austrian section, while the region of the stars and stripes was almost deserted–now the domain of Brother Jonathan is daily filled with crowds of visitors. In the front, trim mercantile men crowd around Hobbs' lock; right opposite the click of Mr Colt's revolvers is unceasing, as the exhibitor demonstrates the facility with which they can be made to perform their murderous task; and in the rear jolly broadshouldered farmers gather about McCormick's reaping machine, and listen in mild stupidity to the details of its wondrous prowess at Tiptree Hall and at Leicester, over rough and smooth land, ridge and furrow.[3]

Punch happily availed itself of the opportunity to lampoon British smugness and vanity by writing 'The Last Appendix to "Yankee Doodle"':

> Yankee Doodle sent to town
> His goods for exhibition;
> Every body ran him down,
> And laughed at his position;
> They thought him all the world behind
> A goney, muff, or noodle.
> Laugh on, good people–never mind–
> Says quiet Yankee Doodle.
> CHORUS–Yankee Doodle, etc.
>
> Yankee Doodle had a craft,
> A rather tidy clipper,

---

[1] Charles T. Haven and Frank A. Belden *A History of the Colt Revolver* (New York 1940) 86. The reasons for the eventual closing of this plant in 1857 are not clear. Haven and Belden simply state that 'He could not prevail upon his American workmen to remain in England and the English ways of doing things did not suit his ideas of mass production' (p. 89). A. Merwyn Carey states of Colt's London activities that 'He developed a successful business and received both British and Russian contracts. His U.S. plant and business required his first attention, with its tremendous development and expansion during this period, and he closed out his English operation in 1857.' (*English Irish and Scottish Firearms Makers* (London 1954) 19.)

[2] It was described and admired also by Charles Dickens in his weekly journal *Household Words* (chapter 15, vol. IX, 1855). This is reprinted in Haven and Belden, op. cit. 345-9.

[3] As quoted in Rodgers, op. cit. 62.

And she challenged while they laughed,
The Britishers to whip her.
Their whole yacht squadron she outsped,
And that on their own water,
Of all the lot she went ahead,
And they came nowhere arter.
CHORUS–Yankee Doodle, etc.

. . . .

Your gunsmiths of their skill may crack,
But that again don't mention;
I guess that Colt's revolvers whack
Their very first invention.
By Yankee Doodle, too, you're beat
Downright in Agriculture,
With this machine for reaping wheat,
Chaw'd up as by a vulture.
CHORUS–Yankee Doodle, etc.

You also fancied, in your pride,
Which truly is tarnation,
Them British locks of yourn defied
The rogues of all creation.
But Chubb's and Bramah's Hobbs has pick'd,
And you must now be viewed all
As having been completely licked
By glorious Yankee Doodle.
CHORUS–Yankee Doodle, etc.[1]

II

The London Exhibition was so spectacular a success, commercially and otherwise, that it not surprisingly soon found imitators.[2] It quickly became the prototype for expositions of truly international scope as well as innumerable more modest undertakings.[3] The idea of an exhibition

---

[1] Ibid. 91. For a choice specimen of transatlantic versification on the same subject, see *Brother Jonathan's Epistle to His Relations Both Sides of the Atlantic, But Chiefly to His Father, John Bull, Brother Jonathan Being a Leetle Riled by the Remarks Made by John Bull at His Small Wares Displayed At the Opening of the Grand Exhibition* (Boston 1852). For some readers the title may be sufficient.

[2] The Exhibition closed with a surplus of over £200,000, which was used as an endowment for what eventually came to be The Victoria and Albert Museum.

[3] 'Between 1862 and 1926 the United States government alone expended approximately $30,000,000 as its part in over forty foreign and national expositions.' Guy Stanton Ford 'International expositions', op. cit. 24. For an interesting account of American participation in international expositions in the second half of the nineteenth century, see Merle Curti 'America at the world fairs, 1851–1893' *The American Historical Review* 55 (1950) 833-56.

on American soil took root immediately, for on 11 March 1852, just four months after the closing of the exhibition at Hyde Park on 11 October, the necessary approval for a New York exhibition had already been granted. On that date the New York State Legislature enacted a charter of incorporation for 'The Association for the Exhibition of the Industry of all Nations.'[1]

The New York Exhibition was not an unqualified success. It was characterized by considerable enthusiasm, excessive haste, and poor organization. When, after numerous delays, the exhibition was formally opened on 14 July 1853, the building itself was not yet completed and only just over half of the articles were properly exhibited.

It was precisely to this delay that we owe the Special Reports of Whitworth and Wallis. Joseph Whitworth and George Wallis were among the commissioners appointed by the British Government to attend the exhibition in New York. In addition to Whitworth and Wallis the commission included Lord Egerton Ellesmere, Sir Charles Lyell, Mr C. Wentworth Dilke, and Professor John Wilson. Each member of the commission had been assigned specific departments of the Exhibition and asked to furnish a report on them upon his return. The assignments were based on the system of classification employed at the London Exhibition of 1851 and reproduced, as was so much else, at the New York Exhibition.[2]

Whitworth's presence among the commissioners is itself a measure of the recently acquired respect for American technology and of a serious desire to acquire greater familiarity with American achievements. For Joseph Whitworth (1803–87) was recognized not only as an engineer of remarkable talents, but also as the world's foremost manufacturer of machine tools. The range of his machine tool display at the London Exhibition of 1851–his lathes, his machinery for planing, shaping, slotting, drilling, boring, punching, and shearing, his standard gauges and measuring apparatus, among others–established his unquestioned authority as a designer and builder of machine tools.[3]

Whitworth's talents were those of a great systematizer rather than inventor. As Roe has stated of Whitworth's machines: 'Their pre-eminence lay not so much in novelty of design as in the standard of accuracy and quality of workmanship which they embodied.'[4] His sys-

---

The impact of American technology in shaping the European image of America is discussed in Hugo A. Meier 'American technology and the nineteenth-century World' *American Quarterly* 10 (1958) 116-30.

[1] Horace Greeley (ed.) *Art and Industry as Represented in the Exhibition at the Crystal Palace, New York, 1853–4* (New York 1853) ix.

[2] For details see *General Report of the British Commissioners to the New York Industrial Exhibition* Parliamentary Papers, 36 (1854) 1-2.

[3] See *Official Description and Illustrated Catalogue* I, 287-91.

[4] Joseph W. Roe *English and American Tool Builders* (New Haven 1916) 99. Roe also states of Whitworth: 'He was a master experimenter. Tests which he made were thorough, conclusive, and always led somewhere. His experiments, whether in machine tools, screw threads, or ordnance, always resulted in a design or process which sooner or later became standard' (p. 106).

tematic approach to machine design was evident in his earlier proposals which brought about the standardization of screw thread practice in England, his innovations in precision measurement which raised standards of accuracy to levels hitherto unattained, and his introduction of the hollow frame design in machinery construction.[1]

Mr Whitworth was a considerable exhibitor of his own celebrated machinery at the New York Exhibition, where his display included lathes, slotting and shaping machines, planing machines, screwing machines, a measuring machine, cylindrical gauges, and even a horse-drawn road sweeping machine.[2]

George Wallis (1811–91) at the time of the New York Exhibition was headmaster of the Government School of Art and Design in Birmingham and had previously held a similar position at the Manchester School of Design. His professional interests were primarily in industrial art and design, and as a result of this work he had been for many years a close observer of the changing industrial scene. He was superintendent of the British textile division and a deputy commissioner of juries at the London Exhibition in 1851, after which he accepted the headmastership of the Birmingham School of Art and Design at the request of the Board of Trade.[3]

When the members of the Commission arrived in New York, it was discovered that the opening date had been postponed from the first of June to 14 July. It was, furthermore, obvious that even by 14 July at least many of the exhibits could not possibly be in a sufficiently presentable state to permit a serious and systematic examination of their respective merits.[4] Accordingly the commissioners decided that the best possible use of their time would be to embark on tours which would enable them

---

[1] Ibid. ch. 9; Joseph Whitworth *Miscellaneous Papers on Mechanical Subjects* (London 1872); *Proceedings of the Institution of Civil Engineers* XCI (1888) 429-46; F.C. Lea *Sir Joseph Whitworth: A Pioneer of Mechanical Engineering* (London 1946). For some amusing observations on Whitworth by an American engineer who went to the 1862 London Exhibition and subsequently had some business contacts with him, see Charles T. Porter *Engineering Reminiscences* (New York 1908) 123-31.

[2] C.R. Goodrich (ed.) *Science and Mechanism* (New York 1854) 117. As its appearance in New York suggests, Whitworth did not regard his road sweeping machine as among his less important constructions.

[3] *Dictionary of National Biography* XX, 597-8. Wallis left Birmingham in 1858 to join the South Kensington Museum as senior keeper of the art collection. He held this post until shortly before his death.

[4] On the deplorable state of unreadiness of the Exhibition the commissioners rather generously observe in their General Report: '. . . it is but justice to those who had undertaken its management to state that the inclemency of the last winter, the great demand for workmen in the building trades, the novelty of construction in certain portions of the building, and the want of any previous experience in a work of such magnitude, may be considered as an excuse for the delay which took place.' *General Report of the British Commissioners to the New York Industrial Exhibition* op. cit. 4.

to observe first-hand the subjects for which they were responsible at the Exhibition.[1] The Special Reports of Whitworth and Wallis are in fact reports of their observations on these tours of the major centres of American manufacturing in the north-eastern states during the summer of 1853. From the point of view of the record which they have left on the state of American manufacturing as they found it, one can only be grateful for the delay in the opening of the Exhibition. No displays which they might have observed within the confines of a single building–even an American Crystal Palace–would possibly compare with the range and richness of detail encompassed in their Special Reports.

The precise responsibilities of Whitworth and Wallis were described as follows:

> In machinery, Mr Joseph Whitworth undertook to report on Class v., Machines for direct use; Class vi., Manufacturing Machines and Tools; Class vii., Civil Engineering; and on the character and action of the Patent Laws, and on the system of Electric Telegraphs so extensively used throughout the United States.
>
> In manufactures, the report on Class xi., Cotton; Class xii., Woolen and Worsted; Class xiii., Silk; Class xiv., Flax and Hemp; and Class xv., Mixed Fabrics; Class xvi., Leather (except Tanneries); Class xvii., Paper, Printing, etc.; Class xviii., Printed Fabrics; Class xix., Tapestry, Lace, and Carpets; Class xx., Clothing; Class xxi., Cutlery and Edge Tools; Class xxii., Iron and General Hardware; Class xxiii., Precious Metals, Jewellery, etc.; Class xxiv., Glass; Class xxv., Ceramic Manufactures; Class xxvi., Decorative Furniture; and on the position of Art Education as applied to Manufactures and the Copyright of Designs, was undertaken by Mr George Wallis.[2]

The members of the Commission re-assembled in New York in time for the official opening on 14 July. President Pierce and several members of his cabinet were present. But although the Exhibition was officially declared to be open, the building itself was still not near completion and many of the displays were not yet assembled. As the General Report explains: 'The continued incompleteness of the Exhibition after the day of opening, suggested the importance of finishing the examination of the various departments assigned to each member of the Commission, as far as possible in their respective centres.'[3]

The members of the Commission were busy men of affairs with press-

---

[1] 'Under these circumstances . . . it was decided, after consideration, that the various localities of the United States in which raw materials were likely to be most abundant, mechanical skill most largely applied, manufacturing industry fairly established, and art and science most perfectly developed, whether educationally or practically, should be first visited as far as the limited time at the disposal of the majority of the members of the Commission would allow.' *General Report*, 1.

[2] Ibid. 2. [3] Ibid.

ing commitments in England, and although they returned at various times between August and October, none of the Commissioners was able to stay long enough to see the exhibits for which he was responsible properly and fully displayed.[1] Wallis and Whitworth embarked from New York on 10 August, at which time the portion of the building which was to house the machinery for which Mr Whitworth was responsible was still unfinished; in the area of Mr Wallis' responsibilities, many of the most important contributions had either not yet arrived or were not yet properly placed on display.[2]

III

The separate reports of Whitworth and Wallis supplement one another in various ways. Aside from the differences in their respective assignments, there were differences in outlook and responsiveness. Whitworth, the consummate practitioner of nineteenth-century engineering skills, described mechanical processes in technical terms and at some length, and showed a persistent interest in questions of manufacturing efficiency – although, it should be added, he was not always clear concerning the distinction between the engineer's and the economist's understanding of efficiency. Wallis, as might be expected, showed a much greater concern with matters of taste and artistic refinement. Although he admired the forthright and 'common sense' American approach to design, his aesthetic sensibilities, alas, suffered frequent violation. Much that he found meretricious in American taste he traced – perhaps somewhat generously – to a European influence.[3] 'The American manufacturer', he states, 'is in some respects wiser than his foreign competitor, and in many instances leaves the ultra-ornate to be supplied from Birmingham and Sheffield, and directs his energies to the development of a better and less exuberant style, which he finds is demanded by the more refined amongst his countrymen' (p. 281). Unfortunately, Wallis did not have the opportunity at the Exhibition to inspect what may have been a prime specimen of American functional ingenuity. 'In the space of an ordinary sideboard, it was stated that a bedstead and complete chamber suite was contained. As it was locked up, and no key to be found, its merits were unascertained, nor could any one connected with the Exhibition state who it belonged to, or where it came from' (p. 294). One wonders what unexpected involutions of Yankee ingenuity lay concealed from Wallis and posterity within that unyielding sideboard.

---

[1] Mr Dilke was called back earlier by the serious illness of his wife, who died shortly after his return to England. [2] *General Report*, 4.

[3] Wallis' generosity extended also to the occasions when he was not permitted to examine machinery of interest to him or when respondents refused to provide him with the information which he sought. Although there were obviously several such instances, the only individual singled out and rebuked for his obduracy in the Wallis report was the English owner of a pottery works in Pittsburgh (pp. 291-2). All subsequent page citations which are not otherwise identified are to the present volume.

Wallis was alert also to the economic importance of taste differences. In his discussion of textile fabrics, for example, he pointed out that the American consumer typically expected an article to last for only a short period, and therefore he was more reluctant than European consumers to pay a premium for either quality or durability. The American consumer had more homogeneous tastes than Europeans; these tastes were, moreover, concentrated at the low quality end of the spectrum. The taste for short-lived articles, Wallis observes, '. . . is said to run through every class of society, and has, of course, a great influence upon the character of goods generally in demand, which . . . are made more for appearance, and less for actual wear and use, than similar goods are in England' (p. 304). A taste bias in favour of homogeneous, low quality goods of short life expectancy was, of course, highly favourable to the introduction of mechanized techniques.

As is evident from a reading of their reports, both Wallis and Whitworth travelled extensively (Wallis stated that, while in the United States, he travelled 'upwards of 5,000 miles'). Although they confined their travels almost entirely to New England and the Middle Atlantic states, this region accounted for 75 per cent of the nation's manufacturing employment in 1850. While their reports abound in insights into the industrialization of the American economy, it is important to remember that, by any index, the American economy was still predominantly agricultural. In spite of a sharp growth in the percentage of the labour force engaged in manufacturing during the decade of the 1840s, such employment still amounted by 1850 to less than 15 per cent of the labour force. By contrast, substantially over half of the labour force in that year was still employed in agriculture.[1]

The reader of these reports is forcefully reminded of the limited impact of steam power upon American manufacturing even as late as the middle of the nineteenth century. Aside from transportation, where the steam engine was already of major importance, steam power had not yet begun to approach the status which it was to achieve toward the end of the nineteenth century. Transportation improvements in turn, by reducing the cost of coal, played an important role in the diffusion of the steam engine.

The use of steam power was especially limited in New England, a region whose many swiftly-flowing rivers offered excellent sites for the utilization of water power. Even as late as 1869, by which time steam had surpassed water as a source of manufacturing power for the entire United States, less than one third of the power supplied to New England manufacturing establishments came from steam.[2] The dependence of

---

[1] Stanley Lebergott 'Labor force and employment, 1800–1960' in National Bureau of Economic Research, *Output, Employment, and Productivity in the United States after 1800*, Studies in Income and Wealth, 30 (New York 1966) 119.

[2] See the estimate of Allen Fenichel 'Growth and diffusion of power in manufacturing, 1839–1919' ibid. 456.

American industry at this time upon water power sources emerges very clearly as the reader follows Whitworth and Wallis through a succession of small New England towns–Lowell, Lawrence, Hadley Falls, Holyoke, Chicopee, Springfield, Waterbury, Manchester–communities all owing their specific locations to unusual opportunities for the exploitation of water power.

The reports of Whitworth and Wallis provide an invaluable still-picture of American manufacturing techniques at a critical phase in the country's history. During this period the factory system, which had begun to emerge in earlier decades in the New England textile mills, was expanding into new industries. The period was one of increasing specialization and localization of production as individual firms undertook to produce for wider and wider markets. This process was, in turn, linked to the rapid growth of these markets and to substantial reductions in transportation costs. Although the United States was experiencing a rapid growth in its railroad network during the 1850s, other transport innovations had been lowering freight costs in the preceding three decades or so. These included the construction of canals beginning with the spectacularly successful Erie Canal, the use of steamboats on western rivers, and a general reduction in ocean and river freight rates.

The impact of this process in driving out older, decentralized forms of production may be seen in particularly dramatic form in an estimate of the decline of household textile manufacturing in the State of New York (see below).

Household textile manufactures in New York

Year	*total yards of all kinds of textile goods*	*per capita yards*
1825	16,469,422	8·95
1835	8,773,813	4·03
1845	7,089,984	2·74
1855	929,241	0·27

Source: Rolla Milton Tryon *Household Manufactures in the United States, 1640–1860* (Chicago 1917) 304-5.

At the time of the visits of the British Commissioners household textile manufacture, which had been a major industry in New York State thirty years earlier, was clearly moribund.

But although rapid strides were being made, the transitional nature of the period is obvious, and the juxtaposition of the old and new appear at many places in the Commissioners' Reports. Wallis observes the lingering pockets of domestic weaving in Pennsylvania, but asserts: 'There can be no doubt . . . that domestic weaving is gradually giving way, and those manufacturers, especially in Pennsylvania, who formerly did a prosperous business as spinners only, now find that the Eastern States supply the piece goods at a rate so little above the cost of the yarn, that it is not worth the while of the farmer to continue this primitive custom of weaving his own cloth' (p. 210). On the same page as he refers to the 200 power looms of the Lowell Manufacturing Company which produce 25,000 yards of carpet per week, Wallis describes also the predominance of the handloom weaver and the domestic system in Pennsylvania, Delaware, and Maryland (pp. 248-9). The ready-made clothing trade in Cincinnati, which catered to the country's growing western market, and into which the sewing machine was 'latterly' being introduced, was organized mainly on the domestic system[1] (pp. 252-3). Wallis' detailed description of the extensive boot and shoe trade emphasizes its elaborate pattern of specialization, but the organization of the industry is along domestic lines. This industry was, as we now know, on the verge of a rapid transformation, which the voracious demands of a civil war were vastly to accelerate.[2]

In interpreting and evaluating the observations in the following reports it will be useful to keep in mind the relative sizes and rankings of individual industries. The data in the table below are from the Census of

United States manufactures, 1860

item	*cost of raw material*	*number of employees*	*value of product*	*value added by manufacture*	*rank by value added*
cotton goods	$52,666,701	114,955	$107,337,783	$54,671,082	1
lumber	51,358,400	75,595	104,928,342	53,569,942	2
boots and shoes	42,728,174	123,026	91,889,298	49,161,124	3
flour and meal	208,497,309	27,682	248,580,365	40,083,056	4
men's clothing	44,149,752	114,800	80,830,555	36,680,803	5
iron (cast, forged, rolled and wrought)	37,486,056	48,975	73,175,332	35,689,276	6
machinery	19,444,533	41,223	52,010,376	32,565,843	7
woollen goods	35,652,701	40,597	60,685,190	25,032,489	8
carriages, wagons and carts	11,898,282	37,102	35,552,842	23,654,560	9
leather	44,520,737	22,679	67,306,452	22,785,715	10

Source: Compiled from *Eighth Census of the United States: Manufactures* 733-42. The Census treated 'wagons and carts' as a distinct category. For present purposes it was thought to be reasonable to combine them with 'carriages'.

[1] 'The invention of the sewing machine in 1846 promoted the growth of the putting-out system in ready-made clothing, for it was found that routine sewing-machine work could be done effectively and inexpensively by women in their own homes. This industry, therefore, provides an interesting exception to the generalization that the mechanization of industry led to the development of the factory system, for in this case an important mechanical invention actually stimulated home rather than shop or factory production.' G.R. Taylor *The Transportation Revolution* (New York 1951) 219-20.

[2] Blanche E. Hazard *The Organization of the Boot and Shoe Industry in Massachusetts before 1875* (Cambridge, Mass. 1921).

1860, the closest year for which figures on American manufacturing of reasonable reliability are available. Industries are ranked by value added in manufacturing.

The reports of Whitworth and Wallis will be found to include materials dealing with each of these ten industries. Not all are treated extensively. 'Flour and meal', for example, is dealt with only briefly by Whitworth. But, for the most part, these reports pay considerable attention to each of the major industries of the United States. They provide a comprehensive picture of the production of cotton goods, including a description of the cotton mill communities of New England which contrasted in so many ways with the industrial cities of Great Britain. Wallis was so interested by the boarding-houses provided for workers at the Bay State Mills in Lawrence, Massachusetts, that he included drawings of these houses in an appendix to his report. He reproduced also some of the codes which regulated the lives of their inhabitants, and which clearly delineate the extent of the paternalism characterizing the employer-employee relationship. The quintessence of a stern and brooding New England Puritanism is captured in a single sentence: 'A proper observance of the Sabbath being necessary for the maintenance of good order, all persons in the employ of this Company *are expected to be constant in attendance at public worship*, and those who habitually neglect this regulation, or who are known to attend improper places of amusement, will be discharged' (p. 325. Emphasis in the original).

Although there is a good deal of overlapping in the coverage of individual industries between the reports of Whitworth and Wallis–especially in the treatment of textile products–the shorter Whitworth Report contains most of the material concerned with machinery, iron and metalworking generally, as well as an important section on woodworking machinery. Whitworth was deeply impressed with American ingenuity in the development of woodworking machines, most particularly with their capacity to save labour. The Wallis Report, by contrast, covers a much wider range of industries and is less concerned than the Whitworth Report with the more technical aspects of machine operation and productive processes generally.

The reports of Whitworth and Wallis contain much more than mere descriptive accounts of American manufacturing. Each author at various points (Whitworth especially in his conclusions and Wallis especially in his introduction) undertakes to place his observations within a larger frame of reference. Each report may indeed be regarded at one level as a contribution to the study of national characteristics. One might easily discern a merely didactic intent on the part of the authors–as, for example, in their laudatory discussion of the American worker, his long hours,[1] vast energy, initiative, adaptability, etc. Such an attitude, however, would do Whitworth and Wallis a serious injustice. For, as they

[1] Wallis states that he found fewer Englishmen employed in American industrial establishments than he had expected. He attributes this, in part, to the long hours commonly demanded of workers in America (p. 207).

both argued, these characteristics were themselves the product of a certain kind of economic environment. Specifically, both Whitworth and Wallis emphasize that the scarcity and therefore high cost of labour in a bountiful resource environment was responsible for the invention as well as the ready adoption of labour-saving techniques. Indeed, these reports may be considered as the *loci classici* of this particular interpretation of the special features of American economic growth.[1] In Wallis' words:

> Thus the very difficulty in procuring human labour, more especially when properly skilled and disciplined, which would assuredly be the greatest drawback to success, appears to have stimulated the invention of the few workers whose energies and skill were engaged in the early development of manufactures; and to this very want of human skill, and the absolute necessity for supplying it, may be attributed the extraordinary ingenuity displayed in many of those labour-saving machines, whose automatic action so completely supplies the place of the more abundant hand labour of older manufacturing countries (p. 203).

Conversely, Whitworth attributed the restrictive practices and resistances to innovation of the English workman to an abundance of labour and comparative scarcity of employment opportunities. 'With the comparatively superabundant supply of hands in this country [i.e., England], and therefore a proportionate difficulty in obtaining remunerative employment, the working classes have less sympathy with the progress of invention' (p. 388).

Whitworth and Wallis write also in terms of highest approbation of the role of public education in the United States–especially in New England and Pennsylvania. For this system has led to the ready application of 'scientific' knowledge to economic problems, to an 'adaptive versatility' and to a degree of occupational and social mobility greater than anything known in Great Britain. The American devotion to education which so impressed the English visitors seems to have been a matter of long standing.[2] According to a rough international comparison prepared in conjunction with the 1850 census, school enrolment figures for the United States were higher than those for any European country except Denmark.[3]

Whitworth and Wallis also argued that the American environment was more favourable to the free and beneficent play of market forces simply because of the absence of arrangements which were all too common in Europe–for example, the organization of labour along craft

---

[1] For a recent penetrating reformulation of this argument see H.J. Habakkuk *American and British Technology in the 19th Century* (Cambridge 1962).

[2] See Albert Fishlow 'The American common school revival: fact or fancy?' in Henry Rosovsky (ed.) *Industrialization in Two Systems: Essays in Honor of Alexander Gerschenkron* (New York 1966).

[3] U.S. Census Bureau *A Compendium of the Seventh Census* J.D.B. DeBow, Supt. of Census (Washington 1854) 148.

skill lines and the ossifying impact of apprenticeship systems. As Wallis saw it, the American environment placed the worker in a situation which made great demands upon his initiative, independence and self reliance; at the same time it placed no obstacles of traditions, attitudes, or institutions in his way. These were the favourable conditions under which '. . . the ingenuity, the indomitable energy and perseverance' of the American worker were nurtured and finally flourished.

IV

The growth of British interest in American technology after, and largely as a result of, the Crystal Palace Exhibition was reinforced by awareness of the serious national consequences attendant upon backwardness in military technology in particular. It is clear that in the early 1850s there was an immense growth of interest in military hardware. Such interest may be measured by a sharp acceleration in patent applications. As one authority has pointed out:

> From the official publication of the British Patent Office, it appears that from the year 1617 down to the end of the year 1852 not more than about 300 patents were granted for inventions relating to fire-arms of all classes. But just prior to the outbreak of the Crimean War the Patent Office was inundated with applications for letters patent, and within a few years the number of patents obtained for inventions relating to this subject amounted to double the number granted in the two and a half preceding centuries.[1]

British officialdom immediately after the Exhibition was painfully aware of pressing problems connected with military supply. Such concern was further exacerbated by increasing unsettlement in the European balance of power and, in particular, by fears of Russian encroachment upon the Ottoman Empire. The euphoria and self-congratulation which were so much a part of the national mood during the Crystal Palace Exhibition, the talk of universal peace and universal brotherhood, all were dispelled soon after the Exhibition's close. Astute observers of the political scene perceived the dangers of war in 1852, and, although Britain did not enter the war against Russia until March 1854, the imminence of such a war was perfectly obvious in 1853.

To appreciate the government's concern over its military supply arrangements, of firearms in particular, it is necessary to look briefly at

---

[1] Edward C.R. Marks *The Evolution of Modern Small Arms and Ammunition* (Birmingham 1898) 63. The author adds that 'Tommy Atkins is to be congratulated that he has never been ordered to fire a weapon made in accordance with some of these specifications'. He might equally be congratulated for the lack of enthusiasm shown by the Machinery Committee (p. 125 below) in reporting on Colonel Porter's revolving rifle: 'This revolving rifle is held in very low estimation in the United States, and is considered dangerous to use, there being three chambers always pointed towards the person who is firing it.'

the conditions under which firearms were produced in the first half of the nineteenth century.

Birmingham and its environs had long been the seat of the British gunmaking trade. The origins of this trade go back to Tudor times, and records indicating government reliance on this district as a source of military firearms go back to the late seventeenth century.[1] Birmingham's competitive rivalry with the London gunmakers was marked by much intrigue and continual jockeying for special privilege. The Birmingham trade grew rapidly in the eighteenth century, especially the African trade, but also with the East India Company and with other overseas colonies such as those of North America. Toward the very end of the century the district undertook the production of sporting guns on a substantial scale.

During periods of war, such as with the American colonies, the gun trade became a scene of frenzied activity. This characteristic instability of the industry was apparent–and much complained of–from a very early date. The high degree of instability in the demand for the final product–especially of the military component–is one of the most important characteristics of the gun trade. Many of the peculiar features of the industry are closely connected with these drastic shifts in the industry's demand curve.

The production of firearms became concentrated more and more in the Birmingham area. Indeed, a progressively increasing proportion of the firearms sold by London gunmakers were made up of component parts which had been produced by the Birmingham trade, the London gunmaker merely fitting the components together and affixing his own mark to the final product.

By the close of the eighteenth century, in the midst of the Revolutionary Wars, Birmingham emerged as the world's major centre of firearms production. At this time, in fact, it was the only large centre of firearms production not under the control of Napoleon's armies. Its output of military firearms–overwhelmingly muskets, but including a small number of rifles, carbines, and pistols–reached enormous proportions in the last decade of the continental wars.

When war with France was renewed in 1803, the urgency of the British position with respect to firearms was so great that the government undertook to purchase all the firearms which were available in foreign markets. These were not only relatively few in number but also of poor quality. Beginning in 1804, however, the number of arms produced by Birmingham for the Board of Ordnance began to rise steadily, and continued to rise until the peak years of 1812 and 1813.

It appears from Board of Ordnance figures that during the period 1804 to 1815, Birmingham produced 1,743,382 complete pieces of firearms for

---

[1] J. Goodwin 'The Newdigates of Arbury, early memorials of the Birmingham gun trade' *Gentleman's Magazine* (February 1869) 289-302; W. H. B. Court *The Rise of the Midland Industries 1600–1838* (London 1938) 142-8.

the Board.[1] This was nowhere near its total output for, in addition, it produced huge quantities of the basic firearms components which were subsequently utilized by the London trade. Over the same period the Birmingham trade produced 3,037,644 barrels for firearms and 2,879,203 locks, all of which were used by the London trade for eventual sale to the Board of Ordnance. In addition to production for the Board of Ordnance, the East India Company was during this period a major purchaser of firearms, and it is estimated that during this period the materials for nearly 1,000,000 guns for this company were manufactured in Birmingham and made up into finished guns in London. Finally, more than 500,000 fowling pieces were produced by the Birmingham District during this period. During these eleven years then, the Birmingham trade produced over 6,000,000 guns or gun components.[2] The number produced in Birmingham for the government alone far exceeded the number produced for the French Government during the same period, when that government controlled the very considerable manufacturing establishments of Belgium and Italy as well as its own government-operated establishments.[3] By way of further comparison, the number of guns actually produced in the years 1803 to 1816 by London manufacturing establishments and by a small royal manufactory which functioned during part of this period amounted to 845,477.

This rapid growth in the output of military firearms was achieved in part by a relaxation of standards of quality, and there were many complaints of inferior workmanship and materials.

> These arms were . . . of an inferior description, the regular army pattern having been laid aside for others more easily obtained, called the India pattern; and great complaints have at times been made of the bad quality of them. The urgency of the demand prevented attention to quality, and the complaints were in many instances well founded.[4]

In fact, the problem of quality control was a major source of contention between the government and the firearms trade and had much to do with the argumentative tone which characterized the trade and its relations with the government.

---

[1] The following figures on firearms production in Birmingham are from John D. Goodman 'The Birmingham small gun trade' in Samuel Timmins (ed.) *Birmingham and the Midland Hardware District* (London 1866) 412-14. It may be noted that, of the total of 1,743,382 completed firearms, 1,635,435 were muskets, 14,695 were rifles, 38,778 carbines, and 54,474 pistols. (Goodman's slightly larger subtotal for muskets (p. 412) is obviously a typographical error.)

[2] Goodman, after presenting these statistics, unaccountably refers to their total as 'nearly five million guns'. Ibid. 413.

[3] Ibid. 413-14. The major French establishments were at St Etienne, Charleville, and Maubeuge, which had been established as 'Manufactures Royales' in 1718.

[4] 'Memorandum relative to small arms' *Report from the Select Committee on Small Arms* Parliamentary Papers 18 (1854) 451.

To some extent the industry responded to large government demands by drawing upon artisans who were locally engaged in other trades. The existence, in Birmingham, of a highly diversified group of metal using industries apparently provided a pool of labour upon which the gunmakers could draw.[1] Similarly, men employed in carpentry could be engaged in producing stocks. Probably more important, however, were sources of labour elsewhere in the firearms trade. This trade had three main segments: (1) sporting guns, many of them custom made and of very high quality, (2) the military trade, and (3) a very cheap gun, which for many years sold at ten shillings, primarily for the African trade. Since the military trade was a 'feast or famine' trade, gunmakers tended to concentrate on sporting guns or cheaper guns. When large government orders appeared they seem to have drawn heavily for labour upon men engaged in producing cheap guns, who ordinarily worked to standards far below those of the military trade. For some parts of the gun the differences in skill may not have been especially important, but in others –for example, in straightening the barrel–it was critical.[2]

As might be expected, the Birmingham gun trade sank into a torpid state after 1815. With the end of hostilities the domestic demand for guns for military purposes collapsed, and large numbers of skilled workers left the trade. Furthermore, with one important exception, no major technical changes were adopted between 1815 and the 1850s which made an extensive demand upon the resources of the military gun trade. The exception was the introduction of the percussion system for discharging firearms. The basic firearm of the British soldier remained the venerable 'Brown Bess', a flintlock musket whose main design features dated back to the reign of Queen Anne, a hundred years earlier.

The use of fulminating powder to achieve ignition by the percussion

---

[1] Birmingham has long been known for the extraordinary diversity of the output of her metal trades, including some items of the most macabre nature. 'Curiosities of the Birmingham heavy steel toy trade have been the secret manufacture, by men known to the police, of house-breakers' implements, such as gouges, ripping chisels, skeleton keys, and pocket jacks; whilst, by way of an antidote to these, handcuffs, manacles, leg-irons, and chains have also been provided from the same source; 'suits' of chains for Brazilian slave-ships; tomahawks for North American Indians, and a mysterious order, executed as late as 1849, for two dozen thumb-screws, for some South American state.' Victoria County History *Warwickshire* II (London 1906) 197.

[2] '. . . The gun makers are divided, after we come to the setting up, into three classes; one class for making superior arms, fowling-pieces, and so forth; next, the persons who stand as skilled workmen on the scale are military gun makers, and then there are gun makers who make a very inferior sort of arm, I believe even as low as 10s. a stand, and I have no doubt that if that class of men be counted upon you might make, perhaps, 4,000 or 5,000 stand of arm in a week. But that is an arm of a very inferior class, and a man so employed is not competent to do the work of a military gun maker.' *Report from the Select Committee on Small Arms*, op. cit. Q. 90, testimony of Captain Sir T. Hastings.

method was introduced in 1807 by the Reverend Alexander Forsyth, who was awarded a patent in that year. His method, however, posed mechanical difficulties which he did not successfully solve, but which were finally solved, by 1816, by the introduction of a copper percussion cap. The percussion-cap system in turn remained the standard method of ignition until the introduction of the breech-loading system of the 1860s, at which time a cartridge containing its own means of ignition was employed.[1]

British Ordnance was slow in adopting the percussion system – some say, perhaps unfairly, as a result of the Duke of Wellington's excessive affection for the 'Brown Bess' flintlock. Tests of the percussion system were finally conducted at Woolwich Arsenal in 1834, and the superiority of the system was finally acknowledged by military authorities. Percussion muskets were eventually issued to British troops and a complete transition from flintlocks to the percussion system was accomplished by 1842.[2]

The distinguishing features of the Birmingham gun trade throughout this period were (1) smallness of scale, (2) dependence on skilled craftsmen, (3) an extensive division of labour in the production of components and their eventual 'setting-up'; (4) localization – most of the Birmingham gun trade was located in the district around St Mary's Church.[3]

At the time of the Crystal Palace Exhibition, the gun trade was still the province of the skilled handicraftsman. Mechanical assistance was virtually non-existent except in a few instances such as the rolling, boring,

---

[1] Goodman, op. cit. 384. It is worth noting that the breech-loading principle was adopted first by makers of sporting guns. The principle was fairly widespread in sporting guns when military guns were still uniformly employing muzzle-loading methods.

[2] Ibid.; A. Merwyn Carey, op. cit. 35, 113; W. C. Aitken 'Guns', in G. Phillips Bevan (ed.) *British Manufacturing Industries* 3 (London 1876) 8 volumes, 6-9. Nasmyth stated that 'The Duke of Wellington to the last proclaimed the sufficiency of "Brown Bess" as a weapon of offence and defence . . .' James Nasmyth *James Nasmyth Engineer: An Autobiography* edited by S. Smiles (London 1883) 362.

[3] These characteristics applied not only to guns but to a considerable extent to most of the trades producing finished goods in Birmingham. G. C. Allen, writing of the first sixty years of the nineteenth century, observes that '. . . there had been no "industrial revolution" in Birmingham and District. Its great economic development was marked by a vast increase in the number of producing units rather than by a growth in the size of the existing few, and the factory still remained unrepresentative of the majority of the concerns producing finished goods. Of Birmingham itself it was said in 1856 that nothing was on a large scale, and that the manufacturing class had not raised itself in any large degree. The concentration of capital and the development of large-scale enterprises had not taken place there as in the nothern centres of industry, and most master manufacturers, it was declared, employed only five or six workers. And this was equally true of the Black Country's small metal industries.' G. C. Allen *The Industrial Development of Birmingham and the Black Country 1860–1927* (London 1929) 113.

and grinding of barrels, or the rolling, grinding, and polishing of bayonets and rammers.[1] Locks, which were produced in large quantities not only in Birmingham but in the neighbouring towns of Darlaston and Wednesbury, were the product of skilled hand labour. The component parts were initially forged on the anvil by skilled workers, put together by the filers, and finished by polishing and hardening. The file was an indispensable tool in adjusting the separate parts to work together in a smooth and harmonious fashion.[2] Similarly, each of the other components was produced by separate groups of artisans – sight-stampers, gun-furniture makers, bayonet-forgers, ramrod-forgers, etc.

A so-called master gunmaker typically possessed neither factory nor workshop. Often he possessed a warehouse, and arranged to acquire components from specialized handicraftsmen who may have worked at home, in garrets or sheds attached to the home, or in more sizeable workshop establishments. The co-ordination of these separate activities in a manner which assured continuity of operation was a formidable task and, as might be expected, led to interruptions whenever the supply of a particular component was disrupted.

When the components had been acquired from the material makers the gun was gradually and painstakingly constructed by a sequence of 'setters-up,' each of whom had some precise responsibility in assembling and finishing the piece.

> To name only a few, there were those who prepared the front sight and lump end of the barrels; the jiggers, who attended to the breech end; the stockers, who let in the barrel and lock and shaped the stock; the barrel-strippers, who prepared the gun for rifling and proof; the hardeners, polishers, borers and riflers, engravers, browners, and finally the lock-freers, who adjusted the working parts. Some of these were individual outworkers, employed by a particular master; others were shop owners working for several employers. It was not uncommon, moreover, for a craftsman who had no work-place of his own to hire shop-room, a bench, a vice, and a gas-jet from one of these workshop proprietors, and to work quite independently for some gun-maker.[3]

Perhaps the most effective picture of the overall operation of the trade is provided in an estimate (see below) which was made, during the mid-1850s, of the number and occupational distribution of workers in the Birmingham trade.

This estimate illustrates the extent of the subdivision of labour; the numbers also give some indication of the relative importance of the separate activities. The extensive nature of the setting-up operations is obvious – since well over half of the workers are engaged there. This reflects not only the fact that many of the materials arrive at the setters-up in a semi-finished state; more important is the fact that a gun is a

[1] Goodman states that 'No barrels are made in England, except in Birmingham, and its immediate neighbourhood.' Goodman op. cit. 388.
[2] Ibid. 390.
[3] G.C. Allen op. cit. 117.

Classified list of workmen employed in the gun trade, with the estimated number in each branch[1]

Material Makers	
stock makers	100
barrel welders; borers; grinders; filers and breechers; rib makers; breech forgers and stampers	700
lock forgers; machiners; filers	1,200
furniture forgers and casters; filers	100
rod forgers; grinders and polishers; finishers	100
bayonet forgers; socket stampers; ring stampers; grinders and polishers; machiners; hardeners; filers	500
band forgers and stampers; machiners; filers; pin makers	300
sight stampers; machiners; jointers; filers	300
trigger boxers	20
oddwork makers	100
	3,420

[1] Goodman op. cit. 392-3. Goodman refers to the inquiry from which these estimates were derived as having taken place 'about ten years ago'. Since Goodman was writing in September 1865 this would place the time of the inquiry at about 1855.

Setters-up

machiners–prepare the front sight and the lump end of the barrel for the nipple	50
jiggers–lump filers and break-off fitters–prepare the breech end of the barrel	200
stockers–let in the barrel and locks, and roughly shape the stock	1,000
percussioners–finish the nipple seat, put in the nipple, and adjust the hammer to the nipple	200
screwers–let in the furniture, and all the remaining pins and screws	1,000
strippers–prepare the gun for rifling and proof	20
barrel borers	50
barrel riflers	50
sighters and sight adjusters	50
smoothers–prepare the barrel for browning	100
finishers–distribute the several parts to browner, polisher, maker-off, and barrel smoother, and when they are returned put the guns together, and finally adjust the several parts; makers-off–file the stocks to give them their proper finish; glass paper and oil them	1,000
polishers–lock and furniture	50
engravers (lock, etc.)	50
browners (barrel)	50
lock freers–finally adjust the working parts of the lock	50
	3,920

summary–material makers	3,420	
setters-up	3,920	
	Total	7,340

delicate and complex mechanism, the separate parts of which must be carefully adjusted to one another if it is to function efficiently. In the absence of a high degree of precision and standardization which would assure such adjustment, the only alternative is skilled and sensitive workmanship lavishly employed to assure appropriate interworking of parts at all stages. Hence the ubiquitous file, the craftsman's 'precision instrument'. A significant fraction of the total cost of a firearm produced in this manner consisted in the patient and painstaking use of the file to adjust the separate pieces to one another.

A further important point illuminated by this classification was the portion of the labour force involved with the gun-stock. The gunstock in fact created one of the most serious bottlenecks of the handicraft method of firearms production. Its irregular form seemed to defy mechanical assistance, and the hand-shaping of the stock was a tedious operation. Furthermore, the fitting and recessing of the stock so that it would properly accommodate the lock and barrel, and the appropriate arrangement of the pins and screws were extremely time-consuming processes, the proper performance of which required considerable experience.[1]

[1] James Gunner, superintendent of the Royal manufactory of small arms at Enfield, appeared as a witness before the Small Arms Committee. Enfield

V

Although the British Government had had frequent cause for dissatisfaction with the gun trade over the manner in which its contracts were fulfilled, it never undertook the production of firearms in public establishments on a scale which offered a serious alternative to the gun trade in time of emergency. Opposition to such a policy was always powerful, especially, of course, among the members of and spokesmen for the trade. Although public commitment to the principle of reliance upon private enterprise was very strong, there was nevertheless ample precedent for government establishments involving products of strategic military importance. This was particularly the case where there were special problems of quality control, and where the competitive process involved a serious danger of quality deterioration. For example, gun-powder was produced at a government establishment at Waltham Abbey as early as 1790, and during the French war public gunpowder establishments were also operated at Faversham and Ballincollig.

> Were the past history of powder contracts, previous to the existence of a Government manufactory, to be fully investigated, many unjust attempts to defraud the Government, not only as regards the quality but also the price of powder, would be found to have been made. It was the extent to which these attempts were carried which determined the Government, about the year 1790, to have an establishment of their own, and after a few years so much was the quality of the manufactured article improved that the charge of a cannon was reduced from one half of the weight of the shot to one third.[1]

---

had produced a small quantity of firearms, not exceeding 6,000 in one year, according to Mr Gunner's testimony. The following is an extract from his testimony: 'Q. 4911. How many stocks can a man make in a day at Enfield?–. . . a man will stock from eight to ten guns a week; that is, he will take the stock in the rough state, and let the barrel in, and the lock, and round the gun up, and put the bands in about eight or ten a week, the present style of gun: it is not finished then; it is then screwed up, and the mounting is then let in by another man; it is distributed amongst a number of hands with the view of facilitating the operation, and that one man should not be employed using many tools; that there should be no confusion of tools.' *Report from the Select Committee on Small Arms*, op. cit. The productivity of the English stocker would seem to have been of the same magnitude as that of his American counterpart before the introduction of stocking machinery. 'Before the introduction of Blanchard's machinery one skilled man was capable of making 1 or 2 stocks in a day, much of the work being in the fitting of the metal parts.' Charles H. Fitch 'Report on the manufactures of interchangeable mechanism' *Tenth Census of the United States* II (1880) 14.

[1] John Anderson *General Statement of the Past and Present Condition of the Several Manufacturing Branches of the War Department* (London, H.M.S.O., 1857) 31-2.

Introduction

In addition to the gunpowder establishments, the vast Royal Arsenal at Woolwich produced a wide range of the heavier machinery of war. Its gun factories turned out cannon; its carriage department engaged in the construction of gun-carriages, transport wagons, and other means and appliances for transport; and the laboratories produced shot, shell, cartridges, barrels, and cases for the conveyance of ammunition, provisions, hospital and commissariat stores, etc. In addition, naval dockyards were institutions of long and respected standing.

One strongly partisan view expressed the need for these publicly-owned establishments as follows:

> It is the *first* Duty of every Department entrusted with the details, to see that our Fleets and Armies be equipped at every point in the *most perfect manner*.–In all the essential parts this has been tried by competition in Private hands, and failed:–1st: Our ships of war, when built by Contract were notoriously unsound!–The Navy Board were obliged to increase the number of Publick Dock Yards.–2nd: Our Gunpowder made by Private hands would not reach our enemies!–The ordnance Department established their own Powder Mills.–3rd: The Carriages of our Field and Battering Guns when made by Private Carpenters were disgraceful!–The Royal Carriage Department was instituted.–4th: The arms of our Soldiers, made by Birmingham Contractors were as proverbially 'bad as a Brummagem Halfpenny' and even of these the supply was deficient!–The Royal manufactory of arms was in consequence established.–
>
> These several Institutions have arisen and increased out of pure necessity:–The Government has positively been driven into the measures, and what are the results?–
>
> Our *Ships*, our *Powder*, our Artillery, our Arms, are acknowledged even by our enemies to be superior to all the world.–
>
> *That System is good which works well!*–[1]

[1] The views are those of George Lovell, who was for many years the government's Inspector of Small Arms, and at the time that he wrote these comments (1830) Storekeeper of Small Arms at Enfield. They appear on p. 158 of his extensive marginal comments on a copy of *Observations on the Manufacture of Fire-Arms for Military Purposes* (London 1829) which is in the British Museum. The author of this extensive pamphlet, which was published anonymously, is identified by the *Birmingham Journal* of 10 April 1830 as a 'Mr Parsons' of Birmingham. The British Museum catalogue identifies the author with an entry consisting only of the same surname. Although Mr Lovell states in his marginalia at the beginning of the pamphlet that 'This pamphlet is written by two Birmingham gun contractors', he quotes a letter from a friend on a later page stating that 'The author is a Wm. Parsons, who is the armourer at the Birmingham Proof-House . . .'. Both the pamphlet and Mr Lovell's comments, in spite of their obvious respective biases, provide much useful information on the industry for the first three decades of the nineteenth century.

Nevertheless, before the 1850s, the government had made only two rather fitful efforts to provide its own firearms. The possibility of a public manufactory was considered in 1804 as a result of dissatisfaction with the high prices requested by Birmingham manufacturers. In 1806 Parliament approved an expenditure of £15,000 for the erection of an establishment at Lewisham to produce locks and barrels. The appropriation was raised to £22,000 in 1807 and in 1808 manufacture began, but output was never more than a very modest proportion of that of the Birmingham trade.[1] Production of locks and barrels at Lewisham was discontinued after a few years, in part at least because of a shortage of water power for the barrel mills.[2] In 1813 the Board of Ordnance began the erection of a manufactory for the production of locks and barrels at Enfield, a few miles north of London, in Middlesex, and near Waltham Abbey. The facilities at Enfield did not come into use, however, until the end of 1815, by which time hostilities had ceased. It was thereupon decided to use the new facilities primarily as a depot for the storage and repair of arms.

The government acquired its firearms through an elaborate contracting system with the gun trade. Having established the number of firearms of a particular model it wished to acquire, it contracted separately with each of the materials suppliers – of locks, barrels, ramrods, bayonets, etc. These components, when eventually delivered, were examined by government inspectors. The government then contracted separately with the setters-up, who received the materials which had passed government inspection, and undertook the processes of fitting, stocking, finishing, etc. The system was a cumbrous one, subject to delays and uncertainty. The government normally made some attempt to maintain a smooth flow of production by stockpiling materials.

The logic of these arrangements was that it made possible a form of quality control. The government was able to inspect all the components of the firearm and therefore to ensure that inferior materials or sloppy workmanship were not used. The workmanship of the setters-up was also subjected to inspection when the finished firearm was delivered to the government. This system had been retained after its introduction during the French wars when the government became seriously concerned over the use of inferior materials. Up to that time, the Ordnance Department had contracted for the delivery of complete firearms with persons who in turn made their own arrangements for materials with the trade.[3]

---

[1] Editorial, 'Birmingham and the war' *Birmingham Journal* (25 February 1854).

[2] Lovell, op. cit. 198. The building and machinery at Lewisham were sold by auction in 1819. 'Birmingham and the war' op. cit.

[3] 'Till some years after the breaking out of the first revolutionary war with France, it was the practice of the Ordnance department, when the government required a supply of arms, to engage with one or more individuals for the number required: this contractor then made his engagements with the persons employed in the trade for the number so required, – the Board of

The deficiencies of this system became strikingly apparent early in the 1850s when the government not only attempted to obtain firearms rapidly in view of an impending crisis but required delivery of a new model with which the gun trade had had no previous experience. For the government had decided to introduce the Minié rifle–a complex and highly finished instrument–into the British service. The Board of Ordnance took steps to procure a supply of Minié rifles in May 1851–the same month as the opening of the Crystal Palace Exhibition. The extraordinary delays in acquiring these new firearms constituted a decisive turning point. For it was as a direct result of delays which were regarded as intolerable that (1) the Board of Ordnance decided to send a commission to the United States, (2) the Small Arms Committee was formed and held extensive hearings on the state of the arms trade, and (3) the Enfield Armoury was established as a major producer of firearms employing machine technology.

Having adopted the new pattern rifle musket in May 1851, the Board of Ordnance obtained tenders for the requisite materials for 23,000 firearms and contracted for these in June.[1] Since most of the parts were of a new design, the government was not able to draw upon its stores, and as a result did not accumulate sufficient materials to begin the setting-up of the muskets until December. At that point the government called for tenders for setting-up and eventually engaged in contracts for that purpose. Numerous delays occurred in the setting-up process and the muskets were not delivered until November 1853. Some two and one-half years had therefore elapsed between the decision to acquire these arms and their delivery. Roughly similar experiences were encountered with a new pattern rifle carbine which the artillery had adopted in January 1853, and with the modified, smaller bore Minié which was introduced in 1853 to replace the pattern of 1851.[2] The latter instance,

---

Ordnance sending down from the Tower viewers to inspect them. The barrels of these arms were either sent up to the Tower of London to be proved, or else they were proved in the private proof-houses of the manufacturer in Birmingham, under the superintendence of the inspector. About the year 1798, the Ordnance department purchased some land in Birmingham, and erected a proof-house and view-rooms, for the purpose of proving the barrels and inspecting the arms. And about this time the Ordnance department entered into engagements directly with the individuals employed in the manufacturing of the different materials–as the barrel, lock, bayonet, &c. and likewise with the gun-maker, to set them up complete.' *Observations on the Manufacture of Fire-Arms*, 6.

[1] *Report of Small Arms Committee* op. cit. Q. 3082.

[2] Ibid. Q. 3082, Q. 3083, and 'Memorandum relative to small arms', 451-2. Whitworth provided the following useful summary of the adoption of the Enfield rifle:

'From the beginning of the present century the British army was armed with a smooth-bore musket, to which the percussion lock was applied in 1842. So inaccurate was the firing of this weapon that at 200 yards it was found impossible to hit a target 11 ft. 6 in. wide with more than 50 per cent of the shots. In 1852 Viscount Hardinge, then Master General of the

moreover, provided strong evidence of collusion among the parties tendering for contracts.

> The offers received were so unsatisfactory as to price, and evinced so perfect a combination amongst the parties, that they were, after some correspondence, declined; and one of the Board officers, Sir T. Hastings, proceeded to Birmingham, and effected agreements with the parties on better terms; but they have not kept their engagements as to time, assigning as the cause the difficulty of obtaining workmen, and a strike for higher wages. The consequence is, that, up to the present time, the Board have not been able to commence the setting up of the muskets; and though they have made a contract for that purpose, it is uncertain, even if the materials should now come in with regularity, when it will be carried out, from the difficulties which the contractors may again encounter from their workmen.[1]

The causes of this breakdown in the delivery of firearms were numerous. Demand was erratic and unstable, and it tended to be treated by members of the industry–both masters and workmen–as highly inelastic in time of national emergency. Therefore the government's stated intention to contract for the purchase of firearms in substantial quantity was taken as a signal to renegotiate all contractual relationships in the industry.

Ordnance, invited some of the principal gunmakers of England to submit pattern muskets for the use of the army in the hope of obtaining a more efficient arm for the service. All the arms submitted were rifles, and experiments for the purpose of testing their merits were carried on at Enfield in the summer of 1852. At the conclusion of the experiments two rifles were made at the Royal Manufactory at Enfield, in which were embodied the improvements and alterations suggested by the experience obtained during the course of the trials with the experimental arms; and a bullet was adapted to these arms by Mr Pritchett, a gunmaker of London. The shooting of these rifles, at distances up to 800 yards, was reported by a committee to be as accurate as that of any that had then been tried.

'The weapon thus produced, known as the Enfield rifle, 1853 pattern, was adopted for the English army in 1855. The length of the barrel of this rifle, was 39 inches, the bore being .577 of an inch in diameter. The rifling was effected by three spiral grooves, making 1 turn in 78 inches, so that the bullet rotated half a turn during its passage along the barrel. The length of the bullet was 1.81 diameters of the bore.' Sir Joseph Whitworth *Papers on Mechanical Subjects* II (London n.d.) 5-6. Cf. also W. Y. Carman *A History of Firearms* (New York 1955) 111-13.

[1] *Report of Small Arms Committee*, op. cit. 452. Collusion among government contractors was a frequent charge in the history of the industry. In 1842, for example, George Lovell, Inspector of small arms, wrote that the gun makers working for Ordnance had a complete organization, including a secretary. 'They determine the prices that shall be given to the workmen, and the price that shall be paid for the arms by the Ordnance, and no one can undertake to work at an under-price so long as he belongs to that body.' Ibid. 453.

In addition the short-run supply schedule for military firearms was certainly inelastic–output in the short-run could be increased substantially only by (1) drawing highly skilled workers away from other pursuits by the inducement of higher wages, or (2) by employing workers of inferior skills, especially those in the cheap sector of the gun trade. Both sources of additional labour tended to raise the price of firearms, and the latter tended to reduce their quality as well. The situation in this case was further complicated by the introduction of a new firearm of very high quality, with which the trade had had no previous experience;[1] furthermore, since it was a new firearm, there were no government stores of component parts upon which it was possible to draw.

The years 1851–4 were punctuated by workers in the industry striking, or threatening to strike, for higher wages, and by complaints of organized conspiracies and price agreements of military gun-makers against the government. Contractors systematically failed to fulfil their contracts and complained bitterly of the unavailability and intractability of their labour force and of the excessively rigid standards imposed by government inspectors.[2] The reaction to these events was, perhaps, characteristic: the makers organized a trade association and the government appointed a Select Committee. The Birmingham Small Arms Trade was formed in 1854 for the avowed purpose of regulating prices and wages in the industry. Goodman states that 'It originally consisted of the twenty firms who, in 1854, were selected by the Government to supply the arms required on the breaking out of the Crimean War. In consequence of the sudden increase in the demand, when the orders were first given, great difficulty arose in arranging (*sic*) workmen's prices, and still further difficulty owing to the competition among the masters for the workmen skilled in the military branch. To remedy these evils, the Association was formed . . . As among its other functions, it regulates the prices to be paid to workmen, it also fixes selling prices.'[3]

The government, in its frustration, appointed a Select Committee on Small Arms with instructions '. . . to consider the Cheapest, most Expeditious, and most Efficient Mode of providing SMALL ARMS for HER MAJESTY'S SERVICE'. The Committee conducted its hearings be-

[1] Apparently the sights on the Minié rifle, which were new and complicated, had been put on wrong. '. . . the back sight had been put on without reference to the fore-sight'. In some cases at least, '. . . the back sight was on the right-hand side, and the fore-sight on the left . . .'. What is perhaps even more remarkable, however, is the evidence of a well-known Birmingham gunmaker that such rifles passed government inspection! *Report of Small Arms Committee*, op. cit. Q. 7666–Q. 7673.

[2] See ibid. 462-8, which includes extracts of letters from materials contractors stating the causes for the delay in completing their contracts. Among the reasons cited are (1) difficulty in procuring workmen, (2) refusal of workmen to make locks at agreed price, (3) continued strikes of workmen, (4) severe illness of key workmen, (5) vexatiousness of government inspection, (6) machinery broken down, (7) frost and lack of coal, and (8) Christmas holidays.

[3] Goodman, op. cit. 430.

tween 10 March and 7 April 1854 and submitted its report on 12 May 1854.

VI

By the time the Committee held its hearings, the possibility of establishing a public manufactory for the production of firearms by machinery had been under discussion for some months. Indeed, it was a major purpose of the Committee to consider the feasibility of such a project. As a result of the failure of the trade to fulfil its contracts, the Board of Ordnance actively undertook to examine the possibilities of machine production. In October 1853 John Anderson was instructed to consider the prospects for producing bayonets (with which great difficulties were then being experienced) by machinery at Enfield. He reported favourably.[1] As the situation deteriorated further, he was asked in December to examine the possibilities of producing the entire musket by machinery. On 14 December the Board of Ordnance sent a letter to Mr Anderson's superior at Woolwich stating that: 'The time is approaching when it should be considered whether it would not be advisable to have the means of manufacturing small arms at Enfield to such an extent as to render the Ordnance in a measure independent of the contractors for the supply. The Master-General and Board are aware that Mr Anderson's time and attention are fully occupied at this moment, but when he has leisure, they would be glad if he would apply his mind to the consideration of what would be required to establish a manufactory of muskets by machinery at Enfield.'[2]

Anderson produced his report on 23 December. In his own words:

> In December he was again called upon for a report on the manufacture of muskets by machinery. On the 23d December a report on that important question was submitted, in which the writer stated, that being entirely without data, he begged to offer his remarks with caution; that he considered the musket eminently adapted to be made by machinery, and that a suitable plant for the purpose, when organized and in full operation, would repay the necessary outlay in two years; that during the first year little good would ensue, as all the tools and gauges would have to be made, and all the hands would be fresh and untaught in their special duty; that looking at each of the 57 parts of a musket as separate articles, each would have to become a separate study, and that for the manufacture of each it would be necessary to construct a series of simple machines each performing a small part, and imparting thorough identity to each article; that this arrangement of manufacture would enable the simple and special machinery to produce the articles accurately and cheaply by the employment

---

[1] Anderson, op. cit. 26.

[2] *Statement of Services Performed by John Anderson, Superintendent of Machinery to the War Department, from the year 1842 up to the present time* (London 1873) 28-9.

> of unskilled labour; that the required machinery did not exist in England at that time; that to employ the machinery used by engineers would ensure certain failure; that there existed abundance of talent in England to produce the required machinery; but that as much attention had been paid to this subject in the United States, and in order that we might be enabled to begin at the point where the American inventors had left off, it would be advisable to obtain a simple set of their machinery adapted to the stock and lock of our musket.[1]

The Board of Ordnance shortly thereafter submitted to Parliament an estimate of £150,000 as the cost of a public firearms manufactory. Parliament, in turn, established the Select Committee on Small Arms to investigate the matter further. The Committee discharged its obligations by conducting exhaustive examinations of a wide range of highly qualified engineers, contractors, and workmen. Anderson was himself a key witness. He was examined at length and described his proposals for the small arms factory. The Committee interviewed Joseph Whitworth and George Wallis, particularly with respect to their observations on America, since their recently published Special Reports had spoken in glowing terms of American manufacturing methods and had made specific reference to the use of machinery in firearms production. The eminent engineer James Nasmyth, inventor of the steam hammer, was also examined at length. But perhaps the most important witness of all was Colonel Sam Colt, although the impact which he had upon the Committee was exerted less through what he actually said and more through the presence of his own pistol factory in London. Several of England's most respected engineers who had visited Colt's plant spoke in the highest possible terms of Colt's machinery and urged the Board of Ordnance to consider the adoption of American methods. A portion of Nasmyth's testimony, for example, ran as follows:

> Q. 1366. Have you ever visited Colonel Colt's manufactory?–I have.
> Q. 1367. What effect did that produce upon your mind?–It produced a very impressive effect, such as I shall never forget. The first impression was to humble me very considerably. I was in a manner introduced to such a masterly extension of what I knew to be correct principles, but extended in so masterly and wholesale a manner, as made me feel that we were very far behind in carrying out what we know to be good principles. Most of the intelligent mechanics of this country are well acquainted with what are the correct principles of carrying out the details of manufactures; but there is a certain degree of timidity resulting from traditional notions, and attachment to old systems, even among the most talented persons, that they keep us considerably behind. What struck me at Colonel Colt's was, that the acquaintance with correct

[1] Anderson, op. cit. 26.

principles had been carried out in a fearless and masterly manner, and they had been pushed to their full extent; and the result was the attainment of perfection and economy such as I had never met with before; one of the important results of this was the employment of unskilled labour as the agent for managing the machines. That last condition was one of the most impressive of all, and certainly the most encouraging result, to show that if we took the right means we do not require to depend upon what are reputed to be the regularly-bred mechanics in the particular line, or what we call skilled tradesmen. . . .

Q. 1441. Would it be advisable, do you think, to get some of the machines that are used in America, in order to see the state to which they have brought musket-making machines there?–I think, from what I have seen at Colonel Colt's, that their system of tool-making, and the manner in which those tools are contrived for the performance of a given purpose, entitle the American tools and machinery to receive our most attentive study and consideration. I did not inquire the prices of these machines. In those American tools there is a common-sense way of going to the point at once, that I was quite struck with; there is great simplicity, almost a quaker-like rigidity of form, given to the machinery; no ornamentation, no rubbing away of corners, or polishing; but the precise, accurate, and correct results. It was that which gratified me so much at Colonel Colt's, to see the spirit that pervaded the machines; they really had a very decided and peculiar character of judicious contrivance.[1]

Anderson, who bore primary responsibility for the Board of Ordnance's proposals, and who described the projected small arms factory in detail,[2] also paid homage to Colt's methods. He had visited the Colt establishment in December 1853.

Having previously heard of American machinery for small arms, I went to Colonel Colt's factory with high expectations, and in the hope of carrying away some of their mechanical notions that might be applicable in our own service, and I did not leave with disappointment; but the strange contrivances were so numerous, and many rather intricate, that, although deeply interested, yet of many ingenious applications I retain but an indistinct remembrance . . . So far as an old building would admit of, the work in this manufactory is reduced to an almost perfect system; a pistol being composed of a certain number of distinct pieces, each piece is produced in proportionate quantity by machinery, and as each piece when finished is the result of a number of operations (some of 20 or 30), and each operation being performed by a special machine made on purpose, many of these machines requiring hardly any skill from the attendant beyond knowing how to fasten

---

[1] *Report of Small Arms Committee*, op. cit.

[2] Ibid. Q. 345-91.

> and unfasten the article, the setting and adjusting of the machine being performed by skilled workmen; but when once the machine is properly set it will produce thousands. Hence there are more than a hundred machines employed, many of them similar in principle to each other, although differing in the form of the cutting instruments. There is also much that is new in England, and abundant evidence of a vigorous straining after a large and accurate result, which is well fitted to inspire us all with healthy ideas; indeed it is impossible to go through that work without coming away a better engineer; or if our gun-makers were acquainted with and adopted similar tools, it would rouse the latent energies of that trade into life and vigour.[1]

A major thrust among those hostile to the establishment of a government manufactory was that none of the witnesses, however competent, possessed the joint qualifications of intimate familiarity with the gun trade plus a detailed knowledge of machine technology. Indeed, Whitworth and Wallis were virtually the only British witnesses who had actually seen the manufacture of guns by machinery in America, and they had not paid special attention to the subject on their American trip. The standard tactic of the spokesmen for the gun trade was to attempt to discredit each witness who asserted that machine technology could be applied to gunmaking–Anderson, Whitworth, Nasmyth, etc.–by establishing that he knew little in particular about either guns or gunmaking. The typical reply, often delivered with growing signs of impatience, was that there was no special mystery about guns, and that a general knowledge of the principles of engineering and machine methods were sufficient qualifications to speak with authority of their application to gunmaking. The hearings provide many examples of a direct confrontation between the old handicraft conception and the emerging conceptions of the production engineer. When Nasmyth, for example, was asked how, in view of his professed ignorance of gunmaking, he could speak authoritatively on the subject of applying machinery to their production, he replied rather tartly: 'A gun is a piece of mechanism; there is nothing very peculiar about the parts of a gun, and the knowledge that enables one to make a steam-engine should enable one to form an opinion upon it.'[2]

One of the sources of contention in the examination of witnesses was the interchangeability of firearms produced by American machine methods; hardly anyone questioned interchangeability as a desideratum. It was generally recognized that it was of critical value under battlefield conditions, by enormously simplifying the repair or reconstruction of damaged weapons. Interchangeability freed the armed forces of their heavy dependence upon skilled armourers and thereby reduced a serious military supply problem to one of much more easily manageable proportions. A musket with a broken lock had to be set aside until the

---

[1] *Report of Small Arms Committee*, op. cit. Q. 341. [2] Ibid. Q. 1661.

services of an armourer were available, at which time it would either be repaired or a new part would have to be specially made for it and fitted properly into place. Some conception of the magnitude of the problem is conveyed by Roe's observation that 'Even in England where skilled workmen were most available, there were not enough armourers to meet the demand, and in 1811 the government had on hand 200,000 musket barrels which were useless for want of men to make or repair the locks.'[1]

Some witnesses claimed that Colt's pistols were not interchangeable in the sense that if a number of parts were thrown into a basket, they could be made up indiscriminately and with no selection. Indeed, one of Colt's former employees testified that separate parts of the gun were punch-marked and that, at the fitting stage, pieces of similar number were assembled together.[2] However, Colt himself did not claim complete interchangeability.[3] Whitworth, when questioned on the subject, went to some pains to establish that interchangeability is not an 'either . . . or' proposition but rather something which admits of degrees, and that the difference in degree between the American machine-made firearm and the British handicraft product was decisive both in terms of its military advantages and economic efficiency.[4] In his Special Report he states that

---

[1] Joseph W. Roe 'Interchangeable manufacture' *Transactions of the Newcomen Society* XVII 165.

[2] *Report of Small Arms Committee* op. cit. Q. 7447–Q. 7449. Cf. also Q. 3975 –Q. 3978.

[3] 'Q. 1107. Could you state what has been the largest number of arms that you ever made in a year in America by machinery?–About 50,000. But we have simplified and improved our machinery, and there is more capital in machinery now than there was then.

'Q. 1116. When you made those 50,000 in a year, could any one part of one gun, for instance, be adapted to another gun; were they so accurately made that you might assemble the parts together, the locks and other parts, indiscriminately?–I should say that they would do that a great deal better than any arms made by hand . . . It sometimes happens, in hardening a piece of metal, that it will change its position, and it requires the delicacy of touch of a polishing machine. If you break 100 machine arms, the barrel of one, the cylinder of another, the main spring of another, etc. etc., break each in its opposite parts, till you break the whole, say some 20 parts, you will then have lost one arm; but you will get of the 19 probably 18 arms that will interchange. I do not think I could now do it exactly to that extent. Sometimes there would be a little spring of the frame in hardening, a little one way sometimes, and sometimes a little the other; but each being made alike, why should they not come together?' *Ibid.* See also Q. 1120 and Q. 1121. A description of Colt's immense Hartford, Connecticut, armoury, published in the *United States Magazine* in March 1857 and reprinted in part in Charles Haven and Frank Belden, op. cit., states (pp. 356-8) that the separate components of the revolver were numbered and later assembled by reference to this numbering.

[4] 'Q. 2121. With regard to Colonel Colt's pistols; do you think that, made as they are now, if you threw a quantity of parts into a basket, they would all make up together without any selection?–I do not think that they would; nor do I think it is possible in an American musket, although there is such

at the Springfield Armory 'The complete musket is made (by putting together the separate parts) in 3 minutes. All these parts are so exactly alike that any single part will, in its place, fit any musket' (p. 365). It should be noted, however, that Whitworth, as well as other engineers, including Nasmyth, did not pay particularly close attention to prices of machinery and labour. Under close questioning he occasionally indicated that he did not know the prices of the machinery whose *economic* efficiency he had extolled. Whitworth may sometimes have implicitly equated engineering efficiency with economic efficiency; the extraordinary success of his own engineering firm, however, suggests that he was probably not simple-minded on the subject. It is, nevertheless, a matter of more than passing interest that one of the most notable achievements in woodworking machinery in England had been devised for decorative purposes, for executing the wood carvings in the Houses of Parliament. Whitworth himself had built these machines.

> Q. 1978. Could machinery similar to that which you saw at Springfield be applied to the stocks of our military muskets?–Undoubtedly.
> Q. 1979. Could you form an approximate estimate of the value of a single set of such machinery?–I scarcely could, but it would not be less than £5,000, I should say.
> Q. 1980. For making stocks?–For making the machinery that made these stocks that I saw; it might be more; it is merely a rough guess.
> Q. 1981. Were you the maker of Mr. Jordan's machinery, by which the carving for the Houses of Parliament was executed?–Yes.

---

care taken, that the parts could be taken indiscriminately and put together without considerable trouble; for this reason, that one stock is of softer wood than another, and it will warp differently; and, therefore, though it may be true today, in a year's time you will have one stock vary in one direction, and another in another.

'Q. 2122. Suppose you took two stocks cut with the American machinery, and two barrels of precisely the same size, one stock being of hard wood and the other of soft, when the barrels come to be stocked up, will they fit?–They could not fit the same.

'Q. 2123. If one fits, the other cannot?–They both might be fits, but the one would fit better than the other.

'Q. 2124. I suppose you mean that all the stories about mixing up the materials, and putting them together indiscriminately, comes to this, that they would fit in a way?–I think this, that on a field of battle, if a great number of muskets were disabled, made by the American mode of manufacture, that the armourer would be able to put a greater number of parts together; whereas by the mode of manufacture in this country he could not do that.

'Q. 2125. Do you think that he could take them without any sort of selection?–No.

'Q. 2126. He would be obliged to sort them to fit them, would he not? Yes.' See also *Report of Small Arms Committee*, op. cit. Q. 2213.

Q. 1982. To what extent did that machinery cheapen the carving?–I am not aware, but I know that Mr. Jordan in some cases told me that the cost was not more than one-fourth; but what it was on the average I am not aware.
Q. 1983. Is such machinery applicable to the making of gunstocks?–It would be applicable, no doubt, for some parts of a stock, but it is not so well adapted for a musket as the machinery which I saw in America.[1]

Whitworth was examined with particular care concerning the American gunstock machinery which he had described in his Special Report. It is made clear in his testimony that what impressed him most were the American methods of producing the gunstock by machinery; he stated that he found nothing remarkably novel in the preparation of the lock or the barrel, although he generally admired the American ability to devise machines for special purposes. 'Altogether in America, you were more struck with the mode of working the wood than their mode of working the iron?–Much more; they are not equal to us in the working of iron.'[2]

In his Special Report Whitworth provided a detailed breakdown of the total labour time consumed in the machine production of a gunstock at Springfield Armory. Whitworth himself observed each operation and reported (p. 365) that all the operations in producing the gunstock involved just over 22 minutes of a man's time.

There was much scepticism in England over the possibility of forming irregular shapes in wood by the use of machinery–indeed, many witnesses were incredulous concerning reports that the Americans were shaping gunstocks by machinery. In fact the machine, Blanchard's stocking lathe, had been in use in the United States for over thirty years. Thomas Blanchard had developed his gunstock lathe as far back as 1818. It was introduced into the national armouries of the U.S. during the 1820s where it replaced the time-consuming hand techniques of shaping the gunstock by whittling, boring, and chiseling.[3] In fact, 'By 1827 Blanchard's stocking and turning machinery had been developed into 16 machines, in use at both national armouries, and for the following purposes: sawing off stock, facing stock and sawing lengthwise, turning stock, boring for barrel, turning barrel, milling bed for barrel-breech and pin, cutting bed for tank of breech-plate, boring holes for breech-plate screws, gauging for barrel, cutting for plate, forming bed for interior of lock, boring side and tang-pin holes, and turning fluted oval on breech.'[4]

---

[1] Ibid. [2] Ibid. Q. 2043.
[3] Charles Fitch 'Report on the manufactures of interchangeable mechanism' *Tenth Census of the U.S.* II (1880) 13-19. Clapham's statement, that it was 'about 1840' when 'Blanchard's lathe for cutting irregular forms had been applied to gun-stocks at the Springfield arsenal' is incorrect. J. H. Clapham op. cit. 77. Clapham apparently derived this piece of misinformation from Goodman, op. cit. 398.
[4] Charles Fitch op. cit. 14.

The copying technique embodied in the stocking machinery has been described as follows: 'A pattern and block to be turned are fitted on a common shaft, that is so hung in a frame that it is adapted to vibrate toward or away from a second shaft that carries a guide wheel opposite and pressing against the pattern, and a revolving cutter wheel of the same diameter opposite the block to be turned. During the revolution of the pattern the block is brought near to or away from the cutting wheel, reproducing exactly the form of the pattern.'[1]

Blanchard's machine was quickly applied to an assortment of items including hat blocks, handles, spokes of wheels, oars, shoe lasts, and even sculptured busts. Yet during the examination of Richard Prosser, a Birmingham civil engineer, at the Small Arms Committee hearings, that gentleman displayed the handle of an American axe which he had owned for many years; he claimed that he could convince no one that it had in fact been turned by machinery.[2]

The Small Arms Committee, in its final report, recognized the numerous uncertainties and problems to be surmounted in connection with a government factory. 'Your Committee have heard all that could be urged in favour of a Government manufactory for muskets, but they have not received evidence sufficient to satisfy them that a Government factory would afford the cheapest, most expeditious, and most efficient mode of providing muskets.'[3] They therefore recommended that the contracting system be continued and suggested some procedures for improving reliability. Most important, they recommended that the Board of Ordnance be authorized to establish its own manufactory, though on a more modest basis than the capacity of 500 firearms a day, which had been proposed, and to be located at Enfield rather than Woolwich. 'This manufactory would serve as an experiment of the advantages to be derived from the more extensive application of machinery, as a check upon the price of contractors, and as a resource in times of emergency, and it should be arranged with a view to its economical working.'[4]

The committee which had been appointed by the Board of Ordnance, before the proceedings of the Small Arms Committee, to visit the U.S., consisted of Robert Burn, Lieut.-Colonel, Royal Artillery, Thomas Picton Warlow, Captain, Royal Artillery, and John Anderson, Ordnance Inspector of Machinery. Since Anderson's testimony was vital to the proceedings of the Small Arms Committee, he and Captain Barlow were detained and were not able to embark until 15 April. Colonel Burn had left England on 25 February and was charged with making all necessary arrangements to facilitate the progress of the committee in the U.S. when the other members arrived. Barlow and Anderson arrived in Boston on 26 April, where they were met by Burn.

The committee had communicated with Whitworth to receive his advice on the establishments which it was important to visit. It was

---

[1] Dwight Goddard *Eminent Engineers* (New York, 1905) 73.

[2] *Report of Small Arms Committee*, Q. 2655–Q. 2657.

[3] Ibid. x. [4] Ibid. x.

authorized to make limited purchases of machinery or models which would be of particular use in the proposed factory. The initial authorization of £30,000 for the purchase of small arms machinery was subsequently reduced to £10,000; when it later requested that this sum be increased to £12,500 it was informed in reply that it could make any purchases which were considered desirable. Of the *Report of the Committee on the Machinery of the U.S.A.* itself, Anderson stated elsewhere that '. . . the part referring to muskets was written by Captain Warlow, R.A., and the whole of the remainder, relating to tools and machinery, was the work of the writer' (i.e., Anderson).[1]

VII

Although the Small Arms Committee had recommended that the Enfield Armoury be undertaken on a more modest scale than the Board of Ordnance had proposed, under the pressures of war these reservations were cast aside. Plans for the creation of a large-scale plant proceeded as rapidly as possible upon the return of the committee from the U.S. The plant was not completed in time to be of service to the British Army in the Crimea–just as the earlier and more modest establishment at Enfield which was completed in 1815 came too late to assist in the war against France. This unfortunate timing on the part of the government's public armouries was precisely what private industry feared, and this fear had acted as a powerful agent against the introduction of machines by private enterprise. From the point of view of the members of the gun trade, the demand for military guns was so notoriously unreliable as to involve risks which were regarded as prohibitive. In the absence of any assurance about the regularity of demand for military guns, the gunmakers were unwilling to assume the burden of fixed costs which machinery would involve.

The following occurs in the testimony of Richard Prosser to the Small Arms Committee:

> Q. 2651. If the introduction of machinery would facilitate the making of guns, why should not private gunmakers have introduced it?–There is not the slightest inducement for them to spend a penny in machinery; I would not do it.
> Q. 2652. Why not?–Because there is no certainty that they would have Government work to make that machinery remunerative.
> Q. 2653. If a manufacturer had a contract for a long time, do you think that machinery could be introduced with advantage by private gunmakers?–No question about it; if you gave them 500,000 to make a year, there would be no difficulty in finding money to erect machinery.
> Q. 2654. If you had that number of guns to make a year, would you introduce machinery?–I could not make them without.

[1] *Statement of Services Performed by John Anderson*, 35.

The existing labour-intensive handicraft system in effect operated by shifting the social cost of this instability onto the workers in the trade, who experienced a high degree of irregularity of employment, often even during periods of peak demand. In spite of the obvious difficulties, the government seems to have made little effort to organize the issuing of contracts in a way which would reduce this source of instability.[1] In reading the testimony before the Small Arms Committee, one gets a strong impression of a lack of any real effort at consultation and discussion between the military authorities and the gunmakers. Of course it is a basic complaint of the gunmakers, continually reiterated, that the government inspection system was excessively and unnecessarily strict, resulting in a rate of rejection which bore no realistic relation to the fitness of the arms rejected. One gets the feeling, also, of a highly officious 'take it or leave it' attitude on the part of the military authorities.[2] From the government's point of view, the system was intolerably unreliable and unresponsive in time of emergency, since its entry into the market during such periods set forces into motion within the trade which generated protracted delays, sharply rising prices, and inferior quality of product. The undependability and erratic work habits of many of the workers in the gun trade had long been a matter of common notoriety. Writing of the gun-lock trade of Wednesbury, a nineteenth-century observer stated:

> It would appear that the habits of lock-filers were never of the thriftiest–forethought for the future never overshadowed their enjoyment of the present. Let trade be ever so brisk, and be the demands of the masters ever so urgent, a lock-filer could not give up his enjoyment of 'Saint Monday.' Throughout the gun-trade, from the forger to finisher, this was also characteristic of every other workman besides the lock-filer. The lock-filer, however, was perhaps the most indispensable workman among the lot. Besides, it took a hard, laborious apprenticeship of at least five years to produce a skilled filer. So that it would seem that the lock-filer not only held the other branches of the gun-trade in his hands, but that upon his willingness to work, and upon the produce of his skill,

---

[1] 'Q. 3539. Do you know that very frequently men who have been employed at the Government work, have had to stand still 15 or 18 weeks together without work, in consequence of the irregularity of the demand?–I frequently hear complaints from the men that sometimes they have nothing to do and are starving, because the Government contracts are so dilatory.' Testimony of John Dent Goodman, Birmingham merchant and partner in a gun manufactory, before Small Arms Committee, op. cit. Prosser in his testimony cited the example of one of his employees, to whom he paid a wage of 35s. a week, turning down a weekly wage of £3 in the gun trade because of the uncertainty of employment prospects in that trade. (Q. 2654 and Q. 2655). Similar citations may be found throughout the testimony. Cf. Q. 5415–Q. 5417.

[2] *Report of Small Arms Committee*, op. cit. Q. 7789–Q. 7808, and pp. 449-68.

> the safety and the very existence of the nation depended. About this time these two facts–the long period it took to produce a skilled lock-filer, and the difficulty of getting government orders executed expeditiously enough when the independent workman was not willing to work–forced themselves very rudely upon the attention of the War Department. The possibly paralysed condition of the country at any sudden threat of danger, made the military authorities extremely apprehensive.[1]

Given the size and irregularity of the government's demand for firearms, much of this may have been unavoidable.

The members of the Small Arms Committee, in their report, showed a clear grasp of some of the underlying problems posed by the irregularity of demand, although they were not equally clear how they ought to be resolved.

> The result of the evidence as to the advantage of a more extensive application of machinery in the gun-trade appears to be, that wherever a large number of articles, exactly similar in pattern, are required, opportunities are afforded for the application of machinery. The expediency of introducing it must however depend on other considerations, which can only be satisfactorily settled by experience. There is, however, reason to believe, that if the gun-trade could have confidence that a large supply of muskets would be purchased by the Ordnance in future years, the manufacturers themselves would be anxious to introduce machinery wherever it could be profitably employed.
>
> The two questions,–namely, the advantage of an increased use of machinery and the expediency of making all muskets in a Government factory, are not, therefore, in any way necessarily connected. . . .
>
> Your Committee would here observe, that the complaint of the contractors, that the orders of the Government for Small Arms have not been continuous, raises a question of great importance. There is no doubt that a large demand for Small Arms, spread over a number of years, would attract to the gun trade a supply of hands sufficient to meet the demand. Even in the first year considerable additional supplies might by this means be attained, and in each successive year the increase would be augmented. On the other hand, it has been observed that it would not be judicious on the part of the Board of Ordnance to pledge themselves to large and continuous orders, for in this age of rapid invention, such a course might be attended with very inconvenient consequences. For instance, the pattern of 1853 has been substituted for that of 1851; therefore, if large orders for the pattern of 1851 existed,

[1] Frederick Wm. Hackwood *Wednesbury Workshops; or Some Account of the Industries of a Black Country Town* (Wednesbury 1889) ch. 8.

> they would be orders for a now obsolete arm. Some new pattern may soon supply the place of that which is now ordered.[1]

The most effective way of assuring a rapid supply of firearms in time of emergency was to have a 'standby' capacity of machinery. This meant the willingness to incur large overhead costs, and the exigencies of war eventually pushed the government in this direction when it undertook to construct a large government armoury instead of one of a 'limited scale' as recommended by the Small Arms Committee. It is clear that at the time the decision was made, it was based primarily upon military and strategic considerations. The costs of gun production by machinery had not been carefully explored and were certainly not understood, as was made abundantly clear during the examination of witnesses by the Small Arms Committee. Even Anderson, on whose evidence and mechanical abilities the Board of Ordnance rested its case for a public armoury, admitted this. While the Enfield Armoury was still under construction he conceded that '. . . there was a difficulty in proving what its precise effect would be, both on the mechanical question of quality, and likewise the no less important one of price, because no data could be obtained on which to form a correct judgment with regard to the manufacture, and the only information on which to frame an estimate, was derived from a dissection of each part of the musket, and an ideal enumeration of the many hundred machines, that would be required to produce it. A manufactory designed for producing 150,000 muskets per annum was approximately estimated to cost about £50,000. '[2] The extreme flimsiness of this estimate is obvious.

The building of the Enfield Armoury was subject to a variety of delays, some foreseen and some unforeseen.[3] When Anderson returned to England in September 1854 he learned for the first time of the decision to locate the armoury at Enfield rather than at the Royal Arsenal at Woolwich. Since his original plans had been organized around a particular location at Woolwich, they had to be redrawn and other arrangements made to suit the new locality. The work of preparing the new site was actually begun in the spring of 1855. The foundations of the buildings were prepared only very slowly and at considerably greater expense than anticipated since, as it turned out, they were located on a peat bog. The work was entirely suspended in the month of August due to some difficulty over the cost of the buildings. By winter the portions of the buildings which were completed were still without any heating which in turn retarded the critical process of preparing the tools. The buildings

---

[1] *Report of the Small Arms Committee*, vii, x.

[2] Anderson, op. cit. 27.

[3] With respect to the delivery of the American machinery, the Committee noted: 'A considerable quantity of it will be delivered in three months, but it will be fifteen months before the whole is finished, and after making allowance for incidental delays, we cannot calculate on having it in England before eighteen months. This delay the Committee believe to have been unavoidable.' (p. 144).

were completed in 1856, and it was then that the difficult task of preparing the machinery, installing it, and arranging all the finer points of eventual operation began in earnest. The form and dimension of each part of the rifle had to be translated into a set of specifications for the appropriate machines, with appropriate tolerances, in order to ensure interchangeability. An elaborate system of gauges had to be prepared to test the accuracy of the parts at each stage in the process of production. In all of these activities, difficulties were encountered with British labour, to which much that transpired was entirely unfamiliar. Anderson stated that 'Great difficulty has been experienced in obtaining workmen of the required ability, capable of executing these instruments. At first, recourse was had to the workmen of the engineering and machine-making trades, but a few only were found to appreciate the extreme accuracy which is required; latterly the numbers of the tool department have been recruited from among the mathematical instrument makers, who after a few months' training are brought up to the high standard which is essential to the true development of the system now adopted.'[1]

Fortunately the Committee, while in America, had had the foresight to anticipate some of the problems involved in introducing American techniques, and had taken the precaution of securing the services of Mr James H. Burton, a man of considerable experience with American small arms machinery, who had served as engineer for some years at the Harper's Ferry Armory. In addition the services of a number of skilled American workmen, familiar with the technical details of organizing firearms production around a sequence of such specialized machines, had also been secured for the Enfield Armoury. They performed indispensable services in preparing and arranging the machinery for the production of the Enfield rifle.[2] Anderson generously acknowledged the technical assistance he had received at this critical juncture. 'The working

---

[1] Anderson, op. cit. 28.

[2] Nevertheless, difficulties were encountered. Some of the initial troubles with the gunstock are described in Thomas Greenwood 'On machinery for the manufacture of gunstocks' *Proceedings of the Institution of Mechanical Engineers* (1862) 328-34. Greenwood was a prominent Leeds manufacturer of machine tools who became a major supplier of machinery to the Enfield Armoury.

Greenwood also took advantage of the opportunity to claim the Blanchard lathe as of English origin. 'The lathe for turning the butt end of the stock was one of the ordinary Blanchard copying lathes, the invention of which, though generally supposed to belong to America, really belonged to South Staffordshire, the lathe having he believed been originally invented by a gentleman named Rigg, living not far from Birmingham, for the purpose of turning shoe lasts; the invention was afterwards taken out to America, whence it returned again to this country under its present name of the Blanchard lathe, and was now employed very extensively in large numbers of manufactures' (p. 340). There is no evidence whatever on the American side to support this contention. It is repeated elsewhere, e.g., in D. K. Clark *The Exhibited Machinery of 1862*, 221.

out of these fixings, instruments, and other tools has devolved upon Mr Burton, and has been carried out with a degree of artistic refinement and precision which it would be difficult to excel.'[1]

VIII

Anderson was determined to establish high standards of interchangeability at Enfield comparable to what he had seen in the American armouries. The Committee's Report indicated perfect satisfaction with the ability to interchange parts in the American musket. The Committee had conducted the experiment of withdrawing ten muskets from the arsenal at Springfield, each made in a different year between 1844 and 1853, and had watched the workman take each one to pieces. The pieces were then mixed up. The Committee '. . . then requested the workman, whose duty it is to 'assemble' the arms, to put them together, which he did–the Committee handing him the parts, taken at hazard,–with the use of a turnscrew only, and as quickly as though they had been English muskets, whose parts had carefully been kept separate.'[2]

This high degree of standardization and interchangeability was observed by the Committee in the actual production of the muskets themselves. The use of the file to achieve a proper fit was required only

---

[1] Anderson, op. cit. 29. At the same time as this document was printed–January 1857–Anderson read a paper to the Society of Arts, 'On the Application of Machinery in the War Department,' describing the recent technical innovations for that select audience. He stated there that '. . . in order to insure perfect success, the details are being carried out by an American gentleman, brought over by the government, who possesses a thorough and practical experience in the working of this system in the United States, and who has the assistance of several of his own countrymen, from the small-arms factories of New England'. *Journal of the Society of Arts* (30 January 1857) 156. Much of the American assistance was in connection with the stocking machinery. After describing, with a considerable display of 'national and local pride', the set of stocking machinery which the English had ordered for their new armoury, the *Springfield* (Massathusetts) *Republican* stated that 'Mr Oramel Clarke, one of the best workmen in the stocking department of the armoury, has been employed to go out to Europe, and take the charge of the machinery and its operation'. (As quoted in *The Engineer* (11 January 1856) 20.) It is of interest to note that, earlier in the century, there was a large number of British workers employed at the two American armouries. Lovell (op. cit. 31-2) states that, on the basis of a document forwarded to him in May 1826 by Sir H. Hardinge, there were over fifty British artificers employed at Springfield Armory. In addition he lists the names of 21 British artificers employed at Harper's Ferry, 12 of whom were lock filers, 5 barrel forgers, 2 barrel filers, 1 lock forger and 1 gun stocker. Apparently the American armouries were heavily dependent upon foreign sources of skilled labour. Fitch states that, in 1819, most of the filers at the national armouries were foreigners. Fitch op. cit. 5.

[2] p. 122. Some interesting material on the Springfield Armory may be found in Jacob Abbott 'The Armory at Springfield' *Harper's New Monthly Magazine* v, no. 26 (July 1852) 145-61.

rarely. The extreme ease with which the muskets were finally assembled from quantities of component parts is carefully noted.

> The workman whose business it is to 'assemble', or set up the arms, takes the different parts promiscuously from a row of boxes, and uses nothing but the turnscrew to put the musket together, excepting on the slott, which contains the band-springs, which have to be squared out at one end with a small chisel. He receives four cents per musket, and has put together as many as 100 in a day and 530 in a week, but his usual day's work is from 50 to 60.
> The time is 3½ minutes.[1]

The discomfort obviously experienced by the Committee in the use of the term 'assemble', which it always encloses in quotation marks, is a matter of considerable interest. For it in fact highlights one of the fundamental differences between the old and the emergent systems of production. The notion of assembling a firearm, with the use of 'nothing but the turnscrew', was completely strange to men who had had previous experience only with craft methods of gun production. Under this system, as we saw earlier, more than half of the workmen employed in the Birmingham gun trade in the mid 1850s were setters-up as opposed to material makers. To dispense with their skills, to put together a finished firearm with the insolent ease implied by the need to employ only a turnscrew, was indeed revolutionary. Much of the cost-reducing nature of the new machine technology in fact lay in this point. The economies made possible by the use of specialized machinery when its costs can be distributed over a very large volume of standardized output are generally obvious and well-recognized. Somewhat less obvious is the point at issue here: when the specialized machinery can produce the component parts of a complex product with a high degree of precision, considerable labour-saving is achieved not just in the production of these components but in their eventual assembly as well. The importance of this basic point is perhaps insufficiently appreciated today because we are relatively far removed in time from a society where craft skills of this sort predominated. To the writers of the Report, however, 'assembling' a firearm was a technical innovation of major proportions.[2]

---

[1] pp. 142-3.

[2] As late as 1902 the point was still apparently sufficiently novel to most British engineers that it was thought worth while to spell it out carefully to a professional audience. In his paper on 'Modern Machine Methods' delivered to the Institution of Mechanical Engineers, Mr Orcutt undertook to specify the role of the 'Erecting Department' in a modern engineering plant as follows: 'Probably in no department of mechanical work are the contrasts between old and new methods more striking than in erecting or assembling. In the new method, machining is done accurately to dimensions; in the old, machine and tools are mainly used for removing metal, and reliance is placed on the fitter for proper working fits. In the new, accuracy and interchangeability of dimensions are maintained by a suitable equipment of gauges and the establishment of limits; in the old, there is a variety of

The Committee's Report provides a detailed statement of the kinds and quantities of machinery for which the Committee contracted while in the U.S.A. While it would therefore be pointless to reproduce this here, certain observations are in order.

What impressed the Committee was not the comparative quality of American machinery, which was often flimsy and not well constructed by English standards;[1] nor were its members much impressed by the more general purpose machines. Rather, the most forceful impact derived from the manner in which it had been adapted and devised for specialized purposes. It is this, it argued, which is truly distinctive of American machinery.

> As regards the class of machinery usually employed by engineers and machine makers, they are upon the whole behind those of England, but in the adaptation of special apparatus to a single operation in almost all branches of industry, the Americans display an amount of ingenuity, combined with undaunted energy, which as a nation we would do well to imitate, if we mean to hold our present position in the great market of the world.[2]

The Committee had come to America expecting to be impressed with the machinery for making stocks, and here its expectations were amply fulfilled. Indeed, America had, by this time, developed not only stocking machinery but a wide range of woodworking machinery of a sort which was entirely unknown in England, which had deeply impressed Whitworth, and which caused the writers of the Report to speak of 'the wonderful energy that characterizes the wood manufacture of the United States'.[3]

---

sizes depending upon the skill and judgment of individuals. In the new, the time necessary for 'setting up' a particular piece of work, or 'lots' of work, is reduced 25, 50 per cent., and even more, below that taken in the old shop. In the new, the number of fitters is strikingly small compared with the abundance of this class of helpers necessary in works running on old lines. The model erecting department, therefore, is one in which fitters' work is reduced to a minimum, and where the least possible amount of time is occupied in assembling. These results are, of course, only possible when the best methods of machining are practised, when jigs and fixtures are used as much as possible, when a proper equipment of gauges is installed and a system of inspection is amplified.' H. F. L. Orcutt 'Modern machine methods' *Proceedings of the Institution of Mechanical Engineers* (1902) Parts 1-2, 37-8. Henry Ford stated the essential point forcefully in his article on "Mass Production" in the *Encyclopaedia Britannica:* "In mass production there are no fitters."

[1] They were, however, quite impressed with its quantity and apparent under-utilization. See p. 99. Cf. also p. 105.

[2] pp. 128-9. The Report later repeated its clarion call to British industry to emulate the energy and resourcefulness of the Americans in devising special-purpose machinery. (pp. 193-4).

[3] p. 171.

> In those districts of the United States of America that the Committee have visited, the working of wood by machinery in almost every branch of industry, is all but universal; and in large establishments the ordinary tools of the carpenter are seldom seen, except in finishing off, after the several parts of the article have been put together.
>
> The determination to use labour-saving machinery, has divided the class of work usually carried on by carpenters and the other wood trades into special manufactures, where only one kind of article is produced, but in numbers or quantity almost in many cases incredible.[1]

The first contract which the Committee entered into was for a complete set of stocking machinery which it referred to as 'absolutely indispensable' after having seen it in operation at the Springfield Armory.[2] This machinery was of improved design, which the Company had been producing only since 1853. Doubts as to the effectiveness of the American machinery upon English walnut were dispelled, and after careful examination the Committee entered into negotiations with the Ames Manufacturing Company of Chicopee, Massachusetts for the preparation of the machinery. This company, which supplied the stocking machinery for the Springfield Armory, agreed to supply a sequence of fifteen machines to prepare the stock of the 'English rifle musket, pattern 1853'.[3] The Committee accepted Mr Ames's tender, amounting to $30,860.[4]

The Committee had expected to receive instruction in the application of machinery to the shaping of wood. It was less prepared, but no less appreciative, when it encountered a good deal that was novel to it in the working of metals. For here, too, its members found an adaptation of machinery to specialized purposes such as had no counterpart in England. Specifically, it discovered that the Americans had developed and

---

[1] p. 167. The Report adds that 'As a general rule the machinery employed is of a coarse description, and the work which it produces may not in every case come up to our notions of finish; but it is produced cheaply and quickly, and the same or similar apparatus may be advantageously employed in England.'

[2] pp. 100-1, 137-42, 180-1. For some estimates of the effect of the stocking machinery upon labour productivity see Fitch, op. cit. 14.

[3] A good deal of information on the history of the Ames Manufacturing Company may be found in Vera Shlakman *Economic History of a Factory Town: A Study of Chicopee, Mass.* Smith College Studies in History, XX (Northampton, Mass., 1935), especially 81-7, 152-3, 162-7. See also Nathan P. Ames, in *Dictionary of American Biography* (New York 1943) I, 249-50 and James Tyler Ames, ibid. 248.

[4] The Committee later ordered additional equipment from Ames Manufacturing Company, including a wide assortment of jigs and gauges which '. . . though very expensive, the Committee considered it absolutely necessary to order, as it is only by means of a continual and careful application of these instruments that uniformity of work to secure interchanges can be obtained'. p. 187. See also pp. 188-90.

applied the milling machine, a device of extraordinary versatility, to a wide range of purposes in shaping the metal parts of the musket. In so doing the machine had, in the United States, progressively displaced the highly expensive operations of hand filing and chiseling of parts. In fact, the development of specialized forms of the milling machine may be considered one of the really distinctive contributions of America to the system of interchangeable manufacture. For it was through the use of these machines that a high degree of uniformity of metal parts was achieved, far beyond what was possible by hand filing.[1] As Fitch stated in 1880:

> A system of interchangeability largely dependent upon hand-filing is difficult to sustain, even with the aid of jigs. In the earlier attempts filing was the principal means of making interchangeable work; but the inspections were not then severe, nor were the pieces required to be so well made as to fit fine gauges. Filing close to hardened jigs is also very destructive of files – an important element of cost. It is scarcely a matter of wonder that the systems of interchangeability so repeatedly introduced were not well sustained until after the introduction of the practices of close forging with steel dies and metalworking, with efficient machinery for making sensibly exact cuts, without dependence upon the craft of the operative. Drilling with jigs is still the common practice, but filing to jigs was superseded by milling and edging with cutters, which were themselves formers, whose exactness was tested and maintained, in case of wear, by careful gauging of the work.[2]

The Report describes in detail the uses of milling machines on firearms,[3] and pays particular attention to the manner in which milling tools are serviced and kept in a satisfactory state of sharpness and hardness. This was a critical problem in the achievement of interchangeability, since

---

[1] Hand filing, however, was not entirely dispensed with. It sometimes remained to perform an essentially 'cleaning-up' operation after the basic shape had been imparted to the metal by the milling machine. 'The hammer, and some of the other parts, require considerable filing by hand, so as to trim off the little corners that have burred up, or places where the milling tool did not reach.' pp. 136-7.

[2] Fitch, op. cit. 3.

[3] pp. 134-7. In Goodman's chapter on the Birmingham gun trade in Timmins (op. cit.) there appears a fairly extensive description of the application of machinery to gun-making (pp. 396-404). Although there are references to the American origin of gun-making machinery, it is easy for the unwary reader to assume that he is reading a description of techniques of production which are employed in the Birmingham region which the book, after all, purports to be about. In fact much of the description is taken directly from these pages of the Report (pp. 134-7) and is really a summary account of the use of machinery in the Springfield Armory in Massachusetts! Several of Goodman's paragraphs are taken *verbatim*, and without acknowledgment, from this Report.

uniformity of the finished metal part was dependent upon uniformity of the tool imparting the shape to the metal. American machine makers had recently made great progress in this direction, and it was a lesson which the Committee realized would have to be learned by British industry.

> It will thus be seen, that to attain the advantage of the parts interchanging, a degree of refinement in the tools and apparatus is imperative, and this cannot be appreciated by those who have been engaged on a ruder system, and hence the assertion of so many gun makers before the Parliamentary Committee, that to have perfect interchange was unattainable.[1]

On the other hand the Committee also observed that the American practice was to give the metal parts only as much finish as was justified by purely functional considerations. A frequent observation is that both American machines and machine-made products appeared less 'finished' than their British counterparts. Americans were more reluctant than the British to incur additional costs in ways which did not improve the efficiency of the final product in some narrowly-defined utilitarian sense.[2]

The Committee made substantial purchases of milling machines and other specialized machinery from the remarkable Robbins and Lawrence Company of Windsor, Vermont, a company which made many seminal contributions to the development of American machine technology in spite of its rather chequered financial history.[3] This was the company whose interchangeable rifles generated so much interest at the Crystal Palace Exhibition in 1851. When the Board of Ordnance informed the Committee in June 1854 that the previous financial limit upon their machinery purchases for the Small Arms Factory had been removed, the Committee went directly to the Robbins and Lawrence machine shop to arrange further purchases.

---

[1] p. 136.

[2] 'Many of the parts, after they are case-hardened, are polished on buffs, in the same manner as practised in England; but on the whole much less attention is bestowed on what is called high finish given to the parts, only to please the eye; but that can be done, if necessary, after the parts are brought to shape by the machines.' p. 136. It may be noted that Whitworth had played a major role in exorcising unnecessary ornamentation and decoration from British machine tool design and reducing these tools to strictly functional lines.

[3] Ibid. 76-9. On the history of the Robbins and Lawrence Company see Joseph Wickham Roe *English and American Tool Builders* (New Haven 1916) ch. 15, and Winston O. Smith *The Sharps Rifle* (New York 1943) ch. 3. This chapter also gives the details of the contract, for the manufacture of 25,000 Minié rifles, between Robbins and Lawrence and the British Government. Robbins and Lawrence in 1852 set up a separate plant in Hartford, Conn., to produce, under contract, the famous Sharps rifle. They bore the primary responsibility for the high degree of perfection to which that rifle was eventually brought.

It was from this firm, then, that most of the American metal-working machinery for Enfield was purchased, and the purchases of milling machines may be said to have constituted the introduction of milling techniques into British industry.[1] These machines were primarily for working the component parts of the lock, and among the machines other than the milling machines were a large number of drilling machines, with which the Committee was also particularly impressed. So struck was it with the 'beauty and efficiency' of the Robbins and Lawrence machinery that it wished to purchase a complete set of their barrel machinery as well. Robbins and Lawrence submitted a tender for such machines which the Committee was anxious to accept. Not knowing whether the Board of Ordnance had in the meantime made arrangements for the purchase of such machinery from Whitworth or other possible British suppliers, and in view of the considerable cost, the Committee decided to await its return to England before acting on the tender.[2] The barrel machinery, which Anderson described as being '... of a very perfect and productive character, and far superior to that used by the gun-makers in England ...' was eventually purchased.[3] The long delay in placing the order was responsible for postponing the completion of the Enfield Armoury.[4] When it was fully equipped, however, it possessed American machinery – lock, stock, and barrel.

Parliamentary returns accounting for the money spent at Enfield indicate the expenditures listed below:

Period	*buildings*			*machinery*		
	£	s.	d.	£	s.	d.
1 Jan. 1854–31 March 1854	85	0	0	0	0	0
1 April 1854–31 March 1855	15,690	13	9	25,478	19	4
1 April 1855–31 March 1856	35,433	18	0	24,082	10	7
1 April 1856–31 March 1857	24,776	16	9	15,105	15	11
1 April 1857–31 March 1858	15,632	7	10	3,985	16	6
1 April 1858–31 March 1859	9,356	6	9	2,908	18	2
	100,975	3	1	71,562	0	6

[1] Nasmyth had introduced, around 1830, a machine which was used primarily for milling the sides of nuts, which might be classified as a milling machine. But the technique was not generalized and had no direct mechanical progeny (Robert Woodbury *History of the Milling Machine*, Cambridge, Mass., 1960, 24-6). In describing the milling machinery displayed at the 1862 London Exhibition, Clark states: 'The application of circular cutters is exemplified in machines for the manufacture of rifle muskets, exhibited by Messrs. Greenwood & Batley. This class of machines has been introduced into this country in a variety of shapes from the Small Arms Factory at Springfield, Massachusetts, U.S.' D.K. Clark *The Exhibited Machinery of 1862*, 130.

[2] p. 191.

[3] Anderson, op. cit. 28.

[4] As of January 1857 Anderson was able to state that 'At the present time the manufacture is all but completed; bayonets of a quality never before

Up to 31 March 1858, 26,739 rifles were produced; between 1 April 1858 and 31 March 1859, 57,256 rifles; between 1 April 1859 and 31 March 1860, 87,405 rifles; and between 1 April 1860 and 31 March 1861, 99,083 rifles.[1]

In this way did the system of interchangeable manufacture cross the Atlantic from the U.S. to Great Britain. The Enfield Armoury quickly became a national showpiece. The engineering literature of the late 1850s and 1860s abounds in enthusiastic descriptions of the Enfield Armoury and its machinery, and the scepticism concerning interchangeability which was so widely and vehemently expressed in the first half of the 1850s was replaced by expressions of national pride and self-congratulation in the workmanship which now made such interchangeability possible. Thus:

> It is only necessary to see a rifle put together at Enfield thoroughly to appreciate the mechanical advantages of the duplicate system. Thus, the separate pieces which form the lock are picked up indifferently from heaps of such pieces, every one fitting into its place most perfectly; and a rifle, the parts of which have never been together before, is formed in a few seconds from a lot of fittings promiscuously thrown together, as any lock will fit any stock, any stock will fit any barrel, and each part, however minute, will fit any rifle. If any part be broken, or requires replacing, there is a new one which will exactly fit it, so that wherever troops are armed with the Enfield rifle, there is not any fear of a mischance through a bayonet not fitting, and the armourer's duty is confined to the exchange of a few delicate cocks or springs; and as it is only a system of exchange, a skilled armourer may be dispensed with . . .[2]

Alternatively, at the time of the London Exhibition of 1862, in commenting upon the milling machines exhibited by Greenwood and Batley of Leeds, and after acknowledging their origin via the Springfield and Enfield Armouries, the following remarks were made upon the rifles produced with these machines:

> When it is remembered that all the parts of a rifle musket are made in perfect duplicate, and that each individual part of a

made in England are made at the rate of 900 per week; the lock and stock machinery is also complete; and the greater part of the barrel machinery has arrived from America; so that there is comparatively little remaining to be done.' Ibid. 30.

[1] *Parliamentary Papers*, 37 (1857–8) 139; *Parliamentary Papers*, 15 (1859) 217; *Parliamentary Papers*, 41 (1860) 457; *Parliamentary Papers*, 36 (1864) 607. There was a parliamentary debate on the subject of the government's manufacturing establishments on 22 July 1864. See *Hansard*, Ser. 3, CLXXVI (1864) especially columns 1916-62.

[2] John Fernie 'On the manufacture of duplicate machines and engines' *Minutes of Proceedings of the Institution of Civil Engineers*, XXII (Session 1862–63) 604-5. The paper is a plea for standardization and 'duplication' of machinery parts in locomotives.

> musket may be interchanged with the same part in any other like musket, making an equally good fit; that muskets are manufactured at the rate of two thousand per week; that upwards of seven hundred operations are required for the production of a single rifle, some idea may be formed of the extreme degree of accuracy with which the machines must be employed.[1]

It would not be fair to leave this part of our discussion with the impression that the movement of new techniques was solely a new-world-to-old-world phenomenon. In the 1840s and 1850s in particular, there was a great deal of international exchange of information involving ordnance work. If it is true, as Deyrup suggests, that the '. . . United States government . . . with what might seem naive helpfulness . . . aided certain countries to modernize their weapons',[2] it is also true that the United States Government received some benefits from what was, after all, a reciprocal relationship. This was particularly the case with respect to the manufacture of barrels. In 1840 the United States Secretary of War dispatched an ordnance commission to Europe to visit arsenals, cannon foundries, and small arms manufactories. Nathan Ames of Ames Manufacturing Company accompanied this commission.[3] In 1848 Brevet Major Hagner, an ordnance officer, was authorized to travel to England and Belgium and instructed particularly to examine the English process of barrel-rolling and the Belgian manufacture of percussion caps.[4] In 1855

---

[1] D. K. Clark, op. cit. 130. One of the best contemporary accounts of the new Enfield Armoury appeared in a series of articles in *The Engineer* (1859) 204-5, 258-9, 294-5, 348-9, 384-5, 422-3. See also the eulogistic account in *Mechanics Magazine* (New Series) II (1859) 44, 69-70, describing the operation of the Enfield machinery, and ibid. VI (1861) 110-11, 127-8, 144-5. Some illustrations of the factory and its machinery appeared in *The Illustrated London News* (21 September 1861) 298 and 304. The British Museum has a set of ten photographs of the original machinery which was sent to the Museum on 28 February 1859. They are in a volume titled *Photographs Representing the Manufacture of the Enfield Rifle*. (Catalogue number 1802 a.24.) Several of the machines (including the stocking lathe) bear the identification of the Ames Manufacturing Company, and on one machine the name of Robbins and Lawrence is legible. One of the original Ames machines, whose function was to cut out the recess in the stock to receive the lock, is now in the possession of the Science Museum in Kensington and is described in K. R. Gilbert 'The Ames recessing machine: a survivor of the original Enfield rifle machinery' *Technology and Culture* IV, no. 2 (Spring 1963) 207-11.

[2] Deyrup, op. cit. 130.

[3] Stearns, op. cit. 7; Shlakman op. cit. 83.

[4] Colonel Talcott, in applying for permission to send two officers on this trip (in the event only Major Hagner was sent) justified it to the Secretary of War as follows:

'There are several matters connected with the manufacture of arms and the preparation of military supplies, in which this department is immediately, and the country generally, interested, to which I desire particularly to direct the attention of these officers during their visit to Europe. Among these I may mention, as of greatest importance, the process of manufac-

and 1856 Major Mordecai travelled about Europe, visited Enfield in March 1856, and wrote a detailed report which also paid particular attention to British techniques of barrel manufacture.[1] The introduction of barrel-rolling was eventually accomplished at Springfield through the efforts of James T. Ames who had been sent to England as an agent of the United States Government. Along with his purchases of barrel machinery Ames brought an expert Birmingham barrel welder in the person of William Onyans, under whose supervision the barrel machinery was introduced at Springfield.[2]

Anderson himself was anxious that the armoury, in addition to producing firearms, should serve an educational function in acquainting British manufacturers with American methods and techniques and thereby encouraging their adoption. Anderson foresaw, perhaps more clearly than any of his contemporaries, the coming American challenge to British industrial leadership. This is clear in the Report reprinted here and in his later writings–e.g., in his capacity as a British juror, at later international expositions. He had effectively understood and grasped the logic of the forthcoming techniques of mass production and the role of production engineering. In speaking of the Enfield Armoury when it was virtually finished, he wrote:

> But there is yet a still higher and a more patriotic view to be taken of this manufactory. Notwithstanding the opposition which has been made to its establishment, the gun makers, if they are wise in their generation, will be the first to avail themselves of the new light which it has thrown upon their trade. Some are already doing so, and the advantages of the system will be recognized by other trades. The American machinery is so different to our own,

---

turing gun-barrels by rolling, known to be in use in England, and understood to combine great economy with good quality of work, and the Belgian method of fabricating percussion caps for small-arms, understood to possess advantages over other methods, particularly in being less dangerous.

'The English process of rolling gun-barrels, it is expected, would effect a saving in this work of 20 per cent. of the whole amount, which is upwards of 25,000 barrels annually at the national armories. Efforts were made to obtain a knowledge of this process (as will be seen by the inclosed correspondence) through our consul-general at London, more than a year ago, but without success.' *A Collection of Annual Reports and Other Important Papers, Relating to the Ordnance Department*, Prepared under the Direction of Brig. Gen. Stephen V. Benet, Washington, 3 vols. vol. II (1878) 242-3. Major Hagner's report, dated 25 October 1849, appears in ibid. 290 *et seq.* His description of English barrel rolling appears on pp. 310-12.

[1] Alfred Mordecai *Military Commission to Europe in 1855 and 1856*, U.S. Senate Executive Documents, 36th Congress, 1st Session, 1861.

[2] Fitch, op. cit. 8; Deyrup op. cit. 150-3. The technique of hook-rifling of barrels, which was widely adopted in America, was introduced from England with some improvements by the Remington firm around 1861. Fitch, op. cit. 11.

> and so rich in suggestions, that when fully organized it should be thrown open to the study of the machine makers of the kingdom, for it is a positive addition to the manufacturing resources of the nation, and shall enable us to keep up with our transatlantic competitors. The knowledge which will thus be gained is the more desirable, as the American machinist takes every opportunity of becoming acquainted with our inventions and scientific literature, while the great majority of persons in this country are comparatively careless on this point. A few hours at Enfield will show that we shall soon have to contend with no mean competitors in the Americans, who display an originality and common sense in most of their arrangements which are not to be despised, but on the contrary are either to be copied or improved upon. Of the many scientific and engineering friends of the writer who have visited the establishment at Enfield, but one opinion has been expressed, of the great good which the country at large will derive therefrom, the musket manufactory being a secondary consideration. It must therefore be a source of satisfaction to all that the Government should thus be outstripping private enterprise.[1]

IX

Throughout the first half of the nineteenth century the making of firearms in the United States decisively affected the development of specialized machinery and techniques of precision manufacturing. The process culminated in 1855 with the opening of Colt's armoury in Hartford, Connecticut, which was at the time the largest and most modern private armoury in the world. The new technology was the work of many men, all of whom perceived the advantages of producing guns on an interchangeable basis. Eli Whitney was an important figure in this process, although his contributions have for long been overstated and the contributions of others badly neglected. What is indisputable is that the new machinery and techniques were the result of persistent efforts to overcome a common range of problems not only by Whitney but also by Simeon North and John Hall and by men employed at such places as Robbins and Lawrence, Ames Manufacturing Company, Colt's armoury and the government's armouries at Springfield and Harper's Ferry.[2]

---

[1] John Anderson *General Statement of the Past and Present Condition of the Several Manufacturing Branches of the War Department* (London H.M.S.O. 1857) 31. Anderson spoke to professional groups on several occasions in an attempt to bring about a wider appreciation of novel American methods and their possible applications in Britain. See, for example, John Anderson 'On some applications of the copying or transfer principle in the production of wooden articles' *Proceedings of the Institution of Mechanical Engineers* (1858) 237-47; John Anderson 'On the application of the copying principle in the manufacture and rifling of guns' ibid. (1862) 125-41.

[2] The case for North's priority is made in S.N.D. North and Ralph H. North *Simeon North, First Official Pistol Maker of the United States* (Concord, New Hampshire, 1913). North's contract with the United States

It is not surprising in retrospect that interchangeable manufacturing should have originated in the production of military arms. For the military firearm is a product which requires to be produced in very large quantities. Moreover, it is a completely standardized product. The personal preferences of individual users are of no consequence, by contrast with, say, the British sporting-gun trade, where guns were often custom built to conform to personal specifications–an English gentleman's arrangement with his gun maker was not totally unlike that with his tailor.[1] The possibility of very large output levels justified the introduction of specialized machines whose cost was very high, but which achieved low per unit cost levels when the volume of standardized output was sufficiently large. From this point of view interchangeability was an indispensable companion to specialized machinery which, once set up, produced large quantities of a uniform output. Precision methods were essential so that the output of each machine could easily be assembled–

Government in 1813 was the first one which specified uniformity of parts. It was stated in the contract that '. . . the component parts of pistols, are to correspond so exactly that any limb or part of one pistol, may be fitted to any other Pistol of the Twenty Thousand' (Ibid. 89). The case for Eli Whitney has been made many times, most recently in J. Mirsky and A. Nevins *The World of Eli Whitney* (New York, 1952) and C.M. Green *Eli Whitney and the Birth of American Technology* (Boston 1956). The history of the federal armouries is yet to be written. The most authoritative account of the techniques developed by John Hall and his success with interchangeability of arms is in *A Collection of Annual Reports and Other Important Papers, Relating to the Ordnance Department*, prepared under the Direction of Brig. Gen. Stephen V. Benet, 3 vols. I (Washington 1878) 150-7. Much useful material on the Springfield Armory will be found in Deyrup, op. cit. which drew extensively on unpublished armoury records and constitutes a serious piece of historical research. See also Charles Stearns *The National Armories* (Springfield, Mass., 1852), which is particularly useful for the twenty years or so before its date of publication. The unsuccessful earlier French attempt to establish an interchangeable system of firearms is treated in W.F. Durfee 'The first systematic attempt at interchangeability in firearms' *Cassier's Magazine* v (April 1894) 469-77.

[1] In describing the shaping of a gun stock for use by 'first-rate sportsmen', a Colonel Hawker is quoted as saying: 'the length, bend, and casting of a stock must, of course, be fitted to the shooter, who should have his measure for them as carefully entered in a gunmaker's books, as that for a suit of clothes on those of his tailor'. *A Treatise on the Progressive Improvement and Present State of the Manufactures in Metal* II (London 1833) 105. By contrast, something which several English observers noted about American products was the manner in which they were often designed to accommodate the needs of the machine rather than the user. Lloyd, for example, noted of the American cutlery trade that '. . . where mechanical devices cannot be adjusted to the production of the traditional product, the product must be modified to the demands of the machine. Hence, the standard American table-knife is a rigid, metal shape, handle and blade forged in one piece, the whole being finished by electroplating–an implement eminently suited to factory production.' G.I.H. Lloyd *The Cutlery Trades* (London 1913) 394-5.

rather than fitted–into a final product. If the components of the firearm were not uniform and interchangeable, the economies of mass production would have been lost in the need to undertake a labour-intensive fitting together of the pieces of separate machines. Finally, the property of interchangeability was important to the producer of the firearm, and to its user as well, since it greatly facilitated the repair and maintenance of firearms in war.[1] Indeed, between 1855 and 1875 American gun-making machinery was purchased by the governments of many countries, in addition to the British, where such techniques were unknown in the production of non-military commodities–including Russia, Prussia, Spain, Turkey, Sweden, Denmark, and Egypt. Many other countries purchased interchangeable firearms from American sources.[2]

It is of more than passing interest that the United States established public armouries with little fuss or protest in the last decade of the eighteenth century. Perhaps the comparative ease with which they were established was due to the fact that no American equivalent to the Birmingham gun makers lobby had had a chance to develop. Nevertheless, the question of whether it was 'proper' for the federal government to produce its own firearms seems not to have been seriously debated–although the question of whether the armouries should be controlled by a civilian or a soldier was a hotly contested issue. In fact, the armouries changed hands several times in the battle between civil and military authority.

In addition to its public armouries the federal government always acquired some of its arms by contract. For a brief period it emulated the British practice of contracting for parts of arms, but rejected the system as unworkable. The usual arrangement was for the government to contract with individual firms for the entire firearm. Furthermore, in the early years the governmental financial advances given to contractors constituted the major source of capital to these firms. The arrangements entered into by the American Government provided arms contractors with an assurance of continuity in their operations and made it possible to plan with a long-term time horizon. The reasonably harmonious relationship between the government and its arms suppliers contributed much to the technological dynamism of the industry in the early decades of the nineteenth century.[3] The contrast with the British experience in this respect is, of course, striking.

---

[1] Joseph W. Roe 'Interchangeable manufacture' *Transactions of the Newcomen Society*, XVII, 165-73.

[2] Fitch, op. cit. 4.

[3] This is also the conclusion of Deyrup in her careful study of the industry. '. . . the contract system was of immense value both to the government and the contractor, for aside from bringing the industry into existence it promoted a spirit of cooperation and mutual aid unique among early American manufacturers, which had much to do with the rapid development of the industry in the first thirty years of the nineteenth century. This spirit of mutual helpfulness, which was most strongly expressed in the relations between the New England contractors and the Springfield Armory, extended

The American firearms industry was instrumental in developing the whole array of tools and accessories upon which the large-scale production of uniform metal parts was dependent: drilling and filing jigs, fixtures, taps, and gauges, and the systematic development of die-forging techniques. Die-forging machines were being employed by John Hall at Harper's Ferry as early as 1827, and received their most significant improvement under the guidance of Colt's ingenious and prolific superintendent, Elisha K. Root.[1]

Thomas Blanchard invented his lathe for the shaping of gunstocks in Millbury, Massachusetts in 1818. Since the machine promised to reduce the considerable labour costs involved in the hand-shaping of gunstocks, it was quickly introduced into the Springfield Armory. Blanchard was placed in charge of the whole stocking process at Springfield and '... proceeded to expand and extend the principle of his machine, first to letting in the barrel, then the mounting, and finally the lock, which the old stockers said could not be done by machinery'.[2] Although the stocking lathe had originally been developed for the shaping of wooden materials, it used a versatile technique which was soon employed for reproducing irregular shapes in a variety of other materials. The machine was applied to such sundry items as tackle-blocks, wheel-spokes, ox-yokes, oars, sculptured busts, etc.

The milling machine, perhaps along with the turret lathe one of the two most important of all modern machine tools, owed its origin in the United States to the attempt to provide an effective machine substitute for highly expensive hand filing and chiselling operations. Earlier attempts to achieve interchangeability in firearms production seem to have been undertaken by the use of hand filing with hardened jigs. Woodbury even advances the conjecture, based on an inventory of Eli Whitney's estate, that '... the large number of files listed as on hand would suggest that for much of his work Whitney used only a filing jig or fixture to guide a hand operated file as his principal means of producing uniform parts for the locks of his muskets'.[3] Of course the degree of uniformity attainable by such crude hand methods did not approach what was possible with the later machine technology. As Fitch observed, comparing the

---

to many aspects of arms manufacture. The Superintendent of the Armory openly aided contractors by advising them of strategic times for applying for contracts. Trade secrets apparently did not as yet exist in the new industry, and the Armory and the contractors exchanged advice and information relative to interchangeability, gun design, manufacturing processes and machine tools.' Deyrup, op. cit. 66.

[1] Fitch, op. cit. 20.

[2] Asa Waters, *Biographical Sketch of Thomas Blanchard and his Inventions* (Worcester 1878) 8; see also Deyrup, op. cit. 97-8.

[3] Woodbury, op. cit. 249. At the Springfield Armory in 1809, files were consumed at a rate slightly exceeding one file for each musket produced. James E. Hicks *United States Ordnance* II (Mt Vernon, New York, 1940) 131. Some interesting material on the two federal armouries for the same period will be found on pp. 129-34.

uniformity of 1880 with the early attempts in the beginning of the century:

> If parts were then called uniform, it must be recollected that the present generation stands upon a plane of mechanical intelligence so much higher, and with facilities for observation so much more extensive than existed in those times, that the very language of expression is changed. Uniformity in gun-work was then, as now, a comparative term; but then it meant within a thirty-second of an inch or more, where now it means within half a thousandth of an inch. Then interchangeability may have signified a great deal of filing and fitting, and an uneven joint when fitted, where now it signifies slipping in a piece, turning a screw-driver, and having a close, even fit.[1]

The exact origins of the milling machine are shrouded in obscurity, but it is known that both Eli Whitney and Simeon North employed crude milling machines in their musket-producing enterprises in the second decade of the nineteenth century, as did John Hall at the Harper's Ferry Armory.[2] These early milling machines were driven by hand. It has been stated that Hall's milling machines 'obtained in 1827 an efficiency of only one-third greater than by filing'.[3] The subsequent improvements in the machine were in great measure achieved at the national armouries and by such gun-producing firms as Robbins and Lawrence. Thomas Warner at Springfield Armory undertook important improvements which included a milling machine which made lock-plates of uniform thickness and the construction of milling machines with spindles which were adjustable by the use of vertical slides.[4] Warner's machines are reported to have drastically reduced the cost of milling operations. Major further improvements in the miller were made by the Robbins and Lawrence Company which was also responsible for introducing the machine into the industrial community of Hartford, Connecticut.[5] The design of the plain milling machine was stabilized in the early 1850s in the form which came to be known as the Lincoln miller. In this form it began to assume a conspicuous place in the whole range of metal trades. It has been estimated that, between 1855 and 1880 '... nearly 100,000 of these machines or practical copies of them, have been built for gun, sewing machine and similar work'. It is interesting to note that the Lincoln miller, which played such a critical role in the manufacture of interchangeable products, was actually *constructed* of interchangeable parts.[6]

---

[1] Fitch, op. cit. 2.

[2] Ibid. 22-6; Joseph W. Roe 'History of the first milling machine', *American Machinist* XXXVI (27 June 1912) 1037-8; Robert S. Woodbury *History of the Milling Machine* (Cambridge, Mass., 1960) chs. 1 and 2. Woodbury's book is an invaluable guide to the technical development of the milling machine.

[3] Fitch, op. cit. 24.

[4] Ibid. 25; Deyrup, op. cit. 153-4.

[5] Fitch, op. cit. 25.

[6] Ibid. 25-6. Fitch cites examples for the year 1880 of individual arms-making plants where milling machines constituted between 25 and 30 per cent of the total number of machines in use (p. 22).

The final major contribution of the American arms makers was the role which they played in the development of the turret lathe which was indispensable to the economical production of all commodities based upon interchangeable parts. The turret lathe, holding a cluster of tools placed on a vertical axis, made it possible to perform a rapid sequence of operations on the work piece without the need for resetting or removing the piece from the lathe. It therefore revolutionized all manufacturing processes requiring large volumes of small uniform components, such as screws, which were produced on turret lathe machines shortly after that machine was developed.[1]

The origin of the turret lathe, initially employing a horizontal axis for the turret, has been attributed to Stephen Fitch, of Middlefield, Connecticut, in 1845, while he was engaged on a government contract for the production of percussion locks for an army horse pistol.[2] The turret lathe principle was employed and improved at the Colt armoury (where Root introduced a double-turret machine in 1852) and by Frederick W. Howe while he was superintendent of the Robbins and Lawrence Company at Windsor, Vermont; and turret lathes were built and sold commercially by that company in 1854. In its early forms the arrangement of parts and the working action of the turret lathe and other machines employed in the production of the Colt revolver bore a striking resemblance to the revolver itself. 'It is noteworthy that the same arrangement of parts characteristic of the Colt revolver seems to have been carried through the principal machines for its manufacture, the horizontal chucking lathes, cone-seating and screw machines, barrel-boring, profiling, and mortising machines, and even the compound crank-drops, exhibiting the same general arrangement of working parts about a center.'[3] A turret screw machine, designed in 1858 by H. D. Stone of the Robbins and Lawrence Company, was sold commercially by Jones and Lamson Machine Company, successors of the now-defunct Robbins and Lawrence firm.[4] From that point on, the turret lathe was adapted and modified for innumerable uses in the production of uniform components for such products as sewing machines, watches, typewriters, locomotives, bicycles and, eventually, automobiles. Its most important subsequent improvement, it might be noted, was also achieved by a person who had

---

[1] Ibid. 28. Anderson stated that the screw-making variant of the turret lathe was introduced into England by Hobbs for use in his lock factory. John Anderson 'Report on machine tools,' in *Reports on the Paris Universal Exhibition, 1867*, III, Parliamentary Papers 30 (1867–8) Part II, 373.

[2] E. G. Parkhurst, 'Origin of the turret, or revolving head' *American Machinist* XXIII (24 May 1900) 489-91. Cf. also Guy Hubbard 'Development of machine tools in New England' *American Machinist* LX (21 February 1924) 272-4.

[3] Fitch, op. cit. 27-8.

[4] Ibid. 28. The Jones and Lamson Machine Company remained one of the leaders in turret lathe production for many years. See James Hartness *Machine Building for Profit* (Springfield, Vermont, 1909). Hartness introduced the flat-turret lathe and was a pioneer in the application of hydraulic feeds to machine tools.

been trained in the firearms industry: Christopher Spencer, a former Colt employee and the inventor of the Spencer repeating rifle. As a result of a machine which he invented for turning sewing-machine spools, Spencer went on to explore methods for making metal screws automatically and, in so doing, invented the automatic turret lathe.[1] This was an innovation of momentous importance, since the self-adjusting feature of the cam cylinder with adjustable strips, through which automaticity was achieved, was eventually to make possible all modern automatic lathe operations. Together with the subsequent perfection of multiple spindle techniques, it was instrumental in a major acceleration in the pace of machine tool operations.

X

We have seen that when the 'American system of manufacturing' crossed the Atlantic during the 1850s, it did so under forced draft conditions. The transfer took place at a time and under circumstances when considerations of national security took precedence over purely and narrowly economic calculations. It would take us far beyond our present scope to consider the subsequent history in Great Britain of the application of interchangeable methods and mass production technology more generally. What must be recorded, however, is that after the system had been planted in Great Britain it failed to flourish as it had in the United States. Its adoption was halting and fitful. Handicraft methods showed a remarkable capacity to survive, methods to assure precision workmanship–such as the use of limit gauges–were adopted very slowly, and in engineering workshops the shift from general purpose machinery to more specialized machines was not nearly as rapid or as extensive as in the United States.[2]

---

[1] Guy Hubbard, op. cit. LXI (21 August 1924) 314; Roe, op. cit. 176.

[2] Much useful material on Anglo-American differences in engineering methods at the beginning of the twentieth century will be found in the paper by H.F.L. Orcutt, 'Modern machine methods' and the subsequent discussion, *Proceedings of the Institution of Mechanical Engineers* (1902) Parts 1-2, 9-112. The greater degree of specialization of American firms and machinery by comparison with their English counterparts received much attention. J.R. Richardson, who operated a general engineering workshop, justified his unwillingness to purchase American moulding machines in the following way: 'It was not that English engineers did not understand American methods, but that Americans did not as a rule understand the conditions which obtained in large engineering works in England having a big general practice. There must be a large run of work. Even the most enthusiastic Americans had told him that a large quantity was not needed, that it could be done perfectly well with a dozen, but very often a dozen was a large quantity. Not only was it necessary to have on his catalogue 500 different types and sizes of steam-engines, but an infinite variety of mining and general machinery; and in addition his firm was expected to do anything required, and had to do it even if it only had to be done once.' (pp. 72-3.) The important unanswered question, of course, is *why* the general engineering workshop persisted as it did in England. Cf. S.B. Saul 'The American impact upon British industry' *Business History* III (1960) 21-7.

Thus the Birmingham gunmakers decided, in 1861, to respond to the establishment of the Enfield Armoury by forming their own mechanized gun-making factory which would produce interchangeable firearms. The result was the founding of the Birmingham Small Arms Company at Small Heath. But although machinery was introduced, many of the processes continued to be hand operations performed by skilled labour. Goodman, who was writing in Birmingham in 1865 and who spoke of the factory from his own observations, referred to 'the hand-filing of the locks, sights, etc.' and commented:

> At Enfield, every part of the gun, from the earliest stage, is produced in the factory. This has not been found necessary in Birmingham; advantage has been taken of having the source of supply so near at hand, to obtain certain parts in an unfinished stage from the manufacturers of the town. By this means great saving has been effected in the outlay for machinery which would otherwise have been required, without prejudice to the quality of the work, as it all passes through the finishing processes in the factory.[1]

More generally, the specialized American machine tools which were so closely associated with interchangeable manufacture achieved only a very gradual acceptance in Great Britain. In 1867 Anderson deplored the failure of British industry to adopt the milling machine[2] which was not, in fact, widely adopted in England until the bicycle craze of the 1890s.[3]

The British experience with the turret lathe, which began to be widely used in the United States in the late 50s and 60s was also one of slow and

---

[1] Goodman, op. cit. 404. A description of the operations of the Birmingham Small Arms Company from the vantage point of 1910 clearly suggests that at least some of the firm's workshops were dominated by handicraft methods. '... workshops formerly occupied by benches and vices are now almost exclusively filled with machinery.' *Proceedings of the Institution of Mechanical Engineers* (1910) Parts 3-4, 1324. In 1862, before the firm had acquired its stocking machinery, it was observed that 'At present the gunmakers of Birmingham had to pay a high price to have their gunstocks made by machinery in London, in order to secure greater accuracy and finish of workmanship than was obtained in hand work.' Such machinery then existed at Enfield and '... a nearly similar set at the London Armoury Company's works in London'. Thomas Greenwood 'On machinery for the manufacture of gunstocks,' *Proceedings of the Institution of Mechanical Engineers* (1862) 335-6. The firm eventually acquired a complete set of stocking machinery from the Ames Manufacturing Company. *The Engineer* XVI (25 December 1863) 375.

[2] 'This class of tool is the great feature of the Enfield small-arms factory, and it is much more used by engineers and machine-makers in America than it is in England. The difficulty of keeping up the sharpness of the cutting instruments is much overrated now that there are proper means for sharpening without reducing the temper.' John Anderson 'Report on machine tools', in *Reports on the Paris Universal Exhibition, 1867* III, Parliamentary Papers, 30 (1867–8) Part II, 373-4.

[3] S.B. Saul, op. cit. 22-3.

grudging acceptance. Moreover, even where it was adopted, principles of machine design which were taken for granted in the United States were often stoutly resisted. The turret lathe does not seem even to have been exhibited at the London Exhibition of 1862. As late as 1902 an eminent British engineer found it necessary to reprimand British manufacturers for using turret lathes which did not incorporate interchangeability in their construction – a feature which was embodied in American turret lathes in the 1850s.

> The introduction of the best class of turret machinery would assist to put on the market a superior article at very little if any advance over the rubbish now sold, where the loss of even the simplest part necessitates the purchase of an entire apparatus, so little is interchangeability practised.[1]

However slow the introduction of the new techniques may have been, and whatever the reasons for the slowness, it is clear that their introduction into British industry was directly linked to the importation of American firearms machinery during the historical episode with which we have been concerned. Allen makes the point forcefully with respect to the Birmingham area.

> It cannot be too strongly emphasized that the Birmingham Small Arms Company was the first of the local factories to turn out highly finished complicated metal articles by mass-production methods. This meant that certain kinds of complex machinery began to appear for the first time in large quantities in a Birmingham factory, while new methods came into existence for the production of standardized parts. For instance, it now paid to sink dies and to stamp out rifle parts which previously had been forged by the smith on the anvil. Thus the coming of the interchangeable rifle brought not only the machine-shop and the tool-room in their modern forms, but also a development of hot stamping. This process, previously confined to such products as keys and edge tools, now began to play a much more important part in Birmingham's manufacturing operations.[2]

---

[1] H.F.L. Orcutt, op. cit. 29. A young American engineer, visiting Bryan Donkin's Bermondsey works in 1832 and examining his complex Fourdrinier paper machinery, later wrote that '. . . this was the first instance I had seen where making the component parts of machinery interchangeable had been reduced to an absolute system, that is now so universally practised by all first-class machinists'. Eugene S. Ferguson *Early Engineering Reminiscences (1815–40) of George Escol Sellers*, Smithsonian Institution, Bulletin 238 (Washington, D.C., 1965) 127. Donkin was pleased to show Sellers his paper machinery but refused to show him the 'tools and various constructions and appliances' employed in the construction of the machines (p. 118). Chapters 14 to 17 of this beautiful volume, although concerned primarily with papermaking, provide many useful insights into Anglo-American differences in the use of machinery in the early 1830s.

[2] G.C. Allen, op. cit. 191.

The explanation for the differences in productive techniques between the United States and Great Britain is a subject which has been warmly debated for over a century.[1] The framework of the discussion and the variables considered have not, however, changed significantly.[2] Indeed, the reports reprinted in this volume are the most authoritative sources of our information on the nature of these differences as they existed in the middle of the nineteenth century. Furthermore, many of the explanations which still loom large in discussions of Anglo-American differences in technology will also be found in these reports. The most pervasive, of course, is the difference in relative factor prices between the two countries. If the relative price of labour, as compared to capital, was significantly higher in the United States than it was in Great Britain, rational economic behaviour would have dictated that American businessmen adopt labour-saving techniques which it might not have paid to adopt in Great Britain.

To what extent differences in adopted technology can be attributed to differences in relative factor prices is still an unsettled matter, primarily because of inadequacy of data, and is deserving of further careful investigation. It should be noticed, however, that this is a potentially powerful explanatory variable. If these price differentials turned out to be sufficiently large, then explanations couched in essentially sociological terms – resistances of workers, lack of entrepreneurial dynamism, irrational decision-making procedures, peculiarities in the structure of national tastes – become at least logically expendable. One point on this subject deserves to be made here, because of its particular relevance to our earlier discussion. The magnificence of the American natural resource endowment is almost always recognized in general discussions. Yet its importance is frequently lost sight of at a critical point in examining technological differences. Technology is typically examined, for analytical simplicity, in a two-factor framework. The usual expository device when exploring the problem of optimal factor combination, is to measure off units of capital on one axis and units of labour on the other. Production possibilities and factor prices being given, the optimal combination is easily established. Thus, it has become common practice to think in terms of a spectrum of possibilities with capital scarce, labour intensive combinations at one end, and capital intensive, labour scarce

---

[1] The most notable recent contribution has been H. J. Habakkuk *American and British Technology in the 19th Century* (Cambridge 1962).

[2] It should be noted that attempts to explain British relative slowness in terms of the more sizeable markets available to American industry are singularly unpersuasive for the techniques with which we have been concerned. It will be recalled that these techniques had all been developed in the United States by the 1850s. Taking size of populations, average income levels, and transport costs into account, the size of the market for American products before midcentury was certainly smaller than that for the British. In the early decades of the nineteenth century, when many of the techniques were first developed, the size of the American market was far smaller. This point is a general one, of course, and does not apply to the size of the market for particular products.

possibilities at the other.[1] In this framework the third dimension, natural resources, and substitution possibilities between resources and the other two factors, are neglected. But it is possible that many observed differences in adopted technology may be accounted for at this frontier. There is evidence, admittedly fragmentary, which suggests that many techniques which were popular in America and neglected in England were not only labour-saving but also resource-extravagant. Their rapid adoption in America and neglect in England was attributable not only or perhaps not even primarily to relative capital-labour prices but to the cheapness of some natural resource inputs in America by comparison with their high price in England. This point applies with particular force to woodworking machines, which were widely utilized in America and equally neglected in England in the first half of the nineteenth century in spite of the extraordinary start which had been made at the beginning of the century in developing woodworking machinery by Bentham and Brunel.[2] One of the major differences between shaping wood by machinery and by skilled workers with tools was that the former method, while substituting capital for labour, also by its extravagance in wood consumption, in effect substituted resources for labour. The latter method, involving intensive doses of labour, in effect substituted relatively cheap labour for relatively expensive wood.

Testimony of some informed observers suggests that American gunstocking machinery wasted a great deal of wood–which, if true, might go a long way toward accounting for differences in wood-shaping techniques between the United States and Great Britain.[3] It should be recalled that the gunstocking machinery acquired by the Enfield Armoury

---

[1] Notice that Habakkuk's book, *American and British Technology in the 19th Century*, is subtitled 'The Search for Labour-saving Inventions.' With equal justice one could speak in nineteenth-century America of 'The Search for Resource-intensive Inventions.' Habakkuk explicitly recognizes the possible importance of this third factor but backs away from it and chooses to concentrate directly on substitution possibilities between capital and labour rather than between natural resources on the one hand and capital and labour on the other. He concludes his discussion of this issue by stating (p. 34): 'We feel justified therefore in proceeding on the assumption that the dearness of American labour is the most fruitful point on which to concentrate in an examination of the economic influences on American technology.'

[2] S.W. Worssam, Jr., claimed in a letter to *The Engineer* that several woodworking machines which were patented in America and brought to England had been invented many years earlier by Bentham and Brunel. He cited Bentham's planer and Bramah's squaring-up machine, but it is not clear how similar these machines were to those of Bentham and Brunel. He also claimed that there were cases of Americans collecting royalties on their woodworking patents in England which were actually of English origin, and which had earlier been patented in England, but apparently never used. *The Engineer* XV (6 February 1863) 80.

[3] *Report of the Small Arms Committee*, Q. 7273–Q. 7281, Q. 7520–Q. 7521; G. L. Molesworth 'On the conversion of wood by machinery' *Proceedings of the Institution of Civil Engineers* XVII (1857–58) 22, 45-6.

from the Ames Manufacturing Company was of a design which incorporated major recent improvements. It is at least possible that the American stocking machinery, vintage 1835 or 1840, would not have been worth adopting in England at prevailing English timber prices.

Finally, it may be appropriate to make some specific observations on the apparent incapacity of the British firearms industry to generate novel productive techniques in any sense comparable to the American firearms industry. Doubtless some of the difficulty lay in the highly unusual nature of the government's contractual relations with individual firms, in its practice of entering into separate contracts with each of the materials suppliers and then with the eventual setters-up. This peculiar arrangement had originated in the government's determination to control the quality of the inputs into English muskets, as a result of which government inspection procedures were interposed between the separate materials producers and the contractors responsible for setting-up the arms. It is difficult to avoid the conclusion that the longer-term cost of these arrangements was very high, for it effectively perpetuated an artificial separation between individual producers in a productive sequence, each of whose activities was closely dependent upon one another. Such fragmentation meant that the ordinary difficulties of innovation or alteration of techniques were multiplied several-fold, because each producer was arbitrarily subject to constraints not inherent in the product or the technology, but resulting from the peculiar way in which the industry was organized for the production of the firearm. No one individual or small group of individuals were in a position to examine the entire sequence of productive processes with a view to making a succession of alterations through a series of interdependent stages, and certainly no individual producers were in a position to examine the whole manufacturing sequence to consider major alterations of productive methods. To some extent the arrangements were obviously wasteful in ways which would have been quickly eliminated had production been centralized with more effective supervision and dovetailing of sequential processes. In some cases one craftsman actually undid the work of a craftsman earlier in the productive sequence–e.g., destroying the finish and smoothness which a worker had been at pains to bestow on a component before it left his hands.[1]

This point, however, should not be pressed too far. The government's contracting system did not create the remarkable subdivision of handicraft skills which were so characteristic of the Birmingham light metal trades. It might be argued that the contracting system contributed to the freezing of the technology and organizational structure of the military firearms industry which, as American experience showed, was a sector where the need for large quantities of a standardized commodity provided fertile ground for the growth of a mass-production technology.

But, after all, much that was wrong with the organization of the supply of firearms was also wrong with handicraft organization in general.

---

[1] G. C. Allen, op. cit. 119.

Introduction

Looked at from one point of view, the Birmingham light metals trades appear to have been wonderfully inventive and adaptive. Throughout its history Birmingham had shown a quite extraordinary capacity to undertake the production of a variety of new products. When England was swept by a fashion for buckles, Birmingham promptly obliged with a profusion of buckles. When the buckles trade was devastated by a growing taste for the lowly shoelace which not even the illustrious House of Hanover could stem, the Birmingham manufacturers went on to buttons.[1] Although the region made no contribution to originating the techniques of brassfounding generally, it rapidly assimilated the techniques and applied them to numerous new uses.[2]

The apparent adaptiveness of the Birmingham metal trades was deceiving, however, for it was adaptive only within the limits of a sharply circumscribed technology. The Birmingham metal trades were capable of producing any of a wide range of articles which could be produced by highly skilled and ingenious craftsmen working only with tools and the simplest machinery such as the stamping machine which was basic to both the buckle and button trades. Birmingham's versatility with stamping and pressing machines enabled her population to undertake production of a wide range of commodities for which these techniques were important. At one time or another these machines were applied to buttons, buckles, coins, watchmaking, various ornaments, and the whole range of light metalwork generally.[3] New products could be introduced with relative ease so long as they involved only relatively small departures from known craft skills and exploited the considerable virtuosity of these skills. But these skills were at the same time the basic limitation of the Birmingham trades as well as their glory.[4] For the craft-based industries

---

[1] 'By 1786, the first note of decay was sounded in the buckle trade by the change of fashion from the buckle to the shoe-string, an innovation which was contemptuously greeted by the press. "The *literary* buckle is now superseded," runs one such comment, "by the black ribbon in the shoes of Sir Fopling's progeny, whose chief employment now is to tie it on with grace." In December, 1791, the buckle-makers petitioned the Prince of Wales to assist in giving employment to more than 20,000 persons "who in consequence of the prevalence of shoe-strings and slippers were in great distress." The prince and Duke of York thereupon ordered their gentlemen and servants to discard the offending shoe-string. A second petition was presented in 1792 to the Duke and Duchess of York, in this case appealing for the extension of the privilege of protection enjoyed by buttons to buckles also. In 1800 the king was memorialized with a prayer that he would set the fashion of wearing buckles, and so "crush the unmanly custom of wearing shoe-strings."' William Page (ed.) *The Victoria History of the County of Warwick* II (London 1906) 241.

[2] Timmins (ed.) op. cit. 228-9.

[3] Richard B. Prosser *Birmingham Inventors and Inventions* (Birmingham 1881) ch. 26. See also Prosser's comments on Birmingham's prominence in the nail trade, pp. 70-77.

[4] A comprehensive study of Adam Smith's treatment of the division of labour which is not confined to the celebrated passages in Book I of the *Wealth of Nations*, will show that Smith was well aware of the restrictive as

such as the Birmingham gun trade or the Wolverhampton lock trade lacked either the internal resources or the connections with the appropriate external sources which would enable them to break outside of the limited framework imposed upon them by their reliance on human skills. In an important sense such craft skills are dead ends, for several reasons. They generated attitudes and traditions characterized by a preoccupation with qualitative aspects of the final product which were at best irrelevant and at worst hostile to the solution of problems associated with raising resource productivity.[1] The accumulated skills and traditions constituted an obstacle to the learning process which is a prerequisite to the acquisition of new and radically different techniques. In this particular and restrictive sense there may have been a 'penalty of taking the lead'.[2] It may have been easier both to develop and to master the mechanical engineering technology of the nineteenth century with no knowledge whatever of handicraft techniques than with such a knowledge.[3] The possession of certain skills may be worse than merely useless when they constitute impediments to the mastery of new and more useful techniques.

The Birmingham metal trades were superbly successful in exploiting human skills. They lacked the capacity, however, for incorporating these skills in machinery. Moreover, the highly self-contained nature of the organization of these trades tended to cut them off from contacts with other industries – particularly the producers of capital goods – from whom they might have learned or borrowed. These trades, by the middle of the nineteenth century, had gone about as far as possible given their reliance upon the speed, strength, precision and dexterity of the human hand. Further technical progress involved a recourse to completely different techniques for ways of achieving form and precision in the shaping of materials. But Birmingham's industrial history had left it ill-prepared in the novel engineering techniques and broader range of mechanical skills upon which further technical progress depended.

---

well as the beneficent aspects of increasing division of labour. Indeed, Smith had some sophisticated views on the social determinants of technological change which have long escaped attention and which are relevant to the present discussion. See Nathan Rosenberg 'Adam Smith on the division of labour: two views or one?' *Economica* (May 1965) 127-39.

[1] The incredible ingenuity lavished by English craftsmen upon locks, keys, and chests may be seen by examining the numerous illustrations in George Price *A Treatise on Fire and Thief-Proof Depositories and Locks and Keys* (London 1856).

[2] Cf. Edward Ames and Nathan Rosenberg 'Changing technological leadership and industrial growth' *Economic Journal* (March 1963) 13-31.

[3] Cf. Habakkuk, op. cit. 116, where it is argued that '. . . the absence in America of European traditions of very high craftsmanship may have been a positive advantage in the development of new mechanical methods.' Somewhat analogous also is S. B. Saul's argument that the British motor car industry was handicapped in its early years because of the baleful influence resulting from its domination by people trained in the methods and outlook of traditional engineering. S. B. Saul 'The motor industry in Britain to 1914' *Business History* (December 1962) 22-44.

Appendix

Both in England and America there has been a deplorable tendency to ignore or neglect entrepreneurial talents, even of a most unusual sort, when these talents find their expression in the public sector. Thus, John Anderson is almost entirely unknown, and Samuel Bentham, in spite of his extraordinary achievements in inventing and introducing machinery and in effecting a major reorganization in British naval yards, has been relegated to an ironic footnote on history where he is identified only as the person who did *not* invent the blockmaking machinery at Portsmouth. Similarly in America, Eli Whitney's name has become legendary, Simeon North is widely known, whereas the major achievements of John Hall at the national armoury at Harper's Ferry are hardly ever referred to.[1] The explanation for this neglect deserves careful investigation, but it is far beyond the scope of the present discussion. Here we may appropriately give some further indication of Anderson's other accomplishments.

John Anderson was born at Woodside, near Aberdeen, on 9 December 1814. He acquired his mechanical experience and training during a thirteen-year period, 1829 to 1842, which included employment in two of the foremost Manchester establishments of his day, that of Sharp, Roberts and Company, and William Fairbairn and Company. Just before he went to Woolwich Arsenal he was employed by the Napier engineering firm of London, which at the time performed a great deal of work for the Board of Ordnance.[2] In 1841 he began his long period of public employment at Woolwich when he was appointed Superintendent Engineer of the Royal Brass Foundry.

The Arsenal at that time was in a backward and neglected condition, and had experienced virtually no infusion of recent mechanical technology. For the most part the departments of the Arsenal had the same equipment as they had had in 1815. There were, in fact, only two steam engines in the entire arsenal, both in the royal carriage department, rated at 20 and 12 horse-power. The gun factories in 1842 were employing the same three gun-boring machines which they had acquired from The Hague sometime around 1780. These machines had been worked by horse-power and had been in constant use for 60 years or so. Further, '. . . the smoothing of the trunnions and all the shaping and finishing of the outside of the guns was performed, slowly and expensively, by hand labour'.[3]

With the strong and influential support of his superior, Colonel Dundas, Anderson initiated a series of sweeping changes in the operation of

---

[1] But see the admirable recent article by Robert Woodbury 'The legend of Eli Whitney and interchangeable parts' op. cit., which goes a considerable way in attempting to redress the balance.

[2] Charles Wilson and William Reader *Men and Machines: A History of D. Napier and Son, Engineers, Ltd., 1808–1958* (London 1958) 28-30.

[3] Anderson *General Statement* 6.

his department. Not only was machinery purchased, but increasingly the Royal Brass Foundry became an engineering workshop, designing and producing its own machinery and, eventually, preparing machinery for other departments–Royal Laboratory, Royal Carriage Department, Waltham Abbey, and Enfield.

By 1848 the government was obviously involved in a large-scale effort to modernize the operation of its manufacturing departments. Expenditures rose much more rapidly with the Crimean crisis, and in the four-year period 1853–4 to 1856–7 inclusive, the manufacturing branches of the War Department expended over £600,000 in mechanizing their operations.[1] The revitalization of the factories, however, had begun well before the Crimean War, and much of this effort had taken place as a result of the initiative and under the personal supervision of John Anderson. Many of the machines, moreover, were of his own design and several of the most important were of his own invention. In looking back from the vantage point of January 1857 Anderson wrote:

> Without prejudice, the United States of America, the continent of Europe, and our own country, have been searched for the most superior appliances, and hundreds of machines have been designed for purposes peculiar to the War Department, but which may be usefully employed in the general manufactures of the kingdom.
>
> From these important changes and improvements the greater proportion of war stores can now be produced with unskilled labour, the form, dimensions, quality, and quantity of the produce being mostly dependent on self-acting apparatus, a system of operation which has been more fully developed in the wood and

---

[1] Ibid. 36. The failure of British Ordnance to keep pace with mechanical developments in the first half of the nineteenth century was a matter of growing national indignation. In an article discussing Whitworth's rifle experiments in 1855 *The Times* sardonically commented: 'As far as the manufacture of ordnance on the present system goes, the Russians are fully our equals, and, though it may be all very chivalrous to fight them with similar weapons, this mechanical nation has surely a right to expect that, on such a point, the British army should have a decisive superiority over that of any other country.' *The Times*, 13 July 1855. The extent to which chivalrous sentiments were in fact declining may be measured by quoting an earlier reaction to the introduction of the rifle, an instrument of death which offered greater accuracy to its user than the smooth bore musket. 'Besides the ordinary musket, there has been of late years introduced into the military service a more curious and murderous firearm called the rifle, and by the French, in whose hands it was formerly turned to dreadful account, *arquebuse rayée*, its use in war being to shoot officers and others picked out of the ranks by the keen eye and unerring hand of men trained to the duty, and designated riflemen. Humanity shrinks at the idea of thus selecting, with personal precision, and for instant death, even the enemies of our country, and surely the practice ought to be condemned and abandoned by every civilised nation.' *A Treatise on the Progressive Improvement and Present State of the Manufactures in Metal* II (London 1833) 108.

> metal manufactures of the War Department than in any private establishment with which I am acquainted.[1]

In renovating the operations of the factories Anderson searched systematically, both at home and abroad, for techniques which had application in the government establishments. He introduced rigorous techniques of observation and experimental methods wherever they were relevant. This was pre-eminently the case in the casting of guns where an understanding of the properties and qualities of metals was critical to the efficiency of the guns and the prevention of explosions.

> In this establishment the great aim will be to obtain metal of the strongest quality; and for this purpose an extensive course of experiments, both chemical and physical, is now being carried out in the Department, not only with the principal irons of the United Kingdom, but also with Swedish, Indian, and Nova Scotia irons, arrangements having been made to procure samples of irons from those countries. It is hoped that the results of these experiments will tend to throw additional light on this important but hitherto neglected subject, and will lead to valuable national results, not only with reference to the casting of guns, but likewise with regard to all cast-iron structures where strength is an object.[2]

In the testing of metals Anderson introduced the tension and torsion machines which the Committee had purchased from Mr Kemble of Cold-springs Foundry, New York. This 'very superior' machine was '. . . similar in construction to the testing machine employed by the gun foundries in America, where great attention has been paid of late to the strength of materials'.[3] This machine, and modifications of it, was very extensively employed in subsequent years in testing the tensile and compressive resistances of metals.[4]

As part of the systematic approach to its problems the gun factory also made use, in the mid-1850s, of photographic equipment to attempt to establish the causes of the bursting of guns.[5] This must surely have been one of the earliest examples of the application of photography to industrial use.

---

[1] John Anderson 'On the application of machinery in the war department' *Journal of the Society of Arts* v, no. 219 (30 January 1857) 155. See also O.F.G.Hogg 'The development of engineering at the Royal Arsenal' *Transactions of the Newcomen Society* XXXII, 29-41.

[2] Anderson *General Statement* 8. Anderson makes the highly interesting observation that standards of cast-iron tenacity were much higher in America than in England. 'It is seldom that our cast iron reaches a tenacity of 20,000 pounds per square inch, the average being about 17,000 pounds, while the American Government insist upon a minimum of strength for their guns equal to a tenacity of 34,000 pounds, and generally obtain 40,000 and occasionally as high as 45,000 pounds.' Loc. cit.

[3] Ibid. 8.

[4] John Anderson *The Strength of Materials and Structures* (London 1872) ch. 3: 'On a Machine for testing the Strength and Elasticity of Materials.'

[5] Anderson *General Statement* 6.

A complete recitation of Anderson's numerous mechanical inventions would be very tedious. His energy and originality were felt in all the government establishments, and the present discussion will be confined to some of his most important and original contributions.[1]

Anderson introduced a variety of specialized machines which replaced hand labour in the preparation of large guns, particularly in connection with the boring of the gun and the shaping of irregular parts. He also devised a rifling machine which was described as '... a first-rate application of exact and scientific principles'.[2]

The introduction of the Minié rifle created a demand for machinery to produce a special, elongated bullet, accurate in weight and dimensions, in very large numbers. Bullets were, at the time, cast by a hand-moulding technique, at a rate of about 500 bullets per man per day. Although it would have been possible to devise casting methods which would have increased output 'at least ten-fold', this could have been accomplished only at the cost of variations in the dimension and specific gravity of the bullets. This, in turn, would have reduced the accuracy of firing.[3] Anderson introduced a new technique, involving the use of hydraulic pressure, which made it possible to produce bullets in enormous quantities and according to precise specifications.[4] Anderson stated, in 1873, that 'These machines were in constant use for 18 years, they were capable of producing 40,000 bullets per hour, at a cost, for labour of 5½d. per thousand, which includes all the cost of maintaining the plant in good order. The cost of casting by the former method was over 5s. the thousand, hence, irrespective of any advantage in point of quality, there is, by this

---

[1] A more comprehensive account of Anderson's mechanical inventions will be found in his *Statement of Services*.

[2] D.K. Clark op. cit. 177. Of the same machine the Jurors at the International Exhibition of 1862 stated: 'The rifling machine exhibited by Smith, Beacock, and Tannett is accurately constructed upon the most ingenious plans of the Superintendent of Gun Factories at Woolwich, Mr Anderson, and the small one exhibited by Greenwood and Batley is a reduced copy of the same. The beautiful action of the rifling-bar grooving the gun along its helical path with the greatest steadiness, its timely withdrawing of the cutters, the exact performance of the split pinion and rack moved by the inclined guide, and all the other numerous and convenient special contrivances can only be appreciated by those who see them at work.' *International Exhibition, 1862. Reports by the Juries* (London 1863) Class VII, Section B, p. 13. Cf. also *The Artizan* (1 December 1857) 265-6.

[3] Anderson *Statement of Services* 11.

[4] 'In this manufacture the lead is first put into a cylinder, from which it is squirted into a long rod by means of hydraulic pressure; the lead rod is then wound upon iron reels, which are transferred to the machinery for compressing the bullet. The bullet machines are entirely self-acting, and unwind the lead rod from the reels as they require it, first cutting off the required quantity, then compressing it into form, and then delivering the bullet ready for the cartridge.' John Anderson, 'On the application of machinery in the war department' 157-8. Cf. also *The Artizan* (1 November 1856) 241-2, and (1 December 1856) 265-6.

arrangement, a clear saving of 4s. 6d. per thousand.'[1]

Anderson's bullet-making machinery was immediately recognized for its qualities of precision and economy. Indeed, the Russian Government promptly requested a set of drawings of the bullet machinery and the Board of Ordnance approved their delivery–in the early months of 1853.[2]

One of Anderson's most important achievements was his responsibility, over a period of several years, for the complete renovation of the gunpowder mills at Waltham Abbey. These mills had been allowed to fall into a state of serious neglect and disrepair, in part due to the large stock of gunpowder which remained on hand at the end of the war in 1815. In fact '... most of the machinery, water-wheels, and gearing being made of wood, the manufactory in 1840 was almost worn out, and incapable of producing more than 3,500 barrels of powder per annum'.[3]

Anderson designed or invented most of the machinery which was installed in the new plant. His most important single invention here was a machine for granulating gunpowder which replaced the old and highly dangerous system of hand-corning. The machine required no human attendance.[4] Throughout the entire plant Anderson substituted metal for wood in the construction of the machinery–including light iron water-wheels for the wooden water-wheels erected by Smeaton many years earlier. He reorganized and relocated the machinery in a way which reduced the possibility of explosions and at the same time minimized the damage from such accidents. He made extensive use of powerful hydraulic presses, and introduced a wide array of newly designed dusting, glazing, mixing, and drying machinery.[5]

One of Anderson's most important contributions to the progress of the war in the Crimea came in the closing months of 1854 in response to an urgent plea from Lord Raglan for Lancaster shells. These shells, produced out of a single piece of wrought iron and shaped much like an oversized champagne bottle, could not be acquired by contract from private sources in time to meet pressing military requirements. Under Anderson's planning and supervision, a government foundry was erected, equipped with the requisite machinery, and was in full operation within a period of two months. Many of the machines introduced into this plant were specifically designed by Anderson for the Lancaster shell. 'One machine which the Writer designed for the Lancaster shell manufacture, viz., for converting the embryo shell into the bottle form, was

---

[1] Anderson *Statement of Services* 12.

[2] Ibid. 13.

[3] Anderson *General Statement* 32.

[4] The machine '... supplied its own wants with pressed cake;–it separated the produce into the several different sorts; as the boxes became filled the machine of its own accord removed them and brought empty ones into their place, and after exhausting the entire supply, it first stopped its own motion, and when all was safe it rang a bell for the attendant.' *Statement of Services* 14.

[5] Anderson 'On the application of machinery in the war department' 161; *General Statement* 31-3; *Statement of Services* 14-16.

one of the most successful of all his contrivances, and reduced the cost from over six pounds to less than one pound per shell. . . .'[1]

One of the most novel contrivances with which Anderson was associated was the so-called Floating Factory which it was determined to send to the Crimea in 1855 in order to assist the military forces with all sorts of essential engineering and carpentry work. To make the appropriate arrangements a committee was appointed consisting of Colonel Tulloh, RA, Captain Collinson, RE, and John Anderson. Although many of the details for the 'Chasseur', as it was called, were worked out by all three members of the committee, all of the internal arrangements of the ship and its machinery were Anderson's responsibility.

> A steam vessel of 600 tons having been procured, a plan was drawn out for converting her into an engineer's workshop, fitted with 4 forges and 28 heavy machines, consisting of different sized lathes, planing, slotting, drilling, screwing and other tools, which were placed on a floor in the hold, and worked with overhead shafting, while galleries ran along the sides of the vessel containing a number of vice benches; a complete saw mill with three benches was placed on the deck, which also carried a cupola, with fan, portable steam engine, and all the requisite plant of a brass and iron foundry, either to be used in the vessel or on shore, as might be found most convenient according to the situation of the army.
>
> The power to work the machinery was obtained either from the 70 horse power engines of the ship or from the portable engine, the gearing of the vessel being arranged either to work the screw or to give motion to the machinery . . .
>
> The factory contained 24 chests of tools of different trades, and also the tools of the brick-maker and well-sinker; it was provided with stores of wood, copper, tin, steel, and iron, sand, fire bricks, fire clay, crucibles, and a general assortment of small stores used in the various branches of handicraft . . .[2]

The vessel, fully equipped and staffed with workmen, was ready in ten weeks after the order for its preparation had been given. It performed indispensable services for the armed forces upon its arrival in Balaclava. It was acclaimed as a great mechanical triumph by all who visited it, not the least of whom were Russian officers who were unstinting in their praise.

Anderson's other technical contributions in the government factories cover the entire range of machines for the production of military supplies, and are described in some detail in his *Statement of Services*. In addition to his concern with substituting machinery for the handicraft skills

[1] *Statement of Services* 38. See also 'On the application of machinery in the war department' 158-9; *General Statement* 16-17.

[2] *General Statement* 34. A complete list of the equipment on board appears on p. 35. Cf. also *Statement of Services*, 38-9, 43-4.

he was responsible for a large-scale reorganization of the government factories, rearranging the sequence of the work process along the lines of a more modern efficiency expert, and introducing a meaningful system of wage differentials and incentive payment schemes.

Anderson's later activities are not of major interest. He was appointed Assistant Superintendent of Machinery under Sir William Armstrong in 1859 when it was decided to adopt the Armstrong gun,[1] and Superintendent of Machinery in 1866. He delivered the Cantor Lectures in 1869.[2] He retired in 1872. He served as a juror at four International Exhibitions, London in 1862, Paris in 1867, Vienna in 1873, and Philadelphia in 1876, and his reports constitute highly astute observations both on recent technological changes and on national differences in machine tool design. Throughout his later years he remained a devoted exponent of American techniques and urged their adoption upon his countrymen. In his last report, from Philadelphia in 1876, he stated:

> To realize the nature of the competition that awaits us, their [American] factories and workshops have to be inspected, in order to see the variety of special tools that are being introduced, both to insure precision and to economize labour; this system of special tools is extending into almost every branch of industry where articles have to be repeated. This applies to furniture, hardware, clocks, watches, small arms ammunition, and to an endless variety of other things. The articles so made are not only good in quality, but the cost of production is extremely low, notwithstanding that those employed earn high pay.[3]

Anderson's book, *Strength of Materials and Structures*, was widely circulated, in both England and America. The degree of Doctor of Laws was conferred upon him by the University of St Andrews. He was knighted and received the French Legion of Honour in 1878, and he died in 1886.[4]

---

[1] This entire episode, the so-called 'Battle of the Guns', which raged in particular over the relative merits of Armstrong and Whitworth improved ordnance guns, was exhaustively investigated by a parliamentary committee. See *Parliamentary Papers* XI (1863) 'Report from the Select Committee on Ordnance'. The very detailed minutes of evidence include statements from Whitworth, Armstrong, and Anderson.

[2] *Journal of the Society of Arts* XVII 711-19, 731-41, 746-56, 763-72.

[3] 'Reports on the Philadelphia International Exhibition of 1876' *Parliamentary Papers* 34 (1877) 235.

[4] He had also received a medal from the French Government in 1855 '. . . for the introduction of the seamless bag for cartridge purposes'. *Statement of Services* 55. Short biographical sketches of Anderson may be found in W.T.Vincent *The Records of the Woolwich District*, 2 vols. vol. I (London 1888–90) 302; *Proceedings of the Institution of Mechanical Engineers* (1886) 460-1; and *Proceedings of the Institution of Civil Engineers*, vol. 86, 346-53.

REPORT OF THE COMMITTEE ON THE MACHINERY OF THE UNITED STATES OF AMERICA. PRESENTED TO THE HOUSE OF COMMONS, IN PURSUANCE OF THEIR ADDRESS OF THE 10TH JULY, 1855. PRINTED BY HARRISON AND SONS, LONDON

Introduction

Before entering upon this Report the Committee feel that it will be necessary for them to explain to the Honourable Board the system they have endeavoured to pursue in order to render it as clear as possible and avoid repetition. The field over which they have travelled is so extensive, and the number of things that have come under their notice so numerous and various, that it is with the greatest difficulty they have been able to reduce them to anything like order for the work done in the various ordnance manufacturing departments at Woolwich embraces almost every branch of trade, and the Committee therefore deemed it advisable to let as little as possible connected with the private manufactures of the country escape them, and visited many works, which at first glance may seem to be totally irrelevant to the branches of manufacture to which their attention was more particularly directed.

After completing their tour through all the principal manufacturing towns of the Northern States, the Committee returned to New York, and at once deliberated on what form they should adopt for their final Report, and came to the conclusion that it would be necessary to divide it into five chapters under the following headings, viz.:

1. Orders and instructions received by the Committee
2. The Journal
3. Fire-arms and their manufacture
4. Labour-saving machinery, relating to the other manufacturing departments
5. List of contracts entered into.

CHAPTER 1. The first of these chapters contains the orders issued to the Committee on leaving England; such communications as they received from the Master-General and Honourable Board, whilst they have been in America, and how their attentions were influenced by these instructions.

CHAPTER 2. The second chapter is made up from the Journal kept by the Committee, and gives the route they followed, with their reasons for laying down this route and adhering to it; all the towns visited by them are given in regular order, and also the different armories, arsenals, or private works gone over in each, with a slight sketch of how they are conducted, and to what extent machinery is applied in them. A list, regularly numbered throughout of such things as struck the Committee as useful or new is given, and opposite each is the number of the page at which a detailed description of it will be found.

CHAPTER 3. The third chapter gives a list and description of the different arms, whether service or private, that have come under the notice of the Committee whilst in the United States, which is followed by a detailed account of their manufacture, as conducted in the best Government and private establishments visited, viz., at Springfield Armory, and at Sharpe's rifle manufactory at Hartford.

CHAPTER 4. The fourth chapter contains descriptions of various branches of manufacture, and of different labour-saving machines and other articles observed as worthy of notice, having reference to the manufacturing departments, and is divided into headings as follows. Inspector of Artillery's Department; Royal Laboratory; Royal Carriage Department; Miscellaneous.

CHAPTER 5. The fifth chapter gives a detailed list of the contracts accepted by the Committee for machinery, etc., with their reasons for entering into them.

By this arrangement the Committee hope that their Report will be found concise, and that any particular machine or branch of trade to which reference may be required will be easily found.

The time the Committee have remained in the United States, so long exceeds what they had anticipated before leaving England, that they think it due to the Master-General and Honourable Board to explain the reasons for this.

As long as the expenditure for the Small-Arm Manufactory, was limited to £10,000; finding that this sum would not go very far in machinery such as they required, some of which is expensive, the Committee were unwilling to give orders to any extent, fearing that, before their tour was completed, they might fall in with machines they liked better, and not be able to procure them, on account of having already entered into contracts for tools of a similar character, and for this reason they did not order anything but the 'stocking machinery' at Springfield, which they saw was indispensable, and knew could not be surpassed anywhere in the country, and some machines from Messrs. Robbins and Lawrence of Windsor, Vermont, whose tender, however, they reduced very considerably, by striking off a large number of machines, and had the sum placed at their disposal not been increased, several valuable machines could not have been procured, and the plant altogether would have been very imperfect. The Committee therefore on their route, took notes of any serviceable-looking machinery they saw, which they thought it advisable to obtain, intending should they see nothing they liked better afterwards, to return and contract for it.

The delay thus occasioned, together with the oppressive heat of the weather in July, prevented the Committee from getting over their business so expeditiously as they had hoped, but they trust this explanation will satisfy the Master-General and Honourable Board, that they have used their best endeavours to urge on the important business intrusted to them, as much as was in their power, and which they think they have brought to a conclusion that will in many respects be beneficial to the service.

1

[Orders and Instructions]

Owing to the delays constantly recurring in the fulfilment of contracts for arms, the high price demanded by contractors, and the inconvenience occasioned to the service by these causes, the Honourable Board of Ordnance, towards the end of the year 1853, considered it advisable, in order to secure a regular supply of them, to take this branch of manufacture into their own hands, and erect a Government establishment capable of producing muskets in large numbers, and at a moderate price, by the introduction of machinery into every part of the manufacture where it was applicable. They were the more disposed to do this by the very low state of the gun trade in Great Britain, owing partly to a great many of the workmen having emigrated and gone into other trades, and partly to the unenterprising way in which this branch of manufacture was in general conducted. Having caused a plan of the building they proposed to erect to be drawn out, and a calculation made of how much the necessary machinery would cost, they included the sum which amounted to £150,000 in the Ordnance estimates, with the intention of getting the establishment erected and set to work as speedily as possible; and hearing from Mr Whitworth and others that machinery was extensively applied to this branch of manufacture in the United States of America, where, on account of the high price of labour, the whole energy of the people is directed to improving and inventing labour-saving machinery, the Honourable Board considered it advisable to send over to that country some of their officers, with the view of obtaining every information in their power connected with the manufacture of arms as there conducted, and with the power of buying such machinery as they might consider would be more productive than that used in England for similar purposes, as well as any good and new machines, tools, etc., they might see, their expenditure being limited to the sum of £30,000.

In pursuance of their views the Honourable Board selected for this mission the following officers, viz.: Lieutenant-Colonel Burn, Royal Artillery; Lieutenant Warlow, Royal Artillery; and Mr Anderson, the

Ordnance Inspector of Machinery, and conveyed them their instructions by the following Minute.

(Confidential) *Copy of the Board's Minute, dated 13th February, 1854, approved by the Master-General*

Submit to the Master-General that Colonel Burn be directed to proceed to the United States of America on the 18th instant, for the purpose of inspecting the different gun factories in that country, and purchasing such machinery and models as may be necessary for the proposed gun factory at Enfield.

That Lieutenant Warlow and Mr Anderson, who are to be associated with Colonel Burn in these objects, proceed to the United States on the 4th March.

That Colonel Burn at once put himself into communication with Mr Whitworth, and ascertain from him the name and residence of the principal makers of machinery in the United States, and the gun factories, whether in the hands of the Government or of private individuals, that it is desirable for him to visit.

That upon his arrival at New York he proceed accordingly to visit them, and so enable himself to direct the attention of Lieutenant Warlow and Mr Anderson to the establishments best worthy of their notice.

That upon the arrival of those gentlemen he proceed in their company to visit these establishments.

That after the inspection Colonel Burn, Lieutenant Warlow, and Mr Anderson proceed to draw up a report describing the machinery or models which they consider it necessary to purchase or to order in the United States, and that after each gentleman has attached his name to this report, Colonel Burn transmit it to the Master-General.

That when this report has been so transmitted, without waiting for any reply or communication from this country, they proceed at once to purchase or to order the machinery and models which in the report they have transmitted they recommend to be purchased or ordered.

That while they make whatever pecuniary arrangements are usual in the United States as to paying or not paying a deposit on ordering any article, they be empowered to pay ready money for anything they purchase, and that for this purpose the Lords of the Treasury be requested to authorize the Paymaster-General to honour any bills they may draw upon him to the amount of thirty thousand pounds.

That while, on the one hand, great care should be taken not to expend money uselessly in the purchase of machinery obviously imperfect and easily to be improved upon, on the other, it is to be remembered that no aim at ideal perfection is to delay the procuring articles, the improvement of which, though possible will obviously take a considerable time, for the saving made by a machine in a few weeks will in many cases repay its full cost.

It is to be borne in mind that while the improvement of the royal laboratory is principally important on the grounds of economy, as by additional expenditure any additional amount of the articles there manu-

factured can be procured, the speedy completion of the Gun Factory is regarded as a matter of the last importance to the public interests, on grounds entirely separate from mere pecuniary saving.
A true copy
signed J. Wood

(Confidential)
Sir, *Office of Ordnance, February 13, 1854*
I have received the Board's commands to transmit for your information and guidance the inclosed copy of a Minute of the Master-General and Board, dated this day; and I am to signify their request that you will proceed to the United States of America on the 18th instant, and act as therein directed.

I am to add that Lieutenant Warlow, Royal Artillery, and Mr Anderson, have been instructed to proceed thither on the 4th March next, and to place themselves under your direction; and I am to request you will give them such instructions previous to your departure as you may consider necessary.
I am, etc.
signed J. Wood
To: Lieut-Col. Burn, Assistant Inspector of Artillery, Woolwich.

The Honourable Board then applied to the Lords of the Treasury requesting that the sum of £30,000 might be placed at the disposal of the Committee, and received in reply the following communication from Sir Charles Trevelyan.

Sir, *Treasury Chambers, February 17, 1854*
I am commanded by the Lords Commissioners of Her Majesty's Treasury, to acquaint you, for the information of the Master-General and Board of Ordnance with reference to your letter, dated 15th instant, that my Lords have desired the Paymaster-General to pay any bills that may be drawn upon him from the United States by Lieutenant-Colonel Burn, Lieutenant Warlow, and Mr Anderson, not exceeding in the aggregate, £30,000, for the purchase of machinery for a new Small-Arm Manufactory, and to charge the amount to the Ordnance grant.
I am, etc.
signed C. E. Trevelyan
To: Secretary to the Ordnance, etc.
I certify the above to be a true copy.
signed J. Wood *Secretary to the Board of Ordnance*
February 18, 1854.

The above was transmitted to Lieutenant-Colonel Burn with the following communication from the Secretary of the Board.

Sir, *Office of Ordnance, February 18, 1854*
With reference to my communication of the 13th instant, I am directed by the Board to transmit for your information and guidance, the inclosed copy of a letter from the Lords Commissioners of Her Majesty's Treasury,

dated 17th instant, authorizing the Paymaster-General to pay any bills that may be drawn upon him by yourself, Lieutenant Warlow, and Mr Anderson, for the purchase of machinery in the United States of America, not exceeding in the aggregate £30,000.
I have, etc.
signed J. Wood
To: Lieut.-Col. Burn, Royal Artillery, Woolwich.

In consequence of the orders received by him, Lieutenant-Colonel Burn proceeded at once to America, leaving England on the 25th of February, in order to obtain from the heads of Public Departments in the United States, all the information he could on the subject of the manufacture of small arms, and also letters of introduction to the different superintendents of Government works, that no unnecessary delay might occur after the other members of the Committee arrived, who it was proposed should follow him in a week's time.

When the Ordnance estimates were laid before the House of Commons by Mr Monsell, MP, Clerk of the Ordnance, the motion for a grant of £150,000 to erect a Small-Arms Factory was opposed on the grounds that the manufacturers of Birmingham and London were quite capable of supplying the Board of Ordnance with any quantity of arms necessary at a lower rate than they could themselves make them, and a committee of the House was appointed to take the subject under their consideration, and report upon it.

As the Honourable Board of Ordnance saw that if the Committee ordered to proceed to America were detained till this report was completed, it would occasion considerable delay in carrying out their plans, they determined to keep to their original intention as regarded sending them to America, but Mr Anderson's evidence being very important to the Committee of the House of Commons on the supply of small arms, that gentleman and Lieutenant Warlow were not able to leave Woolwich till the 13th of April, when they proceeded to Liverpool, and embarked for the United States on the 15th of that month, having received before starting the following Minute of altered instructions, from the Honourable Board of Ordnance, which reduced the sum placed at their command from £30,000 to £10,000, for the proposed Small-Arm Factory, and £5,000 for the Royal Laboratory, and Royal Carriage Department.

(Confidential) *Office of Ordnance, April 11, 1854*
Ordered that Colonel Burn, Lieutenant Warlow, and Mr Anderson, be informed with reference to the Board's Minute of the 13th February last, that the greatest care is to be taken by them not to purchase or order any machinery of the working of which they have not satisfied themselves beyond reasonable doubt.

That their special attention is to be called to the quality of wood manufactured in the American stocking machines, and that they are to take out with them, and to try the effect of those machines on English walnut before they decide on purchasing them.

That they are not, either by purchase or order, to spend, or involve the spending of a larger sum than £10,000 for the proposed Small-Arm Factory, nor more than £5000 on account of the Royal Laboratory and Royal Carriage Department.

That this limitation of expenditure is not to prevent them from putting this Department in a position to procure, with as little delay as possible, additional machinery in the United States if they so think fit.

That they are to report with as little delay as possible, on the cheapness and efficiency of the muskets manufactured in the United States by machinery, and are particularly to express their opinion as to the finish and efficiency of each of the parts of one of the muskets so manufactured as compared with the Minié rifle of 1853.

signed W.R., W.M., T.H.

Lieutenant Warlow and Mr Anderson were met on their arrival at Boston by Colonel Burn, and the three proceeded together to visit the different gun manufactories, etc., of the country. Happening to meet at Springfield Mr Burton, engineer at Harper's Ferry Armory, and thinking that as that gentleman was on the point of giving up his present situation, it was possible he might be willing to tender his services to the Honourable Board, the Committee consulted him as to his feelings on the subject, and finding that he was disposed to do so, and as they thought that it would be most beneficial to the new establishment to have such a man as Mr Burton to assist in setting its machinery in motion and looking after it, Colonel Burn wrote to Mr Monsell on the subject, requesting to know the wishes of the Honourable Board, and on June 15th received at Washington the following minute from their Secretary:

Sir, *Office of Ordnance, May 31, 1854* o/334

I have the Board's commands to authorize you to engage the services of Mr Burton as Superintendent-assistant Engineer, if you consider him to be thoroughly conversant with small-arms machinery, and if Mr Burton is willing to engage himself at a reasonable rate of remuneration.

You will, of course, arrange the rate of remuneration Mr Burton is to receive before he is engaged.

I have, etc.

signed J. Wood

To: Lieut.-Col. Burn, R.A.

In obedience to the instructions contained in the above, the Committee entered into communication with Mr Burton, and ascertained the terms on which he was willing to engage himself, and Colonel Burn wrote the following answer to the Honourable Board's last communication:

To the Secretary to the Honourable Board of Ordnance

Sir, *Washington, D. C., June 16, 1854*

I have the honour to acknowledge the receipt of your letter of 31st ultimo o/(334), containing authority from the Board to engage the services of Mr Burton, Master Armorer at Harper's Ferry, as Superintendent-Assistant

Engineer, provided that I consider him perfectly conversant with small-arm machinery, and if Mr Burton is willing to engage himself at a reasonable rate of remuneration.

In reply I have the honour to state, that with regard to Mr Burton's qualifications for the important trust here referred to, the Board may depend upon every available means being used to ascertain their extent as soon as my health will permit me to leave Washington, where I have been detained for twelve days with severe sickness.

Referring to the question of remuneration I send herewith the copy of a letter addressed to me at Springfield by Mr Burton, from which it will be seen that he stipulates for an annual sum of £400, and expenses paid to England for himself and family.

I am unwilling to offer here my individual opinion upon the amount of remuneration without consulting the other members of the Committee, but I do not hesitate to say, that if the contemplated factory is to be carried on systematically and successfully in manufacturing muskets, the services of a fully competent person are cheaply purchased at almost any price, and particularly at the commencement of a new and great undertaking, where many difficulties and prejudices have to be overcome.

I am sorry to report that Mr Anderson has been confined to his room at Philadelphia for some time, to which place I sent him and Mr Warlow on Saturday, the 10th instant, to make some arrangement for the purchase of cap machines.

I have, etc.

signed ROB. BURN, *Lieut.-Col.*, R.A.

The Committee after having met again and completed their route as far as Springfield, there saw Mr Burton on the subject, but there being no immediate hurry about engaging him, Colonel Burn wrote again to Mr Monsell, MP, for further instructions.

Finding, when they came to examine the tenders submitted to them by contractors, that the sum of £10,000 for the proposed Small-Arm Establishment was not sufficient to enable them to make purchases of several machines which it appeared to them most desirable to obtain, the committee communicated this to the Honourable Board in the following letter, requesting at the same time that the sum placed at their command for the Small-Arms Factory might be increased from £10,000 to £12,500.

Sir, *New York, May 25, 1854*

I have the honour to report for the information of his Lordship the Master-General that the Committee appointed to visit the United States in order to purchase such machinery as might seem to them applicable to producing the parts of the Enfield pattern rifle musket, more especially the 'stock,' have, as mentioned in their report, accepted a tender for a complete set of stocking machinery, submitted by Mr Ames, of Chicopee, which amounts to ($30,860) thirty thousand eight hundred and sixty dollars, and the whole of which machinery is absolutely indispensable, admitting of no deduction in any way.

The Committee since sending their report have received a tender from

Messrs Robbins and Lawrence, of Windsor, Vermont, for machinery to produce the lock, heel-plate, and trigger-guard, etc., which in the whole amounts to ($22,665) twenty-two thousand six hundred and sixty-five dollars; but as this would with other items of expense rather exceed the sum of £10,000, allowed them by the Honourable Board for the purpose of buying machinery for the proposed Small-Arm Factory at Enfield, and as the Committee are anxious to keep within the sum named, they have been under the necessity of reducing this estimate to ($17,515) seventeen thousand five hundred and fifteen dollars, by dispensing with some tools which they consider very efficient and valuable.

They beg therefore to submit to the consideration of his Lordship the Master-General, that could the sum of £10,000 granted be increased to £12,500 they could not only accept the tender of Messrs Robbins and Lawrence for the whole plant of machines, but also, by getting duplicates of two, and triplicate of one of the sixteen stocking machines (making in the whole twenty), the produce would be trebled, on account of the difference of time taken by the different machines in completing their portion of the work, so that instead of sixteen machines producing, say, fifty stocks per diem, the addition of only four would increase the number turned out to 150 daily.

The Committee hope that his Lordship the Master-General will see the propriety of their making this suggestion, it being their endeavour to obtain such perfectly suited machinery as will yield a maximum of produce with a minimum of outlay in money and labour.

I have to request that you will obtain from his Lordship the Master-General the earliest possible answer to this communication.

I have, etc.

signed ROB. BURN, *Lieut.-Col.*, R.A.

In answer to this, the following communication was received by Lieutenant-Colonel Burn whilst at Washington:

m/209

Sir, *Office of Ordnance, June 14, 1854.*

Having submitted to the Board your letter dated 25th ultimo, requesting further authority for the provision of Small-Arms Machinery in the United States, I have their commands to inform you that you are authorized to make any purchases you may consider to be desirable.

I have, etc.

signed J. WOOD

To: Lieut.-Col. Burn, Royal Artillery.

In pursuance of the instructions contained in this letter, the Committee have purchased all the machinery to which they alluded in their letter; and also some others which they considered necessary, taking care, however, to buy or order only such machines as they felt convinced might be advantageously used in producing the different parts of the musket with accuracy and economy, which they know to be the objects aimed at by the Honourable the Board of Ordnance.

2

[Journal of the Committee]

Lieutenant-Colonel Burn, having in pursuance of orders and instructions already detailed, left England on the 25th February, 1854, by the 'Cunard' steamer, arrived in Boston on the 10th of March, and proceeded to Washington and various other places, in order to ascertain what works, arsenals, etc., were best worth visiting; and also to obtain from the heads of Departments in that city the requisite permission to enable the Committee to obtain every information they might require, from the superintendents of the various arsenals, armories, etc. Having accomplished this, Lieutenant-Colonel Burn awaited the arrival of Lieutenant Warlow, R.A., and Mr Anderson, Chief Engineer to the Ordnance, at Boston, which took place on the 26th of April; and on the following day the Committee assembled and discussed what would be the best route for them to pursue in order to see as many of the armories, arsenals, and manufactories as possible, without passing over the same ground twice, and came to the conclusion that it would be advisable, from information gained by them from Mr Whitworth, Colonel Colt, and others, to visit the following towns in the order in which they are here placed, to see the works placed opposite to them.

Boston	Navy yard and private works
Springfield	United States' Armory and private works
Hartford	Gun Manufactories, etc.
Newhaven	Arms Factory and private works
New York	Navy Yard
Westpoint	Military Academy and Iron Works
Philadelphia	United States' Arsenal and Private Works
Washington	United States' Arsenal and Navy Yard
Harper's Ferry	United States' Armory
Wheeling	Private works.
Pittsburgh	Iron Works and United States' Arsenal
Buffalo	Private works
Utica	Arms Manufactory

Albany	United States' Arsenal and private works
Windsor	Arms Factory and Machine works
Boston	Private works in neighbourhood

And the Committee followed this route as nearly as it was possible for them to do, and commenced to visit different works, etc., in the vicinity of Boston.

BOSTON. *Charleston Navy Yard.* This was the first establishment visited by them; and they could not but remark that, though there was not much work doing there was a large quantity of machinery lying idle, ready to be brought into use at any moment should a supply of stores be suddenly required; and also that the store houses for wood, etc., were most extensive, and filled with material. This indeed was the case in all the Government works visited, and is particularly noticed by the Committee in the conclusion of their report. There are in this yard, and in all the others visited, large shops for the manufacture of machinery.

Here the Committee noticed the following things worthy of note, viz.:

1 A peculiar shaped screw auger, for boring hard wood *p.* 172
2 A lacquer, used for iron ordnance *pp.* 161-2
3 Skidding for guns *p.* 162
4 Extensive rope spinning machinery *pp.* 176-7
5 Grape shot *p.* 162
6 Trucks of gun carriages *p.* 174
7 Cap pouches of soldiers *p.* 177

Mr Alger's Iron Works. In these works, which are situated in South Boston, the Committee saw nothing connected with the immediate object of their visit–the manufacture of small arms–but much of great importance to the other manufacturing departments; and they were pleased to observe the manner in which it appeared that everything connected with the casting of iron guns was conducted, every attention being given to the quality of metal used, and the mode in which it was cast. Specimens cut from the dead-head of each gun being carefully tested in a tension and torsion machine are marked and kept, the result being noted down with a view to continually improving the quality of the mixture used.

Mr Alger accompanied the Committee round his works, and showed them every attention in his power, both by giving them information on the plan pursued in his own establishment and branch of trade, and also in assisting them to see other works in its vicinity.

In Mr Alger's works the Committee noted the following things as important, viz.:

8 Excellent mode of casting guns, and making the moulds for them *pp.* 158-9
9 An arrangement of iron furnace *p.* 160
10 An apparatus for cleansing metal *p.* 173
11 A machine for taking tension and torsion resistance to compression and transverse strength *p.* 162
12 A new sort of trip hammer *p.* 176

Mr Babbitt. The Committee visited Mr Babbitt, the celebrated metallurgist, residing near Boston, who introduced the new system of casting, with whom they had a long conversation on the different modes employed in casting ordnance, on which subject they obtained from him a great deal of useful information, which will be found in Chapter 4, page 159 under the heading of 'Inspector of Artillery's Department.'

Mr Alger's Forge Works. The Committee were accompanied round these works by the proprietor, and here remarked the following things, viz.:

13 A new sort of steam tilt hammer *p.* 176
14 An India-rubber diaphragm pump *p.* 177

Last Making Works. The Committee visited a Last-making manufactory, in which a machine on Blanshard's principle is extensively used. This was the first application of this patent that came under their notice.

15 Machine for turning and polishing lasts *p.* 171

They also noticed the following machine as new to them in a woodworking establishment.

16 A vertical saw, for cutting irregular forms *p.* 168

The Committee called at the residence of Mr Blanchard, the inventor of and patentee for machines for turning irregular solid forms, with a view to consulting him relative to the best mode of obtaining such machinery, but found that he was absent at New York on business.

During their stay in Boston they also called on Mr Abbot Lawrence, who gave them such information as to where they would be likely to see good machinery, and also some letters of introduction to proprietors of works. From Boston the Committee proceeded by railway to Springfield.

SPRINGFIELD. Here the Committee first called on Colonel Ripley, Superintendent of the United States Armory, and presented to him the letters of introduction, etc., that Colonel Burn had procured from the heads of Departments at Washington, and passed the first two or three days in looking round the establishment without going into details, in order to get a good general idea of how the work was conducted. Colonel Ripley received them with the greatest kindness and afforded them every facility in his power, both by giving them information himself and directing those under his command to forward in every way the views of the Committee, and by his assistance were enabled to obtain several valuable papers relating to prices paid for work, book-keeping, etc., which they could not have got but for the Superintendent's extreme willingness to aid them in attaining the object of their visit.

The Committee took an early opportunity of having the English stocks which they brought out with them to America put through the various machines in which the 'stocking' is performed, and were gratified to find that they worked as well, and were cut as smoothly, as the stocks of black walnut used in the United States' service, it being the opinion of Colonel Ripley, and some of the subordinate officers, that the English walnut owing to its hardness, and the closeness of its grain, would, if thoroughly desiccated, work cleaner in the machines than that used by them. Unfortunately, those brought over got slightly damp during the passage from England, on which account they did not cut so smoothly as they would otherwise have done, and there was some difficulty in fitting them to the machines as they were shorter than the American stocks (the United States' musket being three inches longer in the barrel than the English). These stocks the Committee had packed up and returned to England as soon as they were ready.

It having been signified to the Committee that the Honourable Board wished them to procure from the United States' Government two muskets as specimens, Lieutenant-Colonel Burn wrote to General Davis, Secretary-of-War, to request permission to be supplied with these arms for the purpose of sending them to England, and received an answer that the British Government had already applied for some arms through their Minister, and that they had been ordered and sent to the British Consul at New York. The Committee therefore supposed that these arms were already on their way to England, and were astonished to find on their arrival at New York, when they called on Mr Barclay, the Consul there, that owing to some delay in the delivery of the arms, he had not yet received them.

Knowing that the Honourable Board were anxious to get these specimens as soon as possible, the Committee proceeded without delay to Governor's Island, where they were informed by Major Thornton, the Governor, that some mistake had occurred relative to the rifle which ought to have been sent to him from Harper's Ferry, but that he was expecting it hourly. The rifle arrived shortly afterwards, and was packed with the musket, and sent at once to Mr Barclay's office for transmission to England.

The Committee having passed some days at the Armory examining the different processes of manufacture, and watching the working of the machines, more particularly those employed on woodwork, waited on Colonel Ripley and consulted with him as to the best plan of obtaining a set of machinery for stocking, similar to that used in the Armory at Springfield, and upon his recommendation visited Mr Ames, the proprietor of an extensive foundry and machine works at Chicopee, about three miles from Springfield, where they observed that the brass ordnance was cast in a manner similar to that in use at Mr Alger's manufactory at Boston, and after going over his works, where they observed the following things worthy of note, viz.:

17 An apparatus for testing the quantity of power required to work a machine *p.* 161
18 A machine for sifting sand *p.* 161

They requested him to examine the stock of the 'Enfield rifle musket, pattern 1853;' and to give them an estimate of the cost of a single set of machinery adapted to produce it. After some conversation on the subject, Mr Ames thought it would be better for him before making out any tender to go over the Armory at Springfield with the Committee, in order that by carefully examining the machinery there used, and consulting Mr Buckland, engineer of that establishment and inventor of many of the machines in use there, as to what modification would be necessary on account of the difference of pattern between the English and American arms, he might be the better able to judge what difficulties he would have to contend with, and obtain data upon which he could make out a formal tender. This was accordingly done, and the matter having been most carefully gone into by all the parties interested, the whole were of opinion that in order to produce the English pattern stock fifteen machines would be necessary, and Mr Ames undertook to make out and submit to the Committee a formal tender for a set of such machines, which he accordingly did. A copy of this tender will be found at Chapter 5, page 181.

In order not to be tied down to one tender for these machines, the Committee considered it advisable to go to Hartford and request Colonel Colt to send them an estimate for a plant of machinery of the same description, which they accordingly did, but found that Colonel Colt had such an immense quantity of work on hand (being about to commence a new armory on a most extensive scale), that it was no object to him to make machinery such as they required, viz., which required a great deal of scheming. On conversing with his engineer it appeared that his prices were higher than those of Mr Ames, and also that the machines could not be produced in the time the latter gentleman had undertaken to deliver them in.

Having carefully deliberated on the matter, and taking into consideration that besides the inducements offered as regarded time and economy by Mr Ames, that gentleman had the great advantage of residing near the Armory, and seeing the machines at work constantly there, besides having made a considerable quantity of the machines used in that establishment, the Committee were of opinion they could not do better than follow Colonel Ripley's advice, and they accordingly accepted Mr Ames's tender, which amounted to $30,860 (thirty thousand eight hundred and sixty dollars). A copy of their acceptance will be found in Chapter 5, page 181.

Mr Ames having previously to tendering told the Committee that unless he could be assured of the co-operation and assistance of Mr Buckland, Engineer at the United States' Armory, in designing the 'stocking' machines he could not undertake to make them at all, they conversed with that gentleman on the subject, and, he being willing to give every possible aid to them in carrying out their plans, Colonel Ripley's sanction to this arrangement was obtained.

As this will entail a great deal of thought and labour on Mr Buckland during the time the machines are in course of construction, the Committee thought they had no right to expect him to do it without any remuneration, and therefore having consulted together, before they left Springfield for good, as to what sum they should offer him, were of opinion that $1,000 (one thousand dollars) would not be more than sufficient to compensate him for his time and work, and accordingly informed him that it was their intention to recommend that this sum should be paid to him by the British Government, as an acknowledgment of his services, as soon as the machinery was completed and delivered.

The Committee passed the remainder of their time at Springfield in visiting the different works of the Armory, and the water-shops connected with it, which are situated on a small stream about one mile from the main establishment, and examined carefully all the various details of the manufacture of the different parts of the musket, a detailed account of which will be found in Chapter 3, page 128, under the heading of 'Manufacture of Small Arms.'

Whilst at the Armory the Committee timed the different machines employed in 'stocking' at their work, and noticing how much longer some were in performing the operations required of them than others; they perceived how very much the produce of a plant like this might be increased by getting duplicates or triplicates of some of the slow machines, and thus making their yield equal to that of those that worked more rapidly. After considering the subject, therefore, they were of opinion that it would be most desirable to increase their order to Mr Ames, by getting two or three of the slow-working machines, but that it could not be done out of the sum to which they were limited.

During their stay at Springfield the Committee, accompanied by Mr Ames, visited Holyoke, where they went over some mills, and afterwards Mr Ames took them to some machine works at Chicopee Falls, where they noticed the following things as worthy of remark.

19 Patent magnetic sewing machine *p.* 165
20 Yankee chaff cutter *p.* 176
21 Tourbine water-wheel *p.* 178

Having made arrangements for procuring copies of various papers, showing the way in which the book-keeping, etc. was conducted at Springfield Armory the Committee proceeded on their route.

A full description of the mode in which muskets are there manufactured, and the machines used, will be found in Chapter 3, page 129, under the head of 'Manufacture of Small Arms.'

HARTFORD. From Springfield the Committee proceeded by railway to Hartford, where they called on Colonel Colt, to see if he would be willing to undertake to make some of the machinery necessary to produce the bands and other parts of the metal work about the musket; but he declined doing so on account of press of business.

Sharpe's Armory. At Hartford the Committee went over an armory belonging to Messrs Robbins and Lawrence, of Windsor, Vermont, in which they are manufacturing Sharpe's Breech-Loading Carbine and Rifle for the company who hold the patent of it. This factory is only just established, and all the machinery was not complete at the time the Committee visited it, but it seemed to be conducted on the best manufacturing principles, machinery being applied to every part of the arm. The Committee were so struck with the beauty and efficiency of the machines here used, that, finding they were made by Messrs Robbins and Lawrence at their machine-shop at Windsor, Vermont, they entered into the subject of the manufacture of fire-arms by machinery with Mr Lawrence, who accompanied them over the works, and who proved to be perfectly conversant with this branch of trade, and having showed him the different parts of the musket required to be produced, he undertook to make out and forward to them a tender for the necessary machinery, on behalf of the firm in which he is a partner.

Mr Palmer, president of the 'Sharpe's Rifle Company,' called on the Committee and took them to see the workshop where the bullets for that arm are manufactured by pressure; but they found that the machine used was very imperfect, and the bullet produced very inferior to that made at Woolwich. He afterwards accompanied them to a house where the cartridges were made up and filled. In these operations some machines are used, but they are not very perfect.

NEWHAVEN. From Hartford the Committee proceeded by railway to Newhaven, where they went over Mr Whitney's armory at Whitneyville (which Captain Warlow had already visited from Springfield), in which works the United States' Rifle is made. Here the same class of machinery is used as at Springfield and the other armories visited, and nothing was seen worthy of special notice.

At Newhaven the Committee also visited the following works, viz.:

Jerome's Clock Manufactory. In this establishment clocks are made in immense quantities for home use and exportation; 600 per diem being the yield, with 250 men employed. Machinery is most extensively used in all parts of the manufacture, and the clocks produced at a very low price, the movements of some costing only $1.

Messrs Davenport and Mallory's Works. This is a manufactory of padlocks and locks; and the same system of special machinery is applied to every particular part; and all the locks of each description produced are identical, and their parts can be interchanged. The work is turned out at a very low cost, some padlocks being made for 5 cents (2½*d.*), and 2000 produced daily.

Messrs Candie and Co.'s Factory. This is a manufactory of india-rubber shoes, in which machinery is applied as far as practicable, and with 175 hands 2000 pair are daily produced.

NEW YORK. From Newhaven the Committee proceeded by railway to New York. On their arrival in that city they proceeded as soon as possible to Mr Gillespie's, in order to ascertain if any letters were awaiting them there; and afterwards they visited the British Consul, Mr Barclay, from whom they ascertained that the arms applied for by the British Government had not yet been received by him (as already mentioned) for transmission to England, and took steps to insure their being sent as quickly as possible.

The Crystal Palace. The Committee visited the Crystal Palace two or three times, in order to examine the machinery there, and see if any of it was suitable to the purposes of manufacture to which their attention had been directed, and noticed the following as new to them:

22 Machine for cutting files *p.* 176
23 Cask making machinery *p.* 174
24 Packing-up machine *p.* 165

Governor's Island. The Committee were accompanied round the stores and works by Major Thornton, who gave them much information on various subjects. They here remarked the following things as new to them, viz.:

25 Percussion caps for spring guns *p.* 162
26 Brass hammer used with these caps *p.* 162
27 Storing flannel cartridges *p.* 178

Rifling Shop. It having come to the knowledge of the Committee, whilst at Springfield, that Mr George Law had purchased a very large number of the United States' Government Muskets, and was having them rifled in New York, they obtained the address of the workshop where the rifling was being performed, and visited it. There were several self-acting machines at work, similar to those they had already seen at Hartford and elsewhere. The muskets were of the old flint pattern, and were being altered to percussion, and rifled with seven grooves.

Novelty Works. This is one of the most extensive establishments in New York for the manufacture of marine engines, etc.; but the Committee, in going over these works, did not see any thing new to them, or any machinery not used in similar works in Great Britain. Near these works were some very large river and sea-going steamers, two of which, the New World and the Georgia, the Committee went over.

Brooklyn Navy Yard. The Committee were accompanied round the works at the United States' Navy Yard, at Brooklyn, by Mr Burnett, United States Civil Engineer, where they could not but observe that, as at Boston Yard, though but little manufacturing work was being carried on or required, most extensive forges and workshops were in course of erection, and a great quantity of machinery lying idle, ready to be brought into use at any time when its assistance might be necessary.

In this yard there is a very fine engine which is used for pumping the water out of the dry dock.

The Ericson caloric ship happening to be lying off the yard the Committee took the opportunity of going over her and examining her machinery, which has been completely changed since her first trips, the motive power now used being a mixture of steam and air.

The Committee noted the following things as worthy of remark:

28 Altered form of muzzle of guns for sea service *p.* 162
29 Plan on which building sheds for ships are constructed *p.* 176
30 Whitening used by the marines in lieu of pipe-clay *p.* 177

Felt hat making. The Committee visited an establishment of this nature, with a view to ascertaining if the machinery used in this branch of

manufacture was applicable to the production of cartridge bags for the Royal Laboratory; but found it was not.
The Allaire works. This is a large establishment for the manufacture of marine engines, and very similar to the 'Novelty Works,' already visited. Here the Committee saw no machinery different from that used in England in works of this nature.
Wood bending machinery. After leaving the Allaire works, the Committee crossed the river to Williamsburgh in order, if possible, to see some machinery lately invented by Mr Blanchard for bending timber for the 'knees' of vessels; but found that the machines used had not yet been perfected, and were not in working order.
'*Great Republic*.' As the remains of the 'Great Republic' were lying close to Mr Blanchard's yard, the Committee took the opportunity of going over her.
Francis's Metallic Life Boat Manufactory. These are very extensive works for the manufacture of boats out of galvanized corrugated iron, and machinery is very extensively applied to their production. The sides of the boats are stamped into form by means of immense dies worked together by hydraulic pressure. The boats thus made are extensively used by both the Government and mercantile marine as ship boats, and are very highly spoken of by officers of the navy and others, who have had the opportunity of judging of their merits. There are two air tight reservoirs in these boats, which prevent the possibility of their sinking if capsized.
Parcel making machine. The Committee having been informed that one of these machines was at work in New Jersey, went there to see if it would be applicable to any purpose connected with the Royal Laboratory. A description of it will be found in Chapter 4, page 165.
Empire Stone Works. This is a very large establishment for cutting and working stone, and machinery has been extensively applied to most of the processes, such as facing, planing, etc.
Franklin Forge. In this forge the Committee saw two English steam hammers at work, one of them a 7½ ton hammer on Condy's plan, and the other a smaller one, on Mr Nasmyth's principle. The former was the largest hammer the Committee had yet seen, either in Great Britain or the United States. Anthracite coal was used for heating iron in these works.

WEST POINT AND COLD SPRINGS. Colonel Lee, Commanding Officer of the Military Academy at West Point, having written and invited the Committee to be present at the public examination of the cadets on the 1st of June, they went there on that day and had an opportunity of going over the whole establishment, which appears to be in every respect admirably conducted.

They also visited an extensive foundry and machine shop belonging to Mr Kemble at Cold Springs, near West Point, in which iron guns are manufactured for the United States' Government, and the tension and torsion machine used for testing the metal.

In these works anthracite coal is used in a furnace of a peculiar construction, a drawing of which was procured. A description of this furnace will be found in Chapter 4, pages 178-9.

Jenning's Rifle and Pistol. The Committee called at Mr Palmer's, agent for the 'Jenning's Arms Company,' at his request, to examine his new pattern repeating rifle, which was on the same principle as one brought out some time ago under the same name; but has been altered and improved. A description of this weapon will be found at Chapter 3, pages 126-7.

Whilst at New York the Committee received a tender from Messrs Robbins and Lawrence of Windsor, Vermont, for machinery to produce the lock, bands, heel plate, etc., of the English pattern musket; but on reckoning up the sum left at their disposal, after deducting the amount of the contract already entered into with Mr Ames for stocking machinery, they found that unless the tender of Messrs Robbins and Lawrence was considerably reduced they could not accept it, and were, therefore, under the necessity of striking out of it several valuable and useful machines, to bring it within the limits of their authorized expenditure.

The Committee, finding that the produce of the whole plant of machinery would be seriously impaired by the reduction they were obliged to make in this estimate, and also considering what benefits would accrue by increasing their order to Mr Ames, and getting duplicates and triplicates of the slow working 'stocking' machinery, which would double the yield of that plant, were of opinion, that it was their duty to write and represent the case to the Honourable Board, and at the same time request that the sum of £10,000 granted for the purchase of machinery for the Small-Arm Manufactory, might be increased to £12,500, to enable them to increase their orders to Messrs Ames, and Robbins and Lawrence, which they accordingly did in a letter, a copy of which will be found in Chapter 1, pages 96-7, and received in answer a letter from the Honourable Board, Chapter 1, page 97, authorizing them to purchase any machinery that might appear to them necessary for the manufacture of small arms.

From New York the Committee proceeded by railway to:

PHILADELPHIA. *Navy Yard.* The Committee having first called at the British Consul's, proceeded to the navy yard, round which they were accompanied by Lieutenant McBlair, of the United States' Navy. In this yard they observed the following things worthy of remark:

31 A floating dry dock *p.* 177
32 Cable nippers *p.* 178
33 Mode of elevating heavy iron guns *p.* 162
34 Form of steps on cheeks of gun-carriages *p.* 174

Deringer's Pistol-making. The Committee having heard that Mr Deringer was celebrated in the United States for the manufacture of single-barrelled rifle pistols, visited his workshop, and were astonished to find that the arms were made entirely by hand, in very small quantities, and the whole conducted in a most rough manner.

Cornelius Baker and Co.'s Chandelier Factory. Having been informed that in this branch of trade machinery had been very successfully applied, the Committee visited this establishment, which is the largest of the sort in Philadelphia, and found all the work carried on on a manufacturing principle, and an immense quantity of chandeliers, lamps, etc., produced at a very moderate cost.

Horstmann's Fringe and Accoutrement Works. This is one of the finest establishments the Committee saw in the United States. The processes of manufacture are conducted in the same manner as in England; and there was nothing particularly worthy of note, except the extent of the premises and work-rooms, and the wholesale way in which the work was conducted.

While at Philadelphia, the Committee noticed the following as new to them, and ingenious.

35 American lifting-pump *p.* 177

From Philadelphia the Committee proceeded by railway to:

WASHINGTON. The Committee, as soon as possible after their arrival in this city, called on the following gentlemen, viz.: Mr Crampton, the British Minister: General Totten, Chief Engineer; Colonel Craig, Chief of Ordnance; General Davis, Secretary of War: and afterwards left their cards at the residence of the President of the United States.

United States' Arsenal. On visiting the arsenal, the Committee were most kindly received by Major Mordecai, who accompanied them round the works, and showed the greatest willingness to give them every information they wished for. They could not but observe the careful way in which, it appears, experiments of every sort, more especially those with powder, are carried on here, and in the other Government establishments visited by them. They noticed the following things worthy of remark in this arsenal, viz.:

36 A machine for making percussion-caps *p.* 163
37 The Ballistic pendulum and musket pendulum under cover *p.* 159
38 The form of heavy gun-carriages *p.* 174
39 A field hand-drill *p.* 178
40 Can-hooks *p.* 177
41 Russian gravitating tangent-sight *p.* 162-3
42 Screw jack *p.* 177
43 A lathe and planing-machine combined *p.* 174

Before they left, Major Mordecai presented to each of the Committee, a copy of his work on 'Experiments with Gunpowder.'

Colonel Burn having been taken ill whilst at Washington, the other members of the Committee visited the following works together.

Navy Yard. On visiting this yard, Lieutenant Warlow and Mr Anderson first called on Commodore Paulding, the Superintendent, who received them most kindly and requested Captain Dahlgren and Lieutenant Landmann, both of the United States' Navy to accompany them and show them the works, etc.

They here observed a very fine ballistic pendulum, and arrangements for testing powder, as well as the following things which they noted, viz.:

44 Bullet press *p.* 164
45 Captain Dahlgren's lock *p.* 163
46 Eccentric shells *p.* 163
47 Gun-carriage for boat service *p.* 174-5
48 Percussion-cap machinery *p.* 163

Lieutenant Warlow and Mr Anderson called a second time on Colonel Craig for some information relative to the manufacture of small-arm ammunition, and afterwards visited the Patent Office.

Colonel Burn being too unwell to move, Lieutenant Warlow and Mr Anderson proceeded from Washington to Philadelphia in order to see Mr Wright, the inventor of the self-acting cap-making machinery, who, they were told, was at Frankfort Arsenal, near Philadelphia, which they had not visited.

FRANKFORT. *Arsenal.* Mr Anderson after his arrival at Philadelphia for a few days not being able to go about, Lieutenant Warlow visited Frankfort Arsenal, where every attention was paid him by Major Hayner and Lieutenant Balch, both of the United States' Army, and he there found Mr Wright, inventor of the self-acting cap-making machinery, and questioned him as to the best method of obtaining it. Mr Wright informed him that he would be willing to tender for it; and Lieutenant Warlow therefore requested him to make out a formal tender, and put himself into communication with Mr Anderson.

Mr Anderson observing that a machine was used in this Arsenal well adapted to making the cup of the Minié bullet in large quantities, Lieutenant Warlow made inquiries about it, and was informed that the patentee was a Mr Bonton of Troy.

Colonel Burn and Mr Anderson being both unable to proceed on the route Lieutenant Warlow hearing that there was to be a rifle match between two of the most celebrated rifle makers in the United States at Fort Plain, thought he might take the opportunity of seeing some of the American match shooting, more especially as on his way to the place where the match was to come off he would pass near Watervliet Arsenal, West Troy, and be able to get further information relative to Mr Bonton's machine for cup making. He accordingly left Mr Anderson at Philadelphia and proceeded by railway to New York and thence to Albany.

ALBANY. Whilst at Albany Lieutenant Warlow visited Watervliet Arsenal, where Major Lymington received him and accompanied him round the works, though it was very late in the day before he arrived there, and on his inquiring about Mr Bonton, Major Lymington informed him that that gentleman had given up making machines, and that if the Committee were desirous of obtaining them, it would be better to request Mr Wright to tender for them, and Lieutenant Warlow communicated this to Mr Anderson at Philadelphia in order that the latter gentleman might make the necessary arrangements with Mr Wright on the subject.

FORT PLAIN. From Albany Lieutenant Warlow proceeded to Fort Plain, by railway, in order to see the rifle match going on there, which was to take three days, but as one day's shooting was sufficient to show him the method in which these matches are conducted, and the accuracy of the rifles used, he went on in the evening to Utica.

The match was between Mr Morgan James, of Utica, and Mr Nelson Lewis, of West Troy, each of whom was, in the course of the three days, to fire five of his own rifles, 20 shots each, at rest from a distance of 40 rods (220 yards). The weather was not favourable, being windy, but some of the shooting of Mr James, who won the match, was very accurate indeed.

Here Lieutenant Warlow noticed the following things as worthy of remark, which he noted down, viz.:

49 Wind-gauges used *p.* 127
50 Rest used by Mr James *p.* 127
51 Mode of recording the practice *p.* 127
52 Careful method of loading *p.* 128

UTICA. Lieutenant Warlow visited the workshop of Mr James, in Utica, who makes the finest sporting and match shooting rifles, from whom he gained a great deal of information on the subject of rifle arms, which will be found in the remarks on match rifles, Chapter 3, p. 127.

GERMAN FLATS. From Utica Lieutenant Warlow visited Mr Remington's establishment, where rifles are made for the United States' Government. Mr Remington was the first person who introduced into the service barrels bored out of the solid bar of steel, for he found that the welded hand-forged barrels were so defective, and that so many were rejected in view for flaws, etc., that it was cheaper for him, on the whole, to use steel instead of iron.

The barrels are bored straight through in a vertical boring machine, in which manner Mr Remington states he has also bored musket barrels, which he has submitted to the consideration of the United States' Government. He is also the largest manufacturer of steel barrels for the private trade in the United States. All the steel used is brought from England, as that made in the United States cannot be depended on.

From Utica Lieutenant Warlow returned by railway to New York, on his way to rejoin the other members of the Committee as soon as they should be able to proceed on their route.

NEW YORK. Lieutenant Warlow having heard that good shooting could be made at 200 yards with the United States Service musket, and knowing that the English musket was next to useless at that distance, so far as accuracy was concerned, went to Governor's Island to find out if Major Thornton could arrange a day on which he might see some practice at that distance, that officer having, at the time of the visit of the Committee to his post, promised that if he could find time he would have a day's shooting for their satisfaction with the weapon named, but he found that Major

Thornton's time was so much engaged in making out a return of stores, etc., after which he was obliged to start on a tour of inspection, that he could not fix any day on which to have the shooting, and which consequently was never made.

Perry's Arms Factory. Lieutenant Warlow visited this establishment, where a breech-loading rifle is made. The works are not on a large scale, nor is machinery applied in anything like so great a degree as in the production of Sharpe's rifle. A description of this arm will be found in the chapter on small arms, Chapter 3, p. 124, as also the quantity of arms on hand and their price.

Marston's Pistol Factory. This workshop is in New York. No special machinery is used, and no particular pattern of pistol adhered to.

During his stay in New York, Lieutenant Warlow also visited the printing press of the 'Daily Times' and Harper's publishing establishment.

Colonel Burn, during the time he was detained at Washington, was enabled to witness an experiment of a very interesting character with some caissons and waggons, made of galvanized corrugated iron, submitted for trial to the United States' Government by Mr Francis, manufacturer of metallic life-boats, the result of which was most satisfactory as they proved to be perfectly water-tight, strong and very buoyant, so much so that the bodies of the waggons might if necessary be used as pontoons. At Washington, Colonel Burn received the two following communications from the Honourable Board, copies of which will be found in Chapter 1, pp. 95, 97, viz.

Relative to engaging Mr Burton, dated 31st May, $\frac{o}{334}$.

Authorizing further expenditure, dated 14th June, 1854, $\frac{m}{209}$.

Whilst at Philadelphia, Mr Anderson visited Frankfort Arsenal, where every attention was shown him by the Superintendent Major Hayner and Lieutenant Balch, and he then saw Mr Wright and directed him to send in a formal tender to the Committee for some machinery for producing caps and iron cups, which it was desirable to obtain for the use of the Royal Laboratory. He there found a new plant of machinery, just commencing work, to make friction tubes which he remarked as being particularly good, so he requested Mr Wright to include them in his tender.

He also noticed the following machinery as worthy of remark, viz.:

53 Machines for working tin plate *p*. 173

and made inquiries as to where they could be best obtained.

At or near Philadelphia, Mr Anderson also visited the following works, viz.:

Chemical works at Frankfort.

Metal works in Philadelphia.

Messrs Morris and Co.'s Engineering works, where he observed the following machines as new, viz.:

54 A new sort of slotting machine *p.* 178
55 A method of welding wrought and cast-iron together *p.* 160

A Manufactory of Presses. A large manufactory in leather in which machinery was extensively applied in cutting and equalizing the thickness of mill bands.

Lieutenant Warlow having heard from Mr Anderson that he and Colonel Burn were ready to proceed on their route, left New York by railway for Philadelphia where he found Mr Anderson, and they proceeded together to Harper's Ferry, where it was arranged that Colonel Burn should meet them.

HARPER'S FERRY. The Committee, as proposed, met at Harper's Ferry and visited the United States' Armory, which is situated on a tongue of land at the junction of the Potomac and another river, and is surrounded by high perpendicular cliffs, which shutting it in make it most oppressively hot during the summer, and for this reason it cannot be a good situation for an armory or manufactory of any description, more particularly for one where forging has to be done.

The Committee were accompanied by the Superintendent, Colonel Bell, round the works, which form two separate factories, in one of which the United States' musket is made, and in the other the rifle. The machinery is similar to that used at Springfield, but the works are not so compact, or in such neat order as those of that armory.

Some extensive buildings are in course of erection, and among them a large rolling mill. The Committee observed the following things worthy of note.

56 Machine for fine drawing barrels *p.* 131
57 Arrangement of forges, anvils, etc. *p.* 178
58 Precautions in case of fire *p.* 161

From Harper's Ferry they proceeded by railway to Wheeling, and thence by steamer up the Ohio to Pittsburgh.

WHEELING. Mr Anderson visited a work for making iron castings, in which he found the composition for dusting the mould differed from what is commonly used.

59 Composition for dusting moulds *p.* 160

PITTSBURGH. The Committee first visited some rolling mills, but found that owing to the heat of the weather they had stopped work.
Arsenal. The arsenal is situated about three miles out of Pittsburgh. The Committee were accompanied round it by the Superintendent, Lieutenant Rodman, United States' Ordnance, and S. Brereton, United States' Army, both of whom showed the greatest attention, and they observed the following things as worthy of note, viz.:

60 Lathe for turning brass studs *p.* 173
61 Machinery for wheel-making *p.* 167-8
62 Machine for turning staves *p.* 169
63 Eight-inch gun, under trial *p.* 162
64 Proof-shot for guns *p.* 162

Rhodes's Biscuit Making Works. In this establishment, biscuits are made in larger quantities, and machinery used in almost all the processes of manufacture.

Novelty Malleable Cast Iron Works. The articles made of malleable cast iron are particularly well executed in the United States. They are annealed with the wrought iron scale of the iron maker, and are subjected to heat for eight days.

Fort Pitt Iron Works. The Committee here met Major Wade, formerly of the United States' Army and now a partner in these works, who has had much experience in the casting of iron and brass ordnance, and made many experiments connected with this branch of manufacture. Extensive experiments have been made in these works in casting iron guns with a core, that can have a supply of water passing into the interior, viz., down the centre and up the sides, the object being to cool the inside as fast as the outside, in order to secure the metal being in a uniform state of contraction. Difficulties had been met with in carrying this into effect; but it was expected that they would be overcome. A piece of metal is trepanned out of the dead-head of each gun, which is tested in various ways. Such a system is much required in England, to insure improvement in the manufacture of an arm so important to the service. Major Wade is of opinion, that great improvement will arise from a more sudden cooling of the metal, either in brass or iron castings.

Machinery for making wrought iron nails is made in this establishment, and Major Wade offered to accompany the Committee to some works where they were in full operation.

Spade and Shovel Factory. The Committee visited a manufactory of this nature where machinery was extensively used, and remarked the following as worthy of note, viz.:

65 Machine for pressing the iron on to the handles of spades *p.* 178

Phelps and Carr–Waggon Building. In this manufactory, machinery is much used, but it is of a very rough description. The following machines were new to the Committee, viz.:

66 Machine for boring a square hole *p.* 167
67 Machine for boring the nave of the wheel *p.* 168-9

Plough Making Works. The Committee went over some works of this nature, but saw nothing bearing on the object of their mission.

Manufactory of Wrought Iron Nails. Major Wade accompanied the Committee to these works, where they saw wrought iron nails produced in immense quantities by self-acting machinery, which worked admirably.

68 Machinery for making wrought-iron nails *p.* 174

Iron Spike Manufactory. Here the Committee saw large iron spikes, which are used in the United States for holding down the rails on railways, produced in very large quantities by a self-acting machine.

From Pittsburgh the Committee proceeded by railway to Cleveland, and thence by the 'Lake' steamer to Buffalo, where they visited several interesting wood works, etc.

BUFFALO. *Whip Manufactory.* The plaiting both of the handle and thong is here performed in a very ingenious machine.

69 Plaiting machine *p.* 175

A Manufactory of Melodions. In these works, machinery was employed for the mortising, planing, sawing, and whole of the wood work.
The Niagara Pail Factory. The manufactory of buckets, tubs, etc., is newly established at a place about three miles from Buffalo; and machinery is used most extensively, by which means 1000 buckets are produced daily at a very cheap rate, and exceedingly well made.
Railway Car Manufactory. An extensive establishment of this sort was also visited by the Committee, in which machinery was applied for almost every purpose, some of which being very neat and well constructed. The Committee took a note of the maker's name, with a view to obtaining them for the Royal Carriage Department.
Door and Sash Manufactory. An extensive establishment of this sort was also visited by the Committee, in which machinery was applied even to sand papering, which was done on a large drum.

At NIAGARA, Mr Anderson went into a manufactory of wooden pegs for shoemaking, in which they were produced in immense quantities and at a very low price by machinery, being sold by the barrel.

After passing some days in Canada, the Committee met again at Albany. Mr Anderson stopped at UTICA on his way, where he visited the following works, viz.:
Mr James's Rifle Maker. Here he observed that malleable cast iron is used in many parts of the rifle where we should usc wrought iron, and it seems sufficiently good for any purpose, at the same time it is closer in the grain and very much more economical.

In his workshop, Mr James uses but little machinery; but he considers his present system only waste of time, and means to enter on the manufacture of rifles on a much larger scale, and with the aid of machinery.
Mr Remington's Barrel and Arms Manufactory, at German Flats.
Mr Grigg's Screw Manufactory. In these works the screws are made by self-acting machinery, which is particularly good, and in first rate order.

Nettlefold's is perhaps the best Screw Manufactory in England, but very inferior to Mr Grigg's; for in the latter the whole work is performed by self-acting machinery, and one person can attend several machines at once.

ALBANY. Mr Anderson, arriving here before the other members of the Committee, visited the following works, viz.:

Messrs Peizely and Co.'s Manufactory of Wooden Bedsteads. Fifteen thousand bedsteads are produced yearly in these works, to the manufacture of which a great quantity of machinery is applied.
Messrs Goule and Co.'s Carriage Manufactory. Here Mr Anderson saw some very excellent mortising machines, and took down the name of the maker of them.
Townsend and Co.'s Engineering Works. Steam engines and tourbines are extensively made in this establishment. The machinery used is similar to that employed in England for like purposes.

The Committee, after meeting at Albany, visited Watervliet Arsenal, round which they were conducted by Major Symington, the Superintendent. Very little appeared to be doing at present, but the shops are large and well built, and the stores excellent.

They observed as worthy of note:

70 The soldier's bedstead *p.* 177
71 State in which gun-carriages are stored *p.* 178

From Albany the Committee proceeded by railway to:

WINDSOR. Here they first visited the machine shop and armory of Messrs Robbins and Lawrence, in company with the proprietors, who at their request submitted a formal tender for the machines which had been struck out of their first tender by the Committee, to which was added an estimate for additional machinery for producing different parts of the musket, which tender the Committee accepted on the 24th July. They also submitted an estimate for some machinery adapted to the work carried on in the Royal Carriage Department, which the Committee caused to be converted into a tender and accepted. Copies of these will be found in Chapter 5, pages 183-4.

The Committee also visited the State Prison, which is situated in the town of Windsor. Here the convicts are employed upon the work of the contractor, who pays the State 33 cents daily for the labour of each man. At present they are employed making the handles of scythes. Should any prove idle, the contractor reports it to the jail authorities, and the man is punished with solitary confinement, the State having to repay the contractor for the time lost to him. By this arrangement the convicts, instead of being a burthen to the State, are a source of revenue to it. A pail manufactory similar to that near Buffalo was visited.

Hearing that there was a firm who manufactured machines used in tin work, at a town called Woodstock, in the neighbourhood of Windsor, the Company, accompanied by Mr Robbins, went there and visited the works, and at their request Mr Whitney, the proprietor, furnished them with a list of prices. On their return to Windsor they asked Messrs Robbins and Lawrence to tender to them for these machines, and procure them from the manufacturer for them, in order to have dealings with as few firms as possible. Messrs Robbins and Lawrence did this, and the Committee accepted the tender, a copy of which will be found in Chapter 5, p. 185.

The same day they went with Mr Robbins to Lebanon to see Messrs Buck and Co., makers of wood machinery, with a view to purchasing some of his machines for the Royal Carriage Department, and at their request a formal tender was made out which reached them at Boston, and which they accepted. A copy will be found in Chapter 5, p. 186.

From Windsor the Committee proceeded by railway to Boston, in order to visit different works they had heard of since their former stay in that town.

BOSTON. *Cutting Machine Company*. The Committee first visited the agent of the Cutting Machine Company and saw some of the machines at work, with which they were much pleased, and requested the agent to draw them out a formal tender for some machines of this description, which they accepted. A copy will be found in Chapter 5, pp. 186-7.

72 Cutting Machine *p.* 173

They next went to the agency of the Wax Thread Sewing Machine, but not finding any of the machines at work and hearing that they would do so at Lynn, near Boston, the Committee proceeded to that place, where they found that Howe's machine was also much used in shoe-making.

Whilst at Lynn they went over some of the shoe-making establishments there, but did not find machinery so extensively used as they had been led to expect.

LOWELL. *Machine Shop*. The Committee visited Lowell, where they went over the extensive machine shop, and found that very much the same machinery was used there as in England.

Cotton Mills. There are very extensive cotton mills at Lowell, which the Committee also visited, and were much pleased with their order and arrangement. A new mill was in course of building, beneath which power was obtained by two large tourbine wheels of 750 horse-power each.

The girls working in these mills are boarded in large houses belonging to the Corporation that owns the work, about forty in each house. These are superintended by matrons.

In Boston the Committee also visited the agent of the Dorcas Sewing Machine Company, and finding that this machine appeared the simplest and most efficient of anything they had yet seen, purchased two of them for the use of the Royal Laboratory and Royal Carriage Department.

They also during their stay in Boston purchased some vices and small tools, etc., which they consigned to the agent of the Cutting Machine Company to be forwarded to England with the machinery purchased from him.

The Committee proceeded by railway to Springfield from Boston.

SPRINGFIELD. After their arrival here, the Committee first proceeded to Chicopee to see how the stocking machinery ordered by them was progressing at Mr Ames's manufactory, but found that gentleman absent. Mr Carter, of Chicopee Falls, happening to be there, requested them to visit his works where some of the machines were in course of construction, which they did.

At the United States' Armory, Springfield, the Committee received from Mr Allen, clerk at that establishment, a number of valuable papers describing the method on which book-keeping, etc., was there conducted, and which he had been kind enough to draw up for them since their first visit. Colonel Ripley, the Superintendent, gave them all the information they required relative to the armory.

Owing to some mistakes in the Post-office, two tenders sent from Mr Wright at Philadelphia had never reached the Committee, and they having written and informed Mr Wright of this, he came to see them at Springfield on the subject and there made out a formal tender for the machines required, which the Committee accepted. A copy of this will be found in Chapter 5, page 187.

Having authority to buy or order such machinery as they considered necessary for the proposed Small Arms Factory, the Committee entered carefully into a calculation as to what additional 'stocking' machines they should get, in order to increase the produce of the plant sufficiently to keep a flow of work through it. In this they were assisted by Mr Buckland and Mr Burton, and they came to the conclusion that it would be advisable to order:

1 extra machine for rough turning,
3 ,, ,, smooth turning the butt,
1 ,, ,, ,, before [*sic*] ;

both which, though not sufficient to make the plant perfect, would more than treble its yield.

The Committee, therefore, requested Mr Ames to draw up a formal tender for these machines, and also the following, which they considered it necessary to get, viz.:

2 self-acting machines for edging the lock-plate.

Apparatus for testing power.

And a set of jigs and gauges, which, though expensive are absolutely necessary as the means by which accuracy and interchange of parts are insured.

A copy of this tender, which the Committee accepted, will be found in Chapter 5, page 191.

Whilst at Springfield the Committee were visited by Mr Lawrence, of the firm of Robbins and Lawrence, who was desirous of accompanying them round the United States' Armory, in order to avail himself of any improvements he might there see in making the machinery ordered from him by the Committee as perfect as possible. They therefore passed a day in going about the various workshops at the Armory with Mr Lawrence.

HARTFORD. From Springfield the Committee visited Hartford, in order to see if Colonel Colt would be willing to make a tender for some drop-hammers similar to those used in his establishment in London, and for which he holds a patent.

Having found that gentleman, they expressed their wishes to him, and he informed them that he could let them have as many as they wanted at two months' notice.

They also went over Sharpe's rifle manufactory again with Mr Lawrence, and were much struck with the neatness and easy working of the rough boring machines as compared with those they had seen in England.

In a long conversation the Committee had with Mr Lawrence on the subject of manufacturing barrels, to which he has given much time and attention, they were so satisfied of his competence to undertake machinery for producing this important part of the musket, that, at their request, he drew up a formal tender for a plant of machinery suitable to produce all the parts of the barrel of the English rifle musket, and included in this estimate some small tools that had been omitted in his last.

This tender the Committee accepted. A copy will be found in Chapter 5, pages 181-2.

At Springfield the Committee saw Mr Burton, and conversed with him relative to engaging his services for the British Government; but, as there was no immediate necessity for coming to a final arrangement with him now, they thought it better to let the matter stand over till they had consulted the Honourable Board again on the subject.

Mr Burton has an engagement now which it will take him some months to fulfil.

Before leaving Springfield for good, the Committee called on Colonel Ripley and the officers of the United States' Armory, to express how sensible they were of all the kindness and attention they had met with during their two visits to that establishment, and with extreme regret parted with the friends who had treated them with so much courtesy, and so materially aided them in bringing the object of their mission to the United States to a satisfactory conclusion. Having done this they set out by railway for New York, where they proceeded at once to draw up their report for his Lordship the Master-General and the Honourable Board of Ordnance.

3

Fire Arms and their Manufacture

The principal object of the Committee in visiting the United States of America being in connection with the proposed armory, they directed their attention more especially to everything relating to small arms and their manufacture.

1st. To the various descriptions of arms that came under their notice–the quality of their workmanship and the condition of their parts interchanging.

2nd. To the system, machinery, and apparatus by means of which they are produced.

A. *The various descriptions of arms that came under the observation of the Committee, their quality, and the condition of their parts interchanging*

In going over the United States' Government Armories at Springfield and Harper's Ferry, and the works of the Government Contractors, Mr Whitney of Newhaven, and Mr Remington of German Flats, near Utica, the Committee had, of course, every opportunity of examining the two principal service arms of the country–viz., the musket and the rifle; and having been in so many of the large manufacturing towns of the Northern States, where almost all the various descriptions of fire-arms are made, and it having soon become known that the object of their mission was chiefly connected with small-arms, they were enabled to see nearly every description of rifle or pistol now made in the United States, partly by hearing of them and partly by the agents or proprietors bringing their arms to the Committee for their examination.

The following weapons thus came under their observation, viz.:

Service Arms

1 The United States' regulation musket
2 The United States' regulation rifle
3 Sharpe's breech-loading rifle carbine

4 Perry's breech-loading rifle carbine
5 Marston's breech-loading rifle carbine
6 Colonel Colt's revolving rifle
7 Colonel Porter's revolving rifle
8 Wesson and Leavitt's revolving pistol
9 Maynard's revolver
10 Allen's pistol–revolving hammer–repeater
11 Jennings's rifle and pistol–match rifles and common rifles
12 Mr Morgan James's rifles

A description of which follows:

[GOVERNMENT ARMS]

1. *The United States' Musket.* The United States' percussion musket is in all respects very like the French, from which it appears to have been copied.

The following are its weights and dimensions:

Barrel	Length	3 ft. 6 in.
	Weight	4 lbs. 9 oz.
	Diameter of bore	·690 in.
Bayonet	Weight	11½ oz.
	Length of blade	1 ft. 6 in.
Weight of musket complete, with bayonet		9 lbs. 13 oz.
Bullet	Weight	·94 oz.
	Diameter	·650 in. or 17 to the lb.
Powder	4 drams, or 110 grains musket powder.	

The barrel is fitted to the stock by means of broad flat iron bands, similar to those of the French musket, and held in the same manner, viz., by a spring let into the wood of the stock. The lock is of the common hook pattern. The mountings are all of wrought iron. There is no back sight on the barrel. The front sight is of brass, and attached to the nose cap band.

The stock is made of black walnut, and is very short from the trigger to the heel plate, as compared with the new English musket.

The bayonet is of steel and iron welded together, and is fixed on the barrel with the French locking ring. It has no shoulder to the blade.

The stock of the American musket is made of black walnut, which grows in very large quantities in Pennsylvania, from which State it is procured by the persons who supply the United States' Armory with stocks in the rough.

The United States' Government do not enter into any contracts to obtain them, but whoever likes to bring a quantity of them to the armories can obtain 28 cents each for them, provided they pass the Government viewer. In the rough and dried, they weigh 7 pounds.

The Committee having heard in England that the reason American walnut was never used for musket stocks in that country, though it was very difficult to obtain sufficient of the English walnut, was because it was very considerably heavier than the last-mentioned wood, considered it advisable to test this as much as was in their power, with which view they caused three of the English stocks that had passed through the American stocking machines and were the full length (the other three were not) to be weighed, and found the following results.

Weight of three English walnut stocks	6 lbs.

They then took at hazard 3 black walnut stocks which had been passed through the same machine, and had them also weighed, finding that

Three stocks of black walnut weighed 6 lbs. 14 oz.,

they then repeated the experiment on a second and third set of three, taken from different parts of the stocking room.

Second set of three black walnut stocks	6 lbs.
Third set of three black walnut stocks	6 lbs. 5 ozs.,

showing that on an average the American black walnut, when brought to the size of the United States musket stock, only weighed 1 ounce heavier than English walnut of the same size.

It having also been said to the Committee that the black walnut varied very much in weight, they tested this by weighing ten stocks taken from one part of the establishment, and then ten taken from another, with the following result.

First ten of black walnut	19 lbs. 14½ oz.
Second ten of black walnut	19 lbs. 6½ oz.

The black walnut is not such a hard and close grained wood as the English, but the Committee are of opinion that it would be quite serviceable for the English arm should the present supply fail, and that it could be procured in any quantities from the United States at the cost of about 28 cents, (or 14*d.*) per stock, in the rough, but that it would be necessary for it to undergo the inspection of some competent person before being sent to England, as a large quantity is rejected in the rough by the United States' Government viewers.

A species of walnut which is abundantly found in Canada has been tried at Springfield, but is not found good, being very woolly in the fibre, and cutting rough in the machines.

Shortly before leaving New York, the Committee received a communication from Mr Joseph S. Munning of Philadelphia, offering to supply the British Government with black walnut musket stocks, to the amount of 20,000 per annum, at 38 cents each, which they declined to accept without authority from the Honourable Board and wrote to Mr Munning to that effect, especially when the above tender so far exceeds that entered into with the United States' Government, viz., at 28 cents each.

With regard to the interchange of parts between the machine-made muskets of the United States' Government, which has caused so much discussion, the Committee particularly interested themselves; and, with the view of testing this as fully as possible, selected with Colonel Ripley's permission ten muskets, each made in a different year, viz., from 1844 to 1853 inclusive, from the principal arsenal at Springfield, which they caused to be taken to pieces in their presence, and the parts placed in a row of boxes, mixed up together. They then requested the workman, whose duty it is to 'assemble' the arms, to put them together, which he did–the Committee handing him the parts, taken at hazard–with the use of a turnscrew only, and as quickly as though they had been English muskets, whose parts had carefully been kept separate.

The only parts of the musket bearing any mark being the barrel and lock, which are stamped with the year in which they were made, and all these tried being of different years, the Committee took care that no barrel and lock bearing the same date, should come together again, and they were put together as follows, viz.:

	Barrel		Lock
The barrel of	1847	with the lock of	1849
,,	1844	,,	1852
,,	1846	,,	1848
,,	1845	,,	1844
,,	1851	,,	1850
,,	1848	,,	1853
,,	1849	,,	1845
,,	1852	,,	1847
,,	1850	,,	1851
,,	1853	,,	1846

The other parts, having no distinguishing mark, were handed out at hazard.

With regard to the fitting of these muskets when thus interchanged, the Committee are of opinion, that all the parts were as close, and the muskets as efficient, as they were before the interchange took place; but they are diffident in expressing any opinion as to the comparative fit of these and the English rifle muskets, as none of them being viewers, they have no experience in examining muskets so minutely, though they have no hesitation in expressing it as their firm conviction, that the American musket is as good and efficient an arm of its sort as can be put into the hands of a soldier, and if the extra finish on military arms now used in England is considered indispensable, that by a small outlay in manual labour, following the machine work, the English musket may be made as perfect and well-fitting as it now is, and yet possess the great advantage of its parts interchanging.

The experiment of interchanging was also tried on three locks with the most perfect success, the parts fitting as closely, and working as freely after as before the interchange had taken place.

The only part of the United States' musket not capable of interchange is the breech pin.

2. *United States' Rifle*. The following are the weights and dimensions of the different parts of this arm:

Barrel	Length	2 ft. 9 in.
	Weight	5 lbs. 6 oz.
	Diameter of bore	·54 in.
Rifling	No. of grooves	7
	Spiral	1 turn in 6 feet.

Bullet	Form	spherical.
	Weight	·5 oz. or 32 to the lb.
	Diameter	·525 in.

Charge of powder, 2¾ drams, fine.
Weight of arm complete, 9 lbs. 11 oz.

The barrel of this arm is of steel, bored out of the solid bar. It is fixed to the stock by brass bands held by springs let into the wood. The barrel of this arm is browned, and the lock plate case hardened. It is furnished with a block sight for 100 yards; but no elevating sight, and there is no bayonet or sword-bayonet attached.

As far as the Committee have been able to ascertain, no experiments of any extent have been made with the improved rifle muskets of Europe; but the United States' Government having a very large number of the rifle above described on hand, have been trying to improve its accuracy and length of range, and facilitate the loading by using an elongated bullet, with which view they have caused experiments to be made at Harper's Ferry, under Colonel —— which resulted in that officer's submitting to the Government for adoption a bullet of which Mr Burton was the originator.

This bullet is precisely the same in form and construction as that of the Belgian rifle musket, experimented with last year at Woolwich, in comparison with the Regulation rifle musket of 1851, and the Enfield rifle musket of 1853; but Colonel —— and Mr Burton had never seen or heard of that projectile as being in use any where, till informed of it by the Committee. No elongated bullet is yet decided on, and there is no rifle musket in the American service.

The ammunition of the United States' rifle is made up with the powder and ball attached, and is loaded by forcing the bullet into the rifling on the old plan. It appears to the Committee, that in its present state it is by no means a good or efficient arm, owing to the difficulty there must be in loading it.

PRIVATE ARMS—LOADING AT THE BREECH

3. *Sharpe's Rifle Carbine*. This is a rifle carbine loading at the breech, which is formed of a sliding steel peg to which the cone is attached. When the trigger guard is pushed forwards as far as it will go, this breech is lowered so as to expose the bore of the barrel into which the cartridge is placed by hand, and the trigger guard on being restored to its first position, raises the sliding breech again into its place. The upper edge of this slide is sharpened so that it acts like a knife, and cuts off the hinder end of the cartridge, leaving the powder exposed.

A percussion cap can be used with this arm, but it is constructed for the application of a new primer invented by Mr Sharpe, which consists of a small detonating wafer, a number of which are placed at once into a little cylindrical receiver in the lock plate behind the hammer, and a spiral spring at the bottom keeps them always pressing upwards. By an arrangement in the lock plate, as the hammer descends it drives out one of these little pellets, and strikes on it as it is passing over the nipple.

When these primers are used, the arm may be loaded and discharged with very great rapidity, and they seem to be very sure in their action, and are thought superior to Maynard's primers, the paper and wax enveloping of which are apt to choke up the nipple, should a miss-fire occur.

From inquiries made of different United States' officers, the Committee think that this is the best breech-loader which has yet been produced in America, more especially as an appropriation of $150,000 has recently been made for the purchase of 5,000, to be issued to the American cavalry on the frontier.

The Committee thought it advisable to inquire from Mr Palmer, the President of the Sharpe's Rifle Company, as to what number of arms he had on hand, how quickly and at what price he could supply them if required, and with what amount of ammunition, etc., and were informed that in three months, 4,000 could be supplied at a cost of $25 each, and any amount of ammunition required at $13 per 1,000; and also that he could continue to supply the carbines at the rate of from 1,000 to 1,200 per month, with ammunition to any extent.

Those manufactured for the United States' Government are of 32 bore rifled with six grooves having an increasing spiral and weigh 6½ lbs.

All the parts of Sharpe's rifle carbine will interchange with the exception of the sliding breech.

4. *Perry's Breech-Loading Rifle*. The works of the Perry Arms Company are at Newark, New Jersey, where rifles, rifle carbines, and pistols, on their principle are made.

The breech end of the barrel is bored out of solid steel, and hung on trunnions so as to form a sort of chamber to receive the charge. This chamber is raised for loading and lowered again after the charge has been put in by means of a backwards and forwards motion of the trigger guard, which is attached to it. All arms made on this plan lose what is a very great advantage in breech-loading guns, such as the needle gun, Sharpe's carbine, etc., viz., the impossibility of the bullet rising in the barrel by the piece being jolted about whilst loaded on account of the breech being chambered, and the shoulder of the bullet resting against the edge of this chamber.

The Perry rifle is made self-capping by a very ingenious contrivance. A hole is bored from the heel plate through the butt to the point where the nipple rests when the chamber-breech is elevated for loading; in this hole a copper tube is placed containing a spiral spring. The caps being placed in this tube are always kept pressing forward by the spring, and in the act of loading the foremost one is placed on the nipple, if the old cap has been previously removed; and if not, the new cap, though it cannot be placed on the nipple, is not wasted, remaining in the charger.

Mr Perry has 5,000 of these arms on hand, and the wholesale cost of them is $25 each.

5. *Marston's Breech-Loading Rifle*. The manufactory in which this rifle was made has been burnt down and, owing to disagreements in the company which own it, its manufacture has not been again commenced. Altogether it does not seem to be highly thought of in the United States,

but the Committee are of opinion that this must be owing to the manufacture of them having been defective, as the results obtained at Woolwich were very good indeed, but the particular arm then tried was not manufactured in the United States but had been carefully made by Lepage Montier, the celebrated gunmaker in Paris, and was very highly finished and carefully put together.

Some of these arms made by Mr Pritchett of London, for the proprietors, are, the Committee believe, now under trial at Woolwich.

REVOLVERS

6. *Colonel Colt's Revolving Rifles.* None of these arms have been as yet manufactured, the buildings in which they are to be made being now in course of erection, and the model is not yet perfected.

It is Colonel Colt's intention, the Committee understand, to manufacture in large quantities two different sizes of revolving rifles, one of the English Government bore–24 to the lb., and the other smaller. The pattern proposed was not complete, but was shown to the Committee. In principle it was the same as the Colt pistol, but in detail varied considerably from that weapon. Colonel Colt has also a new pattern pistol, very small indeed, which he intends making in large quantities, at a very low cost.

Owing to the large numbers in which these rifles are to be produced, Colonel Colt expects to be able to make them at such a price as will defy all competition, even if his patent should not be renewed.

With regard to the efficiency of the Colt pistol as a weapon, the Committee have everywhere heard it most highly spoken of by officers of the United States' Army and others, who have been for long periods on the frontier in Northern Mexico and Texas, and had every opportunity of judging of its merits as an offensive arm.

It is carried by the United States' Dragoons, and is particularly well suited to the description of warfare in which they are constantly engaged, it being the tactics of the Indians, with whom they have frequent skirmishes, to draw their fire and then rush in and overpower them by numbers. As long as the soldiers only carried single arms this was successful; but now that the revolving pistol has been introduced, the Indians do not dare to close, even with a force very inferior in numbers to their own.

The United States' Cavalry, on the frontier, are allowed as much ammunition as they like for practice with these arms, and become very expert in the use of them.

7. *Colonel Porter's Revolver.* This revolving rifle is held in very low estimation in the United States, and is considered dangerous to use, there being three chambers always pointed towards the person who is firing it. A specimen of this weapon is in the Inspector of Artillery's Department at Woolwich, among the small arms.

8. *Wesson and Leavitt's Revolving Pistol.* This revolver is precisely the same in principle as Colonel Colt's, and on this account the manufacture of it has been prohibited in the United States, as an infringement of Colt's American patent.

The chambers revolve by cocking; but the motion is communicated by two bevelled toothed wheels, instead of a ratchet. This part of the patent is Mr Wesson's. Another difference between it and Colt's is that the danger of the fire communicating from one chamber to another on discharge is prevented by the outer surface of the cylinder being slightly conical, which is said to produce the effect Colt obtains by bevelling the mouths of the chambers in his pistol. This is Mr Leavitt's part of the patent.

9. *Maynard's Revolver.* The cylinder of this pistol is made to revolve by hand, and is held in its place by a small detent just in front of the trigger, which detent has to be pressed with the forefinger of the right hand, whilst the cylinder is turned with the left. This is a comparatively slow operation, but cannot be improved on account of Colonel Colt's patent; and, consequently, the pistol meets with a very limited sale–only 3,000 being made yearly.

The principal peculiarity of this pistol is, that one nipple is common to all the chambers, being completely detached from the cylinder. The fire from the primer strikes through the holes made in the cylinder to receive the detent, into the charge.

Maynard's primer is used with this arm, and is worked forward by a small toothed wheel, in lieu of the double spring formerly used. Caps cannot be used on this pistol, as, there being only one nipple, it would be necessary to put on a fresh cap after each discharge.

10. *Allen's Pistol.* This pistol has five barrels fixed together on a centre (similar to the old pattern English revolver), which do not revolve, but are discharged in succession by a revolving hammer, which, by an ingenious arrangement, shifts its position every time it is cocked and discharged. The cocking is effected by means of a sort of double trigger, and two fingers are used, the middle finger working the hind trigger by which the pistol is cocked, and the fore-finger the front trigger by which it is fired when cocked. The caps are on nipples, placed on the continuation of the axes of the bores, and are inclosed inside the pistol. The revolving hammer works horizontally above the trigger.

This pistol is manufactured at the works of Messrs Robbins and Lawrence, Windsor, Vermont.

REPEATERS

11. *Jennings's Rifle and Pistol.* The Jennings Rifle was brought out some years since, but the proprietors, finding it was defective in some respects, ceased making it and prepared a new pattern, which is said to be an improvement on the first.

Beneath the barrel, and running along parallel to it, is a cylinder, into which the charges are placed, and kept bearing towards the butt by a spiral spring. Part of the trigger-guard is formed into a ring, in which the middle finger of the right hand is placed; and by forcing the trigger-guard forward with the middle finger, whilst cocking with the thumb, a charge is removed from the receiver and placed opposite the breech of the barrel, into which it is thrust on the trigger-guard being restored to its place by a cylindrical plug, which forms a breech pin.

The bullet is cylindro-conical, and is very much hollowed out to contain the charge of powder, which is kept in it by a small cork wad, containing a priming wafer, ignited by a small blunt pin, in a manner very similar to that in which ignition is produced in the needle gun. A pistol of this sort is also made.

Whilst at New York, Lieutenant Warlow accompanied Mr Palmer, agent of the Jennings Arms Company, where he saw 200 shots fired from a pistol on this principle, which seemed to work well, and shot with accuracy and force.

MATCH RIFLES

12. *Mr Morgan James's Rifles.* All the rifles used by the American marksmen are very small in the bore, very rarely being as large as 40 gauge, but usually running from 60 to 100 gauge, and sometimes as small as 200 to the pound. The barrels, which are usually octagonal, and bored out of the solid bar of steel, are long and heavy, and are rifled with from six to ten grooves, usually having an increasing spiral.

In fine shooting the size of the bore is usually determined by the distance at which the practice is to be made, and the state of the weather, it being necessary to use a rather larger bore in windy weather than in calm to insure great accuracy.

In matches wind gauges are used to show the marksman the direction and strength of the wind. In the match at Fort Plain Mr James used four of the gauges along the range, on the windward side of it, at intervals of about fifty yards. They consisted of strips of calico, in the shape of a swallow tail, fixed by small cross bars on poles, and free to traverse in any direction by the wind. To the point a weight of six ounces is suspended; and in order to prevent it from twisting, wires are run across it at intervals of about six inches throughout its whole length.

The rest used by Mr James was very small – adapted for accurate shooting. It consisted of a strong table, supported on four legs, bound together by cross-bars, which formed a tray underneath it. This tray was filled with heavy stones to steady the rest. On the fore part of the table was a solid block of wood, on which the muzzle rested; and, commencing from this, running towards the marksman, was an inclined board, working on hinges, and capable of being raised or lowered by an elevating screw. The rifle was laid with its muzzle resting on the block, and the fore part of the trigger-guard pressing against the edge of the inclined plane, which was elevated or depressed by the screw till the sights were brought to bear on the centre of the object.

The practice in matches is recorded by what are called 'strings.' The distance from the centre spot to the centre of each shot-hole is measured and taken down as it is fired, and after the number of shots agreed upon has been fired, the whole of these are added together and called a 'string' of so many inches.

From 10 to 20 shots are the usual number fired, and 40 rods (220 yards) the most common distance. In the match at Fort Plain Mr James made a 'string' of 20 shots which measured 32½ inches, or each shot rather over an inch and a half from the centre of the paper, at 220 yards. This was considered a good 'string,' as the weather was windy and unfavourable.

The loading of these rifles requires to be very carefully done to insure accuracy; a false muzzle, starter, and various implements being used, and the barrel wiped out after every shot. Mr James, after loading, always drew the bullet a few inches, and then pressed it down again, by means of a rod having a head coned out and a very fine screw thread cut in it, which held the bullet without defacing it. If it did not move freely and smoothly he discharged it and loaded again.

The reason why the loading requires to be so carefully done is, that the bullet is nearly purely conical, and it takes great delicacy of manipulation to keep its axis and that of the barrel coincident in pressing it down and starting it. The powder used with these rifles is coarse in the grain and even, and the charge is very large as compared with the weight of the bullet.

On some of these match rifles a telescope is fixed, by which means very great accuracy is obtained. In his workshop at Utica Mr James showed Lieutenant Warlow a 'string' of 10 shots, fired at 220 yards, by his partner, Mr Ferris, and measuring only 7¾ inches. This is supposed to be the best string on record, and was made with a telescope sight.

The fine rifles used in the United States very much surpass in accuracy the Minié and other rifle muskets of Europe, which are all large in the bore; but the delicacy required in loading, and the number of implements necessary, put them out of the question as military arms, even if the smallness of their bores were not a sufficient objection.

B. *The System, Machinery, and Apparatus by means of which Small Arms are produced*

In consequence of the scarcity and high price of labour in the United States, and the extreme desire manifested by masters and workmen to adopt all labour-saving appliances, from the conviction of such being for their mutual interest, a considerable number of different trades are carried on in the same way as the cotton manufacture of England, viz., in large factories, with machinery applied to almost every process, the extreme subdivision of labour and all reduced to an almost perfect system of manufacture. Such establishments are, in the great majority of cases, got up by Corporations to produce a certain class of articles, and in order to obtain them at the lowest cost, a large proportion of special machines are employed, which are frequently made on the premises, such machinery being, as a general rule, more productive than the ordinary tools or machines that are generally used in the same branch of trade when carried on in smaller workshops.

As regards the class of machinery usually employed by engineers and machine makers, they are upon the whole behind those of England, but in the adaptation of special apparatus to a single operation in almost all branches of industry, the Americans display an amount of ingenuity, combined with undaunted energy, which as a nation we would do well to imitate, if we mean to hold our present position in the great market of the world.

Among the many branches of trade to which these remarks apply, that of small arms stands conspicuous. The two national armories of Springfield and Harper's Ferry, the private establishments of Colonel Colt, Robbins and Lawrence, and the Sharpe's Rifle Company, are all conducted on the thorough manufacturing system, with machinery and special tools applied to the several parts, and every part of the whole arm made on the premises from the raw material.

In order to show the manner in which the work is conducted in these establishments, the Committee consider it advisable to give a detailed account of the different processes of manufacture gone through in Springfield Armory and Sharpe's Rifle Factory, the most perfect Government and private establishment visited by them.

The United States' Armory at Springfield is situated on a hill about half-a-mile out of the town of Springfield. The different buildings, which are detached, are situated round a large quadrangular plot of grass, and consist of two very extensive arsenals, the different workshops, offices, etc., and the residences of Colonel Ripley and the other officers connected with the establishment. The main arsenal is capable of containing 200,000 stand of arms, in two large rooms, arranged upon very light, neat-looking racks, which support 14,000 each, in two tiers. There are now 150,000 muskets in this building. The second arsenal contains 100,000 arms, which have been altered from flint to percussion. There is also a third arsenal, but it is very small as compared with the other two.

Part of the workshops belonging to this armory are situated on a small stream, about a mile from the main works, and in them the barrel forging, bayonet making, etc., are performed.

Sharpe's Rifle Manufactory is an establishment just started at Hartford for making these arms on a large scale; the factory is extensive and very well arranged, consisting of a large workshop on the ground floor, and a second on a gallery above, the centre being left open, and having a skylight above it for light and ventilation.

At the back are the forges, etc.

MANUFACTURE OF ARMS

Forging of musket barrels. The system of forging gun barrels in this country by the Trip hammer, worked by a water-wheel or steam engine, is inferior to the English rolling mill as applied to the same purpose; but at the several places visited by the Committee the adoption of the latter method was spoken of as probable.

The proper quantity of material required for a barrel, bayonet, or other article, is determined by a very simple water-gauge. The end of the bar is immersed until the water rises to a known point, this of course gives the exact quantity whatever may be the outward form or dimensions, and the water line on the bar marks the place where it is to be cut off. The iron is heated in a forge with anthracite coal, drawn out, and welded under a tilt hammer, two men being required for the operation, and produce about twenty barrels per day.

This work is usually performed by the piece, the workmen being responsible for the soundness of the weld. When any defect is found after the barrel is bored and turned, the whole expenditure upon it has to be sustained by him; the same principle of responsibility is carried out in the other branches.

At the national armories they make the musket-barrels of Norway iron of a very superior description, but it does not seem better for this purpose than good English iron carefully prepared. In several of the private establishments they make the rifle-barrels of a mild steel, which are bored out of the solid bar; this they consider to be cheaper and better than the welded barrels, there being considerable less waste from defective workmanship. The steel is obtained from Sheffield.

The boring of the barrels consists of three distinct operations, which are performed in differently constructed machines.

The operation of rough boring is done in a self-acting machine, containing four or six spindles, all running in the same direction; the first bit is a screwed auger on the end of a rod, this rod is passed through the barrel from the breech, and bores on the pull, the cutting points being where the auger joins to the rod, which is the reverse of a common auger. During this operation a stream of water passes through the barrels as they lie upon a self-acting saddle on the boring-lathe. One man can turn out from sixty to eighty barrels per day.

In the second boring, the object of which is to straighten the barrel, and to remove the greater part of the material, the lathe is not self-acting, and contains but one spindle, which is driven at a speed of 1,500 revolutions per minute, the bit is square and used with a spill, no water is employed, and only as much oil as will prevent the bit from scratching the barrel; as the bit revolves, the attendant, by turning a handle, moves it on end either way.

In the third boring operation the object is to straighten, smooth, and polish the interior. In this lathe there are two spindles and two separate saddles, both of which are self-acting either way, and perfectly independent of each other. A square bit with a spill is employed, driven at a slow speed. During the second and third borings the same workman straightens the barrels by hand; but in a large manufactory there would be an advantage in having a man to straighten for several borers.

Previous to the third boring the barrels are cut to the exact length by means of a lathe made on purpose, and so contrived that the barrels can be taken out and put into the machine without stopping. The chuck is a conical socket with the angle such that it holds the end of the barrel when it is pressed into it, the other end being cut to the required length by a stationary milling tool, which is moved till it comes into contact with a stop; thus all the barrels must be of precisely the same length.

The lathes for turning the outside of the barrel are self-acting, and trace a copy of the required form. These lathes turn out about twenty-five barrels per day each. At Springfield one man attends to six lathes.

A second lathe is used for turning the irregular part on the breech, all on the copying principle.

The next operation is to re-turn the muzzle at the part where it receives the bayonet, here it is essential that all should be exactly alike, in order that the bayonets may interchange; to secure the exactness, another mode of turning is employed, termed clamp-milling. The cutting instruments consist of a pair of dies, between which the barrel revolves; these dies are pressed together until they come against a stop. Hence all must be the same after the lathe is adjusted.

After the barrels have received a wipe upon a grindstone, they are polished on the outside by means of a machine at a cost of 1 cent (½*d.*) each. This machine resembles a vertical saw-frame, and contains five spindles, which receive five barrels; this frame with the barrels moves up and down at the same time that the barrels have a revolving motion. Independent of this frame there is a set of clamps which grip the barrels, and being supplied with oil and emery, they tend to round and smooth them at the same time.

As the barrels are required to have the appearance of being draw-filed, this is obtained by giving them a very small amount of circular motion towards the end of the operation, this lays the lines nearly parallel with the axis. A very small amount of turn is necessary to prevent the formation of ruts, which would form if the barrel worked up and down constantly in the same place.

The extremities of the barrel, which cannot be got at by the clams of the polishing machine, are completed on revolving buffs, covered with fine emery. The lump and cone seat at the breech are finished with ordinary drills and milling tools.

All rifling is done by self-acting machinery of simple construction. A bar of the required twist is placed behind or under the machine, and on this bar rests a roller connected with a rack, as this rack is drawn along the bar it rises or falls according to the form of twist, and this rising or falling motion given to the rack acts upon the rifling-bar, thus giving it the required twist; the rifling-bar is worked by a crank, which does not appear to be so uniform as a screw would be for the same purpose. The barrel is turned round a division at every cut; this is accomplished by the motion of the rifling-bar, which also puts a detent out and in to prevent the barrel from shifting.

The cutter is set out by a wedge which touches a gradually moving screw at each full stroke, thus making the whole self-acting, and to clean the oil and chips from the cutter, a brush is passed across it at every alternate movement. Self-acting polishing machines are used to smooth and finish the interior of the barrels, both for muskets and rifles.

[*American bayonet*]. The American bayonet is made with the socket of iron and the neck and blade of steel, the bar of iron for the socket is in breadth equal to the length of the socket, and $\frac{5}{16}$ of an inch in thickness, then cut off by a shears into squares which are chamfered on two edges for welding. The blade is made from a square bar of steel, the proper quantity being cut off by means of a water-gauge. As much steel is cut off as will serve for two blades, which by means of a tilt hammer is drawn out at the ends and rounded in the middle for the two necks; by the same means the blade is forged out into the proper shape. The two blades are then parted, the necks turned round, then gripped in a vice while a smith upsets the ends to be welded to the socket, the form given being similar to a boiler rivet. The steel is then welded to the iron, borax being used to prevent the steel from being injured; after which the socket is trimmed off by a shears, then bent round and welded. It then passes on to the annealing furnace to be softened.

The bayonet socket is bored out with a bit which in appearance resembles a screw auger slightly tapered; this bit cuts along the whole screw, and owing to the first being in the opposite direction of the ordinary auger, the chips are pushed before it at the other end of the socket. The hole is completed by having a standard widener passed through every bayonet, thus making all of the same diameter, so as to fit on any musket.

The exterior of the socket is turned in a lathe of special construction, it can turn in either direction at the will of the attendant. It has two slide rests one in front and another at the back, and both together carry the tools necessary to finish the socket. The lathe first turns in one direction and the tools on that side are pushed into contact against a stop, then the lathe is reversed, and the other set of tools served in the same manner. Thus every socket must be alike. The parts of the socket that cannot be finished in the turning lathe are chiselled off in milling machines.

The blade of the bayonet is shaped with milling tools. In cutting the grooves two are placed side by side, and a copious stream of water plays upon the cutters as they revolve with considerable velocity; the flat edges are finished in the same manner.

The process of hardening and tempering the blade of bayonets is performed in a very perfect manner; for hardening, a bath of lead is heated to the required degree of redness, and into this the bayonet is immersed, till it acquires the same colour, then according to the nature of the steel in regard to mildness or hardness, they are dipped in brine, water, or oil.

The tempering is done in a similar way, a second bath of melted lead at the necessary temperature is used to heat the bayonet, a number being thrown in together, there being no risk of overheating during either process, the attendant takes them out of the bath one at a time, and straightens any bend or twist whilst the heat is upon them, and then throws them into a cistern of cold water.

After the blade is tempered, its milled surface is broken on a grinding stone, then a number of them are fixed in a frame over an emery wheel, where they receive their final polish. Those parts of the groove that cannot be got at on the large wheel are finished with small blocks in a line with the blade.

The bayonet locking ring is forged solid; after it is bored and faced it is first slotted, then sawed open, opened out and closed around the socket. [*Ramrod and Lock*]. Ramrods are forged out of a bar of square steel, in size equal to the thickest part of the large end, the small part is drawn out under a tilt hammer and swedged to nearly the required form, the thick end of the ramrod is turned between a pair of dies, by the clamp milling process, which makes them all exactly alike; the remainder is ground on a stone fitted up for the purpose, the rod is supplied with a handle on the turned part and then laid upon a flat rest close to the edge of the stone, upon this rest lies a flat bar for pushing the ramrod against the edge of the grindstone. The operator turns the ramrod with his left hand, whilst at the same time with his right hand he actuates the pushing bar by means of a lever that is attached to it.

The ramrod is cut to the proper length by being laid in a gauge with the turned end in a suitable position, and the other extremity between a pair of cutters, one of which is struck by a hammer to snip off the end.

The screw upon the end of the ramrod is formed by having a hollow spindle to contain the rod; this spindle is capable of turning and moving in a lateral direction; a pair of screwing dies is placed at one end, the end of a ramrod which projects through the spindle is pushed between the dies and the spindle turned round; thus the thread is cut truly and all are alike.

The ramrods are polished on emery wheels about two feet in diameter, the operator holding four in his hand at the same time, provision being made for carrying away the steely particles which would otherwise float in the atmosphere and render the work unhealthy.

At both of the national armories there is a fine set of tools for forging the several parts of the lock, etc., by means of which a few men are able to produce them quickly and with accuracy. The coal chiefly used is anthracite, which affords a clear fire and makes no smoke in the smithy.

Of the various tools that are used to bring out the shape of the several parts, they are in general too complicated to be readily understood (without drawings) from description, but the general principle consists in having accurately-made dies, between which the iron or steel is laid, these dies being arranged so that the workman can open or close them by a treadle with his foot, also of contrivances to hold the article, thus liberating both hands of the smith.

There are levers connected to treadles, by which they can be gripped in various directions, with cutters on the last of every series, so arrange that the article is cut off the bar directly, and drops into a receiver which forms part of the apparatus, etc.

The several parts after they are forged pass to the annealing department, and remain in the fire for about twenty-four hours. To remove the scale oxide from the surface, in order to save the injury to the tools in the finishing operations, the forgings are subjected to a process of pickling, by being immersed in a cistern containing a mixture of sulphuric acid and water, in the proportions of 1 to 3, the water being hot when mixed.

Malleable cast iron is much used for the ornamental and other parts of small-arms, such as the heel-plate, guard-plate, etc. Some of the malleable castings which the Committee saw were annealed in their own sand in which they were cast, some in clay, but mostly in the scale oxide of the iron maker. Malleable cast iron is much more extensively used in the United States than in England for various purposes where we employ wrought iron. The former is much cheaper, and in the majority of cases answers the purpose equally well. These castings are well pickled in a solution of 1 of sulphuric acid to 2 or 2½ of water.

The various parts of a musket-lock and the other smaller articles that belong to fire-arms, after they are forged or cast, annealed and pickled, are then subjected to a series of different kinds of machines according to the form of the article to be produced, and by means of cutting instruments of various shapes, are brought to a definitive form requiring but a small amount of hand-labour to give a slight finish.

As it would lengthen this report too much to describe the processes that each article is subjected to, the Committee will only give a description of the class of machinery that is employed.

[*Milling Tools*]. The instrument most used in cutting the iron-work of muskets is the circular milling tool. The mill is used of every variety of form according to the form to be cut, and the machines in which they are applied are variously constructed, the particular arrangement being dependent on the same cause. In the most common arrangement of milling machines a headstock carries a running spindle, which can be raised or lowered to modify the thickness of the article to be cut; and in those where a heavy cut has to be taken, or where the spindle and cutters would overhang the headstock too far, a second or steadying headstock is used, which is also adjustable. Under the spindle a vice or holder to suit the shape of the article is secured to a self-acting slide-rest, so that fixing a piece of iron in this vice, and having mills on the spindle of the shape required, by passing the former under the latter, the same form will be cut in every case so long as the mill retains its shape.

In the management of an armory one of the chief points to attend to is, keeping such tools for years up to an original form, for if there is any departure however small, the several parts would not interchange. The way in which this identity is secured in the States is very simple and ingenious: a piece of steel is made into a thing the exact shape of the required article, then cut on its surface like a file, and then well tempered. This article is to be kept as a standard to be used as follows:

When a milling tool has been made as near as it can be by ordinary means, and after the teeth are formed upon it, but before it is hardened, it is set to revolve in the milling machine, and the aforesaid standard is fixed in the vice where the ordinary thing is held, then the two are delicately brought into contact, and the standard is carefully passed under the soft mill, just rectifying any of the proud portions from the edges of the teeth, but without wearing the standard in any sensible degree, which is kept for future use to maintain the article produced at a definite form.

The form and conditions relative to the teeth of milling tools is a subject which occupies much of the attention of arms manufacturers, so as to have them to cut smoothly, and to be easily sharpened. The form most easily made is to have the teeth running straight across, and such are frequently employed; but it is found that the long surface taking hold at the same time is apt to produce defective work, and here there are other modifications, one of which is to cut a screw on the toothed surface: this seems to answer the purpose to some extent.

The best condition of mill is with the teeth upon the skew, and where the surface is large, having the skew in different directions: in making such milling tools it is usual to have the mill in a series of rings, the sum of which in a line complete the required instrument.

Until lately the sharpening of milling tools was a very troublesome operation, and they had to be softened and rehardened each time, which soon injured the quality of the steel. The method they employ now does away with this difficulty, and allows them, however complicated, to be sharpened with facility. On a small lathe driven at a high velocity, are placed small circular cakes, composed of glue and fine emery, and which are made by casting this composition in metal moulds. These are of various sizes and forms, to suit different cutters. In sharpening mills the operator sits on a stool with his arms resting on the lathe, so as to steady the hand, and the teeth of the mill are delicately held against the revolving composition, which very quickly whets the cutting edge.

The greater number of milling machines are such as have been described, but for many kinds of work others, or other appliances, have to be used.

When an article has to receive the shape from the milling tool upon an irregular surface, it is necessary to make the article as it moves along under the cutter describe a line of the form required. This is done in a very simple manner, by having the vice for holding the article hinged at one end, with the other resting on a copy of the required shape, so that as the vice is moved along it traces the copy, and so rises and falls under the mill, thus transferring the shape to the object under operation. This method, from the arrangement of the apparatus is termed bridge milling.

There is a class of double milling machines, in which the articles are held between the two sets of mills, each producing the same or a different form. In some machines the headstocks are set out of the straight line, in order to produce the shape required; the milling tools of such machines are rectified on the plan already described, by holding a standard between the two mills.

The clamp milling system has been referred to, under the description of barrel, muzzle, and ramrod. This is a most valuable class of tool, extensively applied to screw making and other purposes. The two dies of the clamp mill are brought to the definite form and shape on the same principle as the milling tools. The dies are brought to almost the finished dimensions, then a standard of the exact form and size required, is set to revolve between them, and they are gently pushed into contact, thus acquiring the original form. The clamps or dies are then tempered, and as they get blunted from time to time are sharpened on a grinding stone, until nearly worn out, when they are again subjected to the action of the original standard. This standard has to be about $\frac{1}{100}$ of an inch smaller than the article to be produced.

There is another extensive class of tools employed upon the lock and other parts for boring and drilling holes and other work. Of such machines also there are many varieties, and much ingenuity has been displayed in having them so arranged as to produce accurate workmanship with facility. Generally each machine contains a number of spindles, with a standard drill for every sized hole in the article; the thing to be drilled is held in an apparatus so contrived that the drill cannot wander in any direction. This secures perfect identity in all, and also allowing of interchange.

In other machines the article revolves, and a succession of drills is brought to bear one after another in such a manner that the most rigid accuracy is attainable. Such machines are used to drill nipples and such articles.

A very important class of machinery called 'edging' machines, are much used. They are employed to trace irregular figures, and to impart an exact outline. For example, it is absolutely necessary that every lock plate should be exactly alike on the edge that fits against the wood of the stock. This rigid degree of accuracy is produced by the 'edging' machine.

The lock plate after being milled close to the edge by a milling machine, is fixed on a frame, in a definite position with regard to its several holes, etc. Alongside is a steel copy of its true shape, above the two there is a sliding frame which carries two spindles both of the same size, except that one revolves and cuts and the other is dumb. The dumb spindle moves in contact with the copy, thus causing the cutting one to move in contact with the lock plate, and to make it precisely the same as the pattern. It will thus be seen, that to attain the advantage of the parts interchanging, a degree of refinement in the tools and apparatus is imperative, and this cannot be appreciated by those who have been engaged on a ruder system, and hence the assertion of so many gun makers before the Parliamentary Committee, that to have perfect interchange was unattainable.

There is yet another and an expensive class of milling machines, termed 'Universal,' from its capability of being twisted about to suit any purpose, or to cut in any direction or form. As a general rule, machines of this description would be unprofitable, hence the milling machines usually employed are simply for one simple operation; but in a manufactory of small-arms the great number of milling tools that are required are of endless variety. The standard patterns, and other tools, etc., that have to be made are all various, these universal milling machines are used for such purposes, together with the ordinary lathes, planing, slotting and other machines of the engineer, and, consequently, form a considerable item in the daily expenditure of an average armory.

Many of the parts, after they are case-hardened, are polished on buffs, in the same manner as practised in England; but on the whole much less attention is bestowed on what is called high finish given to the parts, only to please the eye; but that can be done, if necessary, after the parts are brought to shape by the machines.

The hammer, and some of the other parts, require considerable filing by hand, so as to trim off the little corners that have burred up, or places where the milling tools did not reach. They also employ revolving files in a sort of lathe in the United States' armories. The files are round, and each file is made of a different diameter at different points in its length. It is not a taper file, but is composed of a series of cylinders, according to what it has to do. In finishing the sear, for example, there are two diameters required, one about $\frac{3}{4}$-inch, and the other about $\frac{1}{2}$-inch, or thereabouts. These are turned to the proper diameter, have teeth cut upon them in the ordinary way, and they are then hardened. They revolve with considerable speed, and the workman lays the article to be cut on a rest, and pushes it into contact, at the same time keeping the surface oiled.

In this way a man can do much more work than he can with a common hand file.

An extensive variety of small tools and instruments are also used to reduce the labour on the parts that have to be done by hand; indeed every workman seems to be continually devising some new thing to assist him in his work, and there being a strong desire, both with masters and workman all through the New England States, to be 'posted up' in every new improvement, they seem to be much better acquainted with each other all through the trade than is the case in England.

STOCKING MACHINERY

No part of the machinery used in making fire-arms has been brought to greater perfection than that for the stock. The shape of a gun-stock is so irregular, the form so difficult to produce, and the many obstacles to be overcome so very formidable, that it is no wonder so many practical men, who are connected with the gun trade, are opposed to the notion of such tools being used either with accuracy or advantage; for before the stock machinery now in use at the National Armories of the United States could have been devised and set to work with its present degree of perfection, the machinists of that country required the experience of twenty years in the use of Blanchard's machines or lathes for tracing or copying any pattern, and which may be said to form the basis and to contain the principle of that machinery.

Blanchard's machine has been used very extensively for about thirty years in the turning of shoe-lasts, boot-trees, oars, spokes of wheels, busts, gunstocks, etc. It may be said to consist of two lathes placed side by side, the one containing a copy of the thing required, and the other the material, out of which the thing is to be made.

The sliding rest which holds the turning tool also carries a dumb-tracer of the same form as the turning instrument, which is in contact with the copy. The turning instrument is a wheel containing a series of rounded cutters, which revolve at 2,000 revolutions per minute, while the material and copy turn together slowly. The dumb-tracer is also a round-edged wheel, capable of turning freely on its axis, a spring or weight is employed to pull this tracer into contact with the copy, so that as the copy turns round, this tracer moves out and in according to the form, and as the material to be turned has the same motion as the copy, and the turning instrument the same as the tracer, it of course produces a shape the same as the copy, and by sliding the rest slowly down the double lathe the article is completed.

It is most remarkable that this valuable labour-saving machine should have been so much neglected in England, seeing that it is capable of being applied to many branches of manufacture; its introduction into the armory will prove a national benefit.

The time required to pass a gun-stock through the sixteen different machines varies from twenty minutes to half-an-hour, but this cannot be taken as the time usually occupied during a whole day of ten hours, seeing that the tools occasionally require sharpening, and that there are other incidental hindrances. Some of the machines require but seven or eight seconds to perform their task, while others take four or five minutes; hence to have a complete plant of tools working up to the maximum of produce, a considerable number of the one would be required to one of the other; to produce 500 stocks per day, only one of the former would be required, but ten of the latter. The slow machines are entirely self-acting, so that a single individual can attend to a number of them.

The stocks when supplied to the United States' Armory are roughly blocked out to the shape. The object of the first machine is to turn off all the superfluous wood so as to save the cutters in the following machines. It consists of a circular saw, with different apparatus for holding the stock, and for following a pattern or copy. In this machine the stock is subjected to eight different cuts, viz.:

1*st Cut.* A slab is taken off the upper side where the barrel is to lie.

2*nd Cut.* This slab is detached by an oblique cut at the place of the lock.

3*rd Cut.* An incision is made around the three other sides of the stock at the termination of the smaller part under the barrel, just in front of the lock.

4*th*, 5*th*, 6*th*, 7*th*, *and* 8*th Cuts.* During these five cuts the stock is held in a fixing with a universal joint, and the object is to detach five distinct slabs from the muzzle up to the incision made by the fourth cut, viz., from the two sides, the bottom, and the two lower corners.

At the same time the attendant makes an indentation on the end of the butt, which is the guiding-point in some of the future operations. The time required varies from 3½ to 4 minutes. The same person attends to the next machine, which he keeps in constant operation, the first machine being to it a provider.

The second machine is one of Blanchard's lathes for rough turning the stock. It carries an iron pattern a little larger than the finished stock, so as to leave a small portion for the other machines to finish upon. Before putting the stock into the machine, a plate of steel is put on the end of the butt, which fits into the indentations previously made, and by means of a draw-knife, any proud corner of the butt is chopped off, so as to give the cutters less work to perform. The time required by this machine varies from 4 to 4½ minutes.

The third machine is termed a 'spotting machine,' its object being to cut off certain flat surfaces along the stock, so as to serve as bearings in future operations. The stock is laid in the machine with the upper side downwards, resting on a steel pattern, to which it is clamped; a pair of horizontal cutters at each end and three single cutters along its length, come in contact with the stock, and produce the required effect. The time required by this machine is from 7 to 8 seconds.

The fourth machine is for bedding the barrel, and cutting out the space for the breech-pin. The stock, as it comes from the third machine, is laid upon and against steel plates, with the spotted surfaces in contact with them. Alongside is a steel copy, in the form of what is required to be done to the stock, and above is a sliding frame which contains four revolving tools, with tracing or guide-pins adjusted to them, of the same form as the cutters: these guide-pins are consecutively brought into contact with the copy, which brings the cutters into contact with the wood, one after the other.

The first operation is to chisel out the greater portion of the groove for the barrel, which is done by a circular cutter, with the guide-pin sliding in the groove of the pattern, the slide containing the stock and copy, being traversed by the left hand, whilst with the right the workman pushes the guide-pin against the sides. The second operation is performed with the next cutter in the series, and forms the flat at the sides of the breech for the squares on the barrel, and also a part of the recess into which the breech-pin is to be fitted, every cut being guided by the pin tracing the steel copy. The third operation is to cut the conical recess for the body of the breech-pin, which is done by No. 3 cutter being pushed on laterally against the wood, the guide-pin determining the depth. The fourth operation is to cut out the remainder of the wood for the 'tang' of the breech-pin, which is done by the cutter and guide No. 4. The fifth operation is to give the finishing touch to the groove for the barrel, the cutter, No. 1, having brought it within a shade of the proper size.

For this another arrangement of cutter and spindle is brought into play–a steel rod or spindle which works horizontally, and is capable of moving laterally in any direction, has a cutter on its extremity. The spindle and cutter are in connection with a similar dumb spindle and tracer, which lies in the groove of the copy. During the operation the workman moves the dumb-spindle all over the interior of the copy, while the corresponding cutting-spindle chisels out the groove with the most rigid exactness. The time required to perform these several operations is under 1¼ minute.

A small amount of hand-labour is now given to the stock. 1st. To round off an edge, left by the cutter where the breech-pin and tang fit. 2nd. To finish a small part of the groove, where the flat part of the barrel has to fit–the time occupied being under 1½ minute.

The object of the fifth machine is to finish the extremities of the butt and muzzle, which is done by a circular saw, kept in good condition, so as to cut away very smoothly. During this operation the stock is laid in the machine upon a steel rod, in shape exactly the same as a barrel, to which it is rigidly clamped; the frame that carries the rod is then brought into the oblique position to cut the butt to the proper angle and length, and then the muzzle in a similar manner. Thus every stock is precisely the same length. The time required is from 10 to 12 minutes.

The object of the sixth machine is fourfold. During the first operation the stock is laid in the machine on a solid bar, which is fixed upon a slide; this bar is of the same shape as the barrel. It is clamped upon the bar with the upper side downwards, and by means of a vertical cutter which revolves with great velocity, the sides of the stock in the vicinity of the lock-plate are shaped to their finished dimensions. In the second operation, the stock is clamped to another sliding frame, where it is planed by a circular cutter from the breech of the barrel to the shoulder of the heel-plate, and also on the other side from the first band to the toe of the heel-plate. In the third operation it is fixed upon another slide, and the upper edges of the stock alongside the barrel-groove are planed from breech to muzzle, and finished in beautiful style with well defined sharp edges. During the fourth operation the stock lies with the lock side-downwards, and alongside lies a copy of the work to be done, viz.: to cut out a recess for the screw, which serves the same purpose as the side-cups in the English musket.

On a frame over the stock a revolving spindle carries a cutter, and also a dumb tracer to follow the copy. This cutter is brought down to the stock, the guide determining the depth and direction in which it cuts. The time required to perform these four operations is about 1½ minute; the period of actually cutting is only a few seconds.

The work performed by the seventh machine consists of five operations; the stock is fixed in this machine with the bar, groove uppermost.

In the first operation a revolving spindle, which carries an auger, is brought to bear against the butt, in order to drill the screw-hole, which fastens the butt-plate. In the second operation the spindle, which carries a revolving tap, follows the auger, to form the thread in the hole for the wood screw. In the third operation the recess for the upper part of the butt-plate is cut out by means of a vertical cutter, having a tracer to follow a guide or copy. The fourth operation is similar to the first, only a vertical auger bores the hole for the second screw of the butt plate. The fifth operation is, to cut the screw of this hole, which is done by a tap in the vertical spindle; the time required for the several operations being less than half a minute.

The object of the eighth machine is to cut that part of the stock which has to receive the bands. This machine is on Blanchard's principle, but rather differently arranged from the one described.

In this machine the cutters are of the same form as the stock is when finished; and hence, as the copy and material turn round, a perfectly smooth surface is obtained, which is not exactly the case with the round-edged cutters that have to slide along and trace an irregular surface. In the ordinary Blanchard's machine there is always a certain amount of roughness that requires to be hand finished; but those parts of the stock, from the first band to the muzzle, are produced with perfect smoothness, and with a degree of accuracy that no hand could imitate. The time required to cut for the bands is less than thirty seconds.

The object of the ninth machine is to cut off the material intervening between the several bands; it is also on the Blanchard principle, and very similar to the eighth machine, but with the cutter arranged on cylinders of the exact length, so as to bring out the requisite lines, the oval form of that part of the stock being dependent on the iron copy, which is traced in the usual manner. The time required is under 30 seconds.

The tenth machine is an ordinary Blanchard machine, containing an accurate iron copy of the stock from the butt to the lock; and in order to have the work as correct and smooth as possible, the cutter or tracer passes along slowly, so that the time required to pass along from the butt to the lock is between 8 and 9 minutes; hence, in a large establishment a number of such machines would be required so as to keep pace with the others.

The eleventh machine is similar to the tenth, and turns the stock from the point where the other left off to the first band, and requires between 5 and 6 minutes to perform the operation.

The object of the twelfth machine is the important one of cutting out for the lock plate. The stock is laid in a suitable holder with the steel copy of the recess to be made placed along side. Over the stock and copy hangs the side of a circular reel or frame, round the periphery of which is placed the series of five spindles; this frame turns on an axis, so that any of the spindles can be successively brought to bear upon the stock. Each of the spindles referred to is in a separate slide upon the reel, and each has a dumb tracer the same shape and length as the cutter with which to follow the steel copy. The driving band runs upon a loose pulley over the frame, which is in a line with the fixed position of the different spindles, and as the several spindles are brought round, the band is drawn down to a fixed pulley on the top of each spindle. The bedding of the lock may be divided into five operations:

The first operation is to cut out the space for the lock plate, which it does with unvarying accuracy, both as to figure and depth. For the second operation another drill cutter is brought round, and five incisions made for the lock preparatory to the work of the followers. In the third operation another cutter of large diameter is brought round to enlarge some of the parts cut in the second operation. In the next a small cutter is brought to bear, and gets into remote points not reachable by the others, the true form depending on the copy, which is traced by the point of the same size as the bit, and can pass into spaces over which larger tracers pass without entering. In the fifth operation a very minute cutter comes into play and delicately touches off certain corners, and the result of the whole, if performed with any degree of care, is, as it ought to be, most rigidly correct, and the whole is accomplished in considerably less than a minute of time, which includes the several shiftings, without any hurry on the part of the attendants; and the removal of certain burrs that are thrown up on some of the corners is all that has to be done to make the recess ready to receive the lock-plate.

The object of the thirteenth machine is to cut out the recess for the trigger-guard, the ramrod stops, and the trigger. In construction this machine resembles the twelfth (for bedding the lock), only that the reel carries four spindles instead of five. The stock is fixed in the machine with its lower side upwards, having a steel copy of the various recesses placed alongside, and the several insertions are made by turning round the several cutters and bringing them to bear in succession; the time required is under one minute.

The object of the fourteenth machine is to cut out the groove for the ramrod, the recess for the ramrod spring, slotting for bearing springs, and boring several pin-holes. In this machine the stock is laid upon the left side, and fixed to a pattern barrel.

The first operation is to cut out the openings or slots for the hand-springs, which is accomplished by a revolving cutter of peculiar construction, consisting of four radial arms, each arm being in the form of an involute. In the second operation the stock is pierced for the band-spring pins, by means of two drills, the one working upwards and the other downwards, so arranged that they cannot come into contact or injure one another. The third operation is to cut out the groove for the ramrod, which is done by a revolving horizontal cutter tracing a copy. By the fourth operation the hollow for the ramrod-spring is cut out by means of a circular cutter, which also is guided by the pattern; time required, about one minute.

The fifteenth machine is an appendage to the lock machine, but only for economy. It consists of a solid barrel for holding the stock, and an apparatus for drilling and screwing several holes.

The first operation is to drill out the hole for the breech-nail. The stock is placed in an inverted position at the requisite angle under the vertical drills which bore the holes. In boring for the side screws another adjustment is applied to the same machine; the time required to perform their several operations being under half a minute. The rest of the work upon the stock consists of boring for the ramrod, filing off the sharp edges, and a general smoothing, which takes altogether nearly an hour.

In some private establishments the arrangement of the stock machinery was different from that described, in some respects more simple; but, on the whole, none was so perfect as the plant in Springfield Armory.

The lock fitter takes the parts of the locks from heaps of each promiscuously, and often making up 40 or 50 without having to use a file.

The part that most frequently requires filing is the main spring, which occasionally sets in tempering, but when once fitted to any lock-plate will fit another equally well; the parts of the lock are not numbered or marked in any way.

The workman whose business it is to 'assemble,' or set up the arms, takes the different parts promiscuously from a row of boxes, and uses nothing but the turnscrew to put the musket together, excepting on the slott, which contains the bandsprings, which have to be squared out at one end with a small chisel. He receives four cents per musket, and has put together as many as 100 in a day and 530 in a week, but his usual day's work is from 50 to 60.

The time is 3½ minutes.

Besides the machinery and tools which have been enumerated, there are hundreds of valuable instruments and gauges that are employed in testing the work through all its stages, from the raw material to the finished gun, others for holding the pieces whilst undergoing different operations, such as marking, drilling, screwing, etc., the object of all being to secure thorough identity in all parts, and facilitate the work. The Committee also saw much that was new to them, which will be valuable to the service, in drills for metal, bits for chiselling wood, modes of cutting screws, turning of nipple cones, and in the tempering of milling tools and cutters, etc., all of which, though apparently trifling in themselves, yet form an important item in the management of an armory.

The Committee also observed that everything that could be done to reduce labour in the movement of material from one point to another was adopted. This includes mechanical arrangements for lifting material, etc. from one floor to another, carriages for conveying material on the same floor, and such like, all of which tends to reduce the price of the musket.

The Committee paid particular attention to the manufacturing capabilities of the United States as compared with England, and they are strongly of opinion that with the same facilities and energy the advantages are all on the side of our own country.

In the national armories of America, the same class of workmen as are employed in our Government works earn much higher wages, although absent from their employment to an extent which could not be tolerated with us. The high price of labour and material and the extreme heat of the weather during summer, all operate powerfully against the Americans; and the only reason why they should produce arms cheaper than in England is altogether owing to the productive capabilities of the machinery and tools that are employed; and it is the opinion of Mr Lawrence, a celebrated manufacturer of arms (including Sharpe's breech-loading rifle), that we shall produce our arms under £1 10*s*. when once in full operation, and at the same time be able to pay good wages to those employed.

The American musket complete costs the Government $8·89. The accompanying tariff shows the value of every separate part, and the price paid to the workman for its production. (The tariff will be found at page 146.)

The prices in America are higher than would have to be paid in England under the same circumstances, and the more so if the proposed armory is to be on a large scale, when the work can be subdivided into minute operations, which will admit of their being performed by unskilled labour, which is not so much the case in a smaller establishment, where one person has not only to perform operations, but also to keep the tools in order.

After having heard what was said at the Parliamentary Committee, by those who were opposed to the New Armory, and having now seen the American system in operation, the Committee are of opinion that the expectations which were held out to result from the proposed Armory will be more than realized.

The Committee, whilst in America, were informed by Mr Robbins, the contractor who makes Sharpe's breech-loading rifle for the Company to which the patent belongs, that this arm, which is much more expensive to make, and for which the American Government pay $30 is, by their agreement with the proprietors, supplied to them at the following rates:

First 10,000 for $16½
Second 10,000 for $12½

and that the makers cleared $5 on every arm turned out, an important and encouraging fact in connection with the establishment of the proposed Armory.

With regard to the time when the machinery, which the Committee have contracted for in America, will be completed; they regret that it is much greater than they had anticipated before leaving England.

A considerable quantity of it will be delivered in three months, but it will be fifteen months before the whole is finished, and after making allowance for incidental delays, we cannot calculate on having it in England before eighteen months. This delay the Committee believe to have been unavoidable.

TARIFF, OR PRICES FOR LABOUR ON THE MUSKET AND APPENDAGES, AT THE UNITED STATES ARMORY, SPRINGFIELD, MASS. SPRINGFIELD, MAY, 1854

Cat. of Components		Branch of Labour	Price per operation
$	c.	Musket	c.
8	89	assembling *by hand*	04
		Ramrod Spring	
	03	forging *by hand*	015
	031	annealing *by hand, by day*	
	032	drilling	0015
	068	filing *by hand*	03
	071	tempering *by hand*	001
		Rod Spring Wire	
	001	cutting off *by hand*	0003
	003	pointing	0007
		Ramrod Stop	
	003	punching	
	008	milling	
	012	filing *by hand*	
		Stock	
	37	first turning	022
	38	spotting	005
	39	grooving for barrel	023
	41	fitting in barrel *by hand*	009
	43	planing	011
	44	bedding for butt-plate	004
	47	bedding for lock	013
	51	bedding for guard	008
	52	fitting-in lock *by hand*	005
	56	fitting bands (3)	016
	58	finish turning	015
	60	bedding for band-springs and rod and spring	01
	62	boring for screws	004
	76	shaping butt *by hand*	} 24
	82	fitting stop and spring *by hand*	}
	86	filing chase *by hand*	}
	96	completing *by hand*	}
		Butt Plate	
	10	forging *by hand*	02
		helping forge *by hand*	013
	102	annealing *by hand, by day*	
	106	trimming	0032
	13	first filing *by hand*	015
	135	drilling	004
	145	first milling	01
	155	second milling	01
	17	second filing *by hand*	006
	18	countersinking	002
	20	polishing	014

Cat. of Components	*Branch of Labour*	*Price per operation*
$ c.	Butt Plate Screw	c.
015	forging *by hand*	0022
016	annealing *by hand, by day*	
02	milling	002
022	slitting	002
025	cutting	0025
026	tempering *by hand*	00018
028	polishing	002
	Guard Plate	
06	forging *by hand*	015
	helping forge *by hand*	01
063	annealing *by hand, by day*	
068	first straightening *by hand*	003
07	first milling	0016
072	second milling	0024
076	trimming	0024
08	second straightening *by hand*	0025
085	third milling	0025
094	drilling	006
10	grinding	0035
105	fourth milling	0047
106	first countersinking	0008
11	punching	0036
1138	third straightening *by hand*	0035
12	drilling stud	003
13	turning	0052
135	fifth milling	003
145	filing mortise *by hand*	0075
17	edging *by hand*	02
22	filing *by hand*	045
23	second countersinking	0018
25	polishing	014
	Guard Bow	
05	forging *by hand*	0275
051	annealing *by hand, by day*	
071	first milling	0188
074	second milling	003
084	edging *by hand*	0035
09	grinding	003
125	filing *by hand*	035
13	polishing ends	004
20	swivelling *by hand*	015
22	polishing	013
	Swivel	
025	forging *by hand*	0125
026	annealing *by hand, by day*	
04	punching	01
055	finishing	012

Cat. of Components	Branch of Labour	Price per operation
$ c.	Guard Bow Nut	c.
002	forging *by hand*	001
0021	annealing *by hand, by day*	
01	finishing	0045
011	tempering *by hand*	0005
012	polishing	0012
	Trigger	
024	forging *by hand*	012
025	annealing *by hand, by day*	
028	trimming	002
04	first milling	0083
05	second milling	0042
078	filing *by hand*	025
08	tempering *by hand*	0006
085	polishing	004
	Trigger Screw	
004	forging *by hand*	0015
004	annealing *by hand, by day*	
01	finishing	004
011	tempering *by hand*	00018
018	polishing	001
	Guard	
62	completing *by hand*	008
	Guard Screw, long	
007	forging *by hand*	002
008	annealing *by hand, by day*	
01	milling	015
012	slitting	0017
015	cutting	0015
016	tempering *by hand*	00018
018	polishing	0012
	Guard Screw, short	
007	forging *by hand*	002
008	annealing *by hand, by day*	
01	milling	0013
012	slitting	0017
015	cutting	0014
016	tempering *by hand*	00018
018	polishing	0012
	Bayonet Stud	
003	forging *by hand*	001
003	annealing *by hand, by day*	
	Breech Screw	
06	forging *by hand*	0125
	helping forge *by hand*	0075
066	annealing *by hand, by day*	

Cat. of Components		Branch of Labour	Price per operation
$	c.	Breech Screw, *contd.*	c.
	074	milling	0045
	086	cutting	01
		polishing (for issue)	002
		Cone	
	004	forging *by hand*	002
	005	annealing *by hand, by day*	
	01	first milling	004
	013	second milling	0012
	015	squaring	0018
	02	cutting	005
	033	drilling	012
	04	filing	0015
	046	tempering *by hand*	0005
		Barrel Plate	
	75	cutting off *by day*	
	78	drawing	02
		helping draw	013
	80	curving	011
		helping curve	0065
		Barrel	
1	16	welding	12
		helping weld	065
1	20	cone seating	02
		helping cone seat	011
1	22	annealing *by day*	
1	26	first boring	04
1	34	second boring	075
1	38	third boring	025
1	44	counter-boring	006
1	50	turning	03
1	51	fourth boring	024
1	51	first milling	004
1	71	grinding	002
1	72	seating	002
1	73	studding *by hand*	0098
1	77	tapping *by hand*	01
1	81	fifth boring	028
1	83	second milling	0125
1	86	proving *by hand, by day*	
2	11	breeching *by hand*	01
2	15	third milling	02
2	17	fourth milling	0125
2	21	fitting *by hand*	033
2	22	drilling for cone	005
2	23	tapping	01
2	27	jigging *by hand*	037
2	31	filing breech *by hand*	0367

Cat. of Components $	c.	Branch of Labour	Price per operation c.
		Barrel, *contd.*	
2	50	filing butts *by hand*	185
2	51	drilling tang and vent	004
2	52	fifth milling	001
2	58	finish boring	005
2	64	polishing	05
2	84	polishing butt	027
3	00	completing *by hand*	008
		Lower Band Spring	
	02	forging *by hand*	001
	021	annealing *by hand*	
	023	first milling	001
	046	fitting *by hand*	02
	048	second milling	0009
	056	filing *by hand*	005
	06	tempering *by hand*	0005
	064	polishing	0025
		Lower Band Plate	
	018	punching	001
		Lower Band	
	032	forging	01
	036	annealing	
	046	first mandrelling	0079
	048	turning	002
	054	first milling	004
	06	second milling	005
	006	second mandrelling *by hand*	0037
	076	filing *by hand*	008
	086	polishing	005
	09	bevelling *by hand*	004
		Tang Screw	
	014	forging *by hand*	0025
	015	annealing *by hand, by day*	
	02	milling	0036
	023	slitting	0017
	026	cutting	0015
	028	tempering *by hand*	00018
	032	polishing	0015
		Middle Band Spring	
	02	forging *by hand*	011
	021	annealing *by hand, by day*	
	023	first milling	001
	046	fitting *by hand*	02
	048	second milling	0009
	056	filing *by hand*	005
	06	tempering *by hand*	0005
	064	polishing	0025

Cat. of Components	*Branch of Labour*	*Price per operation*
$ c.	Middle Band	c.
05	forging *by hand*	0263
053	annealing *by hand, by day*	
063	first mandrelling *by hand*	008
068	turning	0026
078	milling	0065
081	drilling	002
085	second mandrelling *by hand*	0037
105	filing *by hand*	014
111	polishing side	005
186	swivelling *by hand*	013
191	polishing	003
195	bevelling *by hand*	004
	Upper Band Spring	
02	forging *by hand*	011
021	annealing *by hand, by day*	
023	first milling	001
047	fitting *by hand*	02
049	forming pivot	0013
051	second milling	0009
06	filing *by hand*	008
065	tempering *by hand*	0005
07	polishing	0026
	Upper Band Plate	
04	punching	001
	Upper Band Sight	
004	casting *by hand*	001
	Upper Band	
07	forging *by hand*	025
075	annealing *by hand, by day*	
08	first mandrelling *by hand*	0086
09	first filing *by hand*	01
098	turning	005
103	first milling	003
127	sighting *by hand*	011
135	second mandrelling *by hand*	0037
138	second milling	0035
154	third milling	001
164	drilling	001
204	second filing *by hand*	03
218	polishing	021
225	bevelling *by hand*	008
	Lock Plate	
095	forging *by hand*	0175
	helping forge *by hand*	012
098	annealing *by hand, by day*	
122	first milling	0175

Cat. of Components	*Branch of Labour*	*Price per operation*
$ c.	Lock Plate, *contd.*	c.
127	stamping *by hand*	003
14	drilling	013
16	second milling	008
17	lapping *by hand*	0075
175	buffing	0025
21	third milling	012
215	filing *by hand*	01
26	edging	01875
28	bevelling	01875
30	finishing *by hand*	0175
	Tumbler	
06	forging *by hand*	01
	helping forge *by hand*	007
062	annealing *by hand, by day*	
092	first milling	013
14	second milling	004
19	filing *by hand*	116
20	drilling	007
21	tempering *by hand*	004
	Hammer	
07	forging *by hand*	02
	helping forge *by hand*	0125
072	annealing *by hand, by day*	
0765	swedging	0025
	helping swedge	0018
079	fitting *by hand*	0025
086	punching	0005
088	first milling	001
09	first drifting	0009
10	straightening *by hand*	0025
103	second milling	00125
105	third milling	00125
107	fourth milling	002
11	fifth milling	001
113	sixth milling	00125
116	seventh milling	00125
12	eighth milling	00125
123	drilling	0025
13	slitting	00125
135	ninth milling	00125
139	second drifting	0005
142	tenth milling	005
16	eleventh milling	008
172	turning	065
26	filing *by hand*	005
30	checking	005
32	polishing (for issue)	013

Cat. of Components	Branch of Labour	Price per operation
$ c.	Tumbler Screw	c.
007	forging *by hand*	0022
008	annealing *by hand, by day*	
017	finishing	007
019	tempering *included in locks*	
022	polishing	002
	Bridle	
02	forging *by hand*	01
021	annealing *by hand, by day*	
022	trimming	001
029	milling	0035
045	drilling	007
065	shaving	014
125	filing *by hand*	045
	Sear	
03	forging *by hand*	012
032	annealing *by hand, by day*	
034	buffing	001
039	drilling	002
043	first milling	0025
048	reaming	001
058	second milling	006
068	third milling	002
14	filing *by hand*	067
	Sear Spring	
023	forging *by hand*	0085
028	curving *by hand*	0025
03	annealing *by hand, by day*	
034	jointing	004
04	drilling	0018
07	filing *by hand*	025
075	tempering *by hand*	0008
085	polishing	003
	Main Spring	
06	forging *by hand*	026
07	curving *by hand*	0098
073	annealing *by hand, by day*	
08	buffing	005
11	first milling	01
12	first filing *by hand*	01
126	second milling	0044
132	third milling	004
175	second filing *by hand*	042
18	fourth milling	004
22	tempering *by hand*	009
24	polishing	005

Cat. of Components $	c.	Branch of Labour	Price per operation c.
		Lock Screw, long:	
	004	forging *by hand*	0015
	004	annealing *by hand, by day*	
		Bridle Screw	
	012	finishing	004
	015	polishing	001
		Lock Screw, short	
	004	forging *by hand*	0015
	004	annealing *by hand, by day*	
		Bridle Screw	
	012	finishing	004
	015	polishing	001
		Sear Spring Screw	
	012	finishing	004
	015	polishing	001
		Main Spring Screw	
	012	finishing	004
	015	polishing	001
		Lock	
1	20	assembled *by hand, by day*	
1	22	tempering *by hand*	01
1	30	polishing	05
1	80	complete *by hand*	07
		Side Plate	001
	012	punching	001
	031	straightening *by hand*	0008
	018	drilling	002
	024	milling	003
	05	filing *by hand*	018
	053	polishing	0025
		Side Screw	
	001	forging *by hand*	0037
	011	annealing *by hand, by day*	
	018	milling	0063
	022	slitting	0025
	025	cutting *by hand*	0015
	027	tempering *by hand*	00018
	032	polishing	002
		Ramrod	
	16	drawing	001
		helping draw	009
	19	rounding	011
		helping round	009
	23	tempering *by hand*	034
	24	annealing *by hand, by day*	
		milling	004

Cat. of Components	Branch of Labour	Price per operation
$ c.	Ramrod, *contd.*	c.
	grinding	012
	polishing	014
	finishing *by hand*	005
	Bayonet Clasp, Screw	
004	forging *by hand*	0015
004	annealing *by hand, by day*	
012	finishing	004
013	tempering *by hand*	00018
015	polishing	0012
	Bayonet Clasp	
02	forging *by hand*	005
	helping forge *by hand*	0042
021	annealing *by hand, by day*	
027	boring	0055
032	first milling	0036
034	punching	002
04	second milling	005
045	drilling	003
051	third milling	005
056	polishing stud	002
058	slitting	0033
09	filing *by hand*	03
093	reaming	0025
11	polishing	004
	Bayonet Socket Plate	
06	cutting off	0007
	Bayonet	
16	drawing	007
	helping draw	0047
19	swedging	0165
	helping swedge	01
21	heading neck *by hand*	004
	helping head neck *by hand*	003
26	welding on socket *by hand*	027
	helping weld on socket *by hand*	015
32	welding socket	027
	helping weld socket	015
351	swedging neck *by hand*	0045
	helping swedge neck *by hand*	0028
36	annealing *by hand, by day*	
39	first boring	017
41	first turning	008
47	milling	047
50	filing neck *by hand*	02
525	tempering *by hand*	0175
535	edging	005
555	buffing	014

Cat. of Components	*Branch of Labour*	*Price per operation*
$ c.	Bayonet, *contd.*	c.
64	grinding	20
655	grooving	01
67	finish boring	004
69	milling socket	01
71	finish turning	0125
84	filing socket *by hand*	10
	drilling	003
	polishing	058
	completing *by hand*	006
	Wiper	
035	forging *by hand*	02
036	annealing *by hand, by day*	
045	fitting	01
05	drilling	002
065	filing *by hand*	01
09	tempering *by hand*	002
	Screw Driver Wrench	
03	forging *by hand*	015
031	annealing *by hand, by day*	
033	fitting *by hand*	001
037	drilling	004
039	first milling	0015
04	second milling	0009
044	punching	003
046	trimming	001
048	straightening *by hand*	001
049	third milling	00085
054	filing *by hand*	0045
057	polishing	003
06	finishing	0015
064	tempering *by hand*	0015
	Screw Driver Blade	
01	forging *by hand*	015
011	annealing *by hand, by day*	
013	drilling	0075
014	trimming	0022
016	straightening *by hand*	0173
018	milling	003
02	filing *by hand*	0015
029	polishing	0027
032	finishing	0015
036	tempering *by hand*	0005
	Screw Driver Collet	
004	forging *by hand*	0005
004	annealing *by hand, by day*	
005	drilling } finishing	002125
006	milling }	

Cat. of Components	*Branch of Labour*	*Price per operation*
$ c.	Screw Driver	c.
12	completing *by hand*	00575
	Ball Screw	
06	forging *by hand*	015
062	annealing *by hand, by day*	
072	milling	0075
08	drilling	0022
10	finishing	0173
11	tempering *by hand*	003
	Spring Vice Bolster	
021	forging *by hand*	018
021	annealing *by hand, by day*	
025	punching	0035
037	milling	012
04	drilling	0028
044	tapping	0043
054	filing *by hand*	0096
056	polishing	0039
058	tempering *by hand and day*	
	Spring Vice Slide	
018	forging *by hand*	019
019	annealing *by hand, by day*	
03	milling	012
084	drilling	0028
038	tapping	0035
047	filing *by hand*	0087
048	tempering *by hand*	0015
	Spring Vice Slide Screw	
004	forging *by hand*	0015
004	annealing *by hand, by day*	
012	finishing	004
015	polishing	001
	Spring Vice Thumb Screw	
02	forging *by hand*	0085
021	annealing *by hand, by day*	
03	trimming	0015
037	milling	0066
041	cutting	0036
049	filing *by hand*	0083
051	polishing *by day*	
052	tempering *by hand, by day*	
	Spring Vice	
18	completing *by hand*	006

4

Inspector of Artillery's Department

Although the primary object of the Committee in visiting the United States was in connection with small arms and their manufacture, still, in compliance with the instructions they received, their attention was also directed to anything at all likely to benefit the public service.

[CASTING IRON AND BRASS]

The quality of the brass and iron ordnance used in the United States seems to be of a very high order, both as regards the soundness and strength of the material.

The improvement of the material for guns is a subject which occupies much attention, and the metal of every gun is subjected to a series of testing experiments in a machine constructed for the purpose, a specimen of each being kept, and the result of the trial duly registered. The effect of this, continued from year to year, has so quickened the energies of the several contractors, that the strength of the mixture used has been greatly increased, and is now much greater than in England, where nothing is done to ascertain the precise conditions of the metal in tenacity, transverse strength, compressibility, and torsion. The average tenacity of our cast iron per square inch, is under 17,000 pounds, in the United States it varies from 25,000 to 40,000, and some has been got up to 45,000; this, however, is a rare exception.

An extensive series of experiments have been going on for some time at Pittsburgh, under the direction of Major Wade, in connection with the casting of iron guns, having for their object to obtain a more uniform condition of the metal with regard to contraction of the particles during the process of cooling; it being the opinion, that in the cooling of such masses of metal, when cast in the present mode, the exterior is under a certain amount of tension from being brought to a cool state before the interior.

The direction in which these experiments have been hitherto conducted, has been casting the guns upon a core instead of solid, as hitherto practised; and having this core made in such a manner that a stream of cold water flows down its centre to the chamber of the gun, then up an annular space on the outside; but within the core, the flow of water being regulated so that the interior and exterior surfaces shall cool regularly.

From iron guns which have been cast in this way, most surprising results have been obtained; but the obstacles to be overcome before it can be safely employed in practice, are found to be very formidable.

The same principle is carried out in the casting of hydraulic cylinders and other work.

The system of moulding and casting brass guns, the Committee consider to be very much superior to that practised in the Royal Arsenal, both as regards accuracy, economy, and the soundness of the metal.

The inaccuracy of our system consists in having to make a loam model for every gun that is cast, the correctness of which depends on the workman, not only with regard to the diameter, but also to the position of the trunnions, dispart, and other projections. In the United States, a metal or wooden standard pattern is invariably employed, and the gun is moulded and cast in a double cast iron flask.

The material used for moulding guns in England is loam, but owing to the liability of this material to contain vegetable or mineral gaseous generating substances, and its porosity in allowing the liquid metal to penetrate its surface, the American founders have abandoned it for other materials.

That adopted by Mr Babbitt, of Boston, is pounded fire bricks, reduced to a powder, and made adhesive by the addition of pure pipe-clay water. The mould is burnt in a kiln, and then washed with a glaze of the same material, reduced to a creamy consistency.

Mr Ames, of Chicopee, one of the United States' Government contractors, employs for the same purpose decomposed granite, commonly called 'kaolin.' The crude kaolin is made adhesive for moulding by the addition of a small quantity of pure pipe-clay water.

This material is used in a double flask in the same manner as sand, and after being dried, is coated with a thin wash of pure kaolin.

Either of these materials, viz., brick dust or kaolin, yield most perfect castings, by which–together with the mode of running the metal into the mould at the lower end, invariably used, and leaving a dead head of large diameter–a soundness of texture and freedom from porosity towards the muzzle is obtained, which is not usually met with in bronze ordnance.

A piece of metal is trepanned from the dead head of every gun, both of bronze and iron, to be subjected to the testing machine, to ascertain the several conditions of the metal, which is in addition to the usual proof with hydraulic pressure and gunpowder, etc.

The 'Ballistic Pendulum' is much used both in the Arsenal and Navy Yard at Washington, for experiments both from cannon and small arms. These machines are got up in the very best style of workmanship, at a great expense, but must prove of great value to both branches of the service.

The great influx of workmen from France and Germany to America, from districts which excel in the production of fancy metal work, has given a very high tone to the fineness and smoothness of general castings in this country; some of the ornamental work and bronze statuary, that came under the notice of the Committee, seem equal to the fine productions of Berlin or Munich. Even in castings for ordinary purposes, such as door-locks, stoves, etc., they greatly excel in lightness, sharpness of figure, and an extreme smoothness of surface, which the Committee have rarely seen equalled. Cast iron is more extensively used in furniture and for other household purposes in the New England States than with us; and the variety of uses to which it is put, where we employ wood, brass, or stone, is endless.

The fine surface on their castings is partly due to the use of pipe-clay on the surface of the mould, a material, which is not to the knowledge of the Committee, employed for this purpose at Woolwich, or anywhere in England. The thin castings are made of Scotch hot-blast pig-iron, mixed with a portion of Pennsylvanian iron.

Malleable cast iron is also much used for a variety of purposes where we employ wrought iron or brass, and, as a general rule, it answers the purpose equally well and is much cheaper.

The cores for pipes and many other sorts of castings are made of green sand. In moulding small articles, only one double flask is used, which is made with hinges; so that when the mould is finished, the flask is opened up and the mould of sand laid upon the floor. In a lock manufactory in Connecticut, where work was conducted on this plan, a man will lay down 100 moulds per day from one flask, thus saving 99 of that article.

All through the United States, the cupolas for melting metal are made with a hinged bottom, which is kept up by a prop; and when the work is over, the prop is knocked away, the bottom of the furnace drops down, and the whole of the white hot fuel, slag, or metal in the interior runs out upon the floor, thus saving that hot and most laborious operation of raking out a cupola furnace.

A revolving cylinder is everywhere used for cleansing small and middling sized castings. The cylinder is filled from time to time, and set to revolve without any attendance, the workman, meanwhile, being employed in other ways. In the foundries generally, any labour-saving appliances that may be suggested are more readily adopted by the workmen than is the case in England.

Among the many things that came under observation, may be mentioned a method of welding cast and wrought iron together. Filings of soft cast iron are mixed and melted with calcined borax, and sprinkled on the part to be united. They are then heated and welded together.

A method of reducing castings both in thickness and weight which is pursued by stove makers, in order to alter patterns, is to mix equal parts of nitric acid and water, and in this mixture the castings are immersed until of the required dimensions.

With regard to the machinery employed in finishing ordnance, the Committee found them considerably behind those used in the Royal

Arsenal; but at the same time there are several simple contrivances which may be adopted with advantage, one of which is a mode of connecting a long sliding rest with its screw. For this, a kind of screwed pincers is employed, which has only to be opened to allow the workman to change its position by hand, thus saving the time and labour of winding it by the screw and handle.

A method employed for straightening long bars in a lathe consists of a bridle, which takes hold of projections cast upon the bed, and which is furnished with a screw and lever, with a bolster on each side to take the thrust. By this simple arrangement no hammering is required.

In Jerome's clock manufactory the metal-turners have a simple method of making articles the size without gauging.

The tool-holder is a bar of iron that lies in a line with the article to be turned, and rests upon two hooks, one upon each head-stock. For every diameter in the required article there is a tool set at the proper position in the holder, and there pinched with a set screw. The lathe is set to revolve, and the workman employs tool after tool on the several parts, which produces definite dimensions with surprising facility: these tools are laid upon a rest, as in ordinary hand-turning.

In all the national establishments ample provision is made for the prevention of accident by fire, by having pumps which can at any moment be set in operation by the prime mover.

In each of the manufacturing arsenals there is a proper grinding and polishing room, in which the things that are made have a smooth tidy appearance given them at a trifling expense, as compared with hand-finishing, and yet quite sufficient for such articles.

In all the United States' arsenals there are machine shops for the construction of special machinery, ordinary machines being purchased.

A very simple but accurate apparatus is employed for testing the amount of mechanical power that is required to work any machine, or to accomplish any object. The driving power is communicated through the testing-apparatus, and the precise amount of strain is determined by a steel-yard. The amount of motion is ascertained by counting the revolutions per minute, with a stop watch.

An excellent hand-sifting machine is used in foundries. The sieve works on rollers, on a frame. The motion is obtained by a crank-handle acting on a cam, which gives three shakes during every revolution.

The method of joining leather bands in factories is simple. Two plates, at the joint, are gripped together by a screw or screws, having their inner surfaces cut like a blunted ratchet. These teeth are forced into the leather, and so retain a firm hold.

An improved smith's vice has come into general use, in which the two jaws are always parallel to each other. In this vice there are two screws, the motion being conveyed from the one to the other by means of an endless vertebrated chain.

The iron guns in the navy-yard at Boston were very neatly kept, and looked particularly well. Instead of being lacquered they are black-leaded, in the following way:–a coat of black paint is first given them, and before this is quite dry–whilst yet sticky–a coat of black lead is sifted on, and rubbed in with a rag.

The skidding used in this yard was also very simple, consisting of a round bar of wrought iron, laid on long blocks of granite.

The grape-shot used in the United States' service is made up with iron rings encircling it and holding the balls together, in lieu of the perforated iron plates used in England. Through the nut at the top of the spindle a wire is passed and twisted, which acts as a handle.

A large percussion cap, made of card-board, has been tried in the United States' service, for firing heavy guns, in lieu of tubes, etc., and is found to answer well. It is fitted on to the hammer, and on the lanyard being pulled, strikes on the vent, its outside upper surface downwards, and the fire is driven into the charge.

The hammer used with these caps is of brass and very heavy, weighing one pound. It has under its head a large iron nipple, on which the caps are placed.

The form of the muzzle of guns used in the United States' Navy, has been altered to that proposed by the late Colonel Colquhoun, R.A., some time since, the mouldings at the muzzle being entirely removed.

At all the gun-foundries and arsenals an excellent machine is employed to test the several qualities of the metal. It is a system of levers, mounted on knife-edges, with arrangements for holding the material to undergo a trial of its tenacity, power of resisting compression, transverse strength, and resistance to torsion, and having a delicate steel-yard, by means of which the exact result is easily determined.

At Pittsburgh Arsenal the Committee were shown an 8-inch iron gun, which had been cast on a hollow core, and was under extreme proof, having been fired 2500 times.

Instead of bushing the vent when it became worn, it was plugged up altogether, and a new one bored in another part of the breech.

A proof-shot for the 32-pounder gun was shown to the Committee. It was double, being somewhat the form of a dumb-bell, and weighed twice as much as the service-shot of the same gun.

The heavy iron guns of the United States' service are elevated in a manner different from that used in the British service. A segmental ratchet passing through the centre of the caseable is cast upon the breech, each tooth of the ratchet being equal to 2½ degrees of elevation. The teeth of the ratchet-breech rest upon a pawl, which is attached to the carriage. In raising or lowering the gun an iron crowbar is passed through an opening in the pawl, by means of which, acting on the teeth of the ratchet, the gun may be readily lowered or raised, any number of degrees. To obtain intermediate degrees of elevation, the pawl is jointed to a wedge, which rests on an inclined plane, so that by means of a horizontal screw, the wedge is pulled backwards or pushed forwards; thus raising or lowering the pawl, and the breech of the gun that rests upon it.

In Washington Arsenal the Committee saw a gravitating sight, used by the Russians in artillery practice. It consists of a bar of gun-metal, with a slot up the middle, and a weight at the lower end, and is slung by means of two small trunnions, about one-third of the way up, on to a metal crotch, which is screwed on the breech. A slide, similar to that on the Minié rifle sight, traverses up and down, and the different elevations are marked along the sides of the sight. On the muzzle a small dispart pin is screwed. By this arrangement the gun may be laid, without any allowance being made for the wheels not being on level ground, and accuracy obtained, as the tangent sight being suspended is always vertical.

A gun-lock has been proposed by Captain Dahlgren, United States' Navy. The head of the hammer is hollow throughout, so as to permit of the escape from the vent passing through it. With this lock a tube is used, having a circular head to it.

The shells used in the United States from the 'Columbiads,' are cast eccentric, the fuze hole passing through the centre of the eccentric lump. They are very thick in the metal.

THE ROYAL LABORATORY

In the several arsenals which the Committee visited, they observed that whilst there was comparatively little work going on in any of them, each had a good supply of machinery, and in most of them they found workmen engaged in constructing machines to be applied to manufacturing Ordnance stores.

The machinery used for making percussion caps in the several laboratories are of a very perfect character. A single machine, which is entirely self-acting, is supplied with the sheet of copper which it converts into caps, charges the caps with composition, then compresses it, and finally delivers them into a box ready for the varnish.

One of these machines is said to produce from 50,000 to 60,000 per day, and one person can look after two of them, as they only require to be supplied with the necessary material. To complete a cap requires four operations.

1. To cut the blank.
2. To form the blank into a cap.
3. To fill the cap with composition.
4. To compress the powder.

As the machine operates only on a single cap at a time at any of the operations, and as the several operations are performed at some little distance from each other, the risk of accident is very small, and during the experience the United States' officers have had with them, there has not been an explosion of more than a single cap at a time. That portion of the percussion cap machine that forms the cap, is equally applicable to the manufacture of cups for the Minié bullet.

At Frankfort arsenal there is a beautiful set of machinery for the manufacture of friction tubes.

The first is a machine similar to that for the cup of the Minié bullet. In the second machine this cup is elongated into a capsule of about $\frac{3}{4}$-inch in length. The third machine converts this capsule into a tube of the proper length, with one end open and the other closed. In the fourth machine these tubes are put into a hopper, which delivers them into recesses on the periphery of a revolving cylinder, with an intermittent motion. During the passage round, the tubes are trimmed at the end by a revolving cutter, and have the hole drilled in them for the branch tube, the whole being completely self-acting.

Round bullets are made by compression from a cast rod. Two machines are required, the first to form the bullet, and the second to detach it from the rod. For the first machine, which is self-feeding, the rod of lead is passed through a horizontal tube, the dies being vertical. A self-acting ratchet motion pushes the rod through the tube; the front end of the tube is depressible as the upper die closes, but it is counterbalanced sufficiently to lift the newly made bullet out of the lower die, which is then pushed forward so as to make room for the next bullet. They are then detached by hand in a machine that is moved by steam power.

In a private establishment at Hartford, there is an extensive manufacture of elongated bullets carried on, which in the opinion of the Committee is very defective.

The bullets are first cast in a mould, and then compressed in dies, which meet and part on the longitudinal section, and consequently leave a small fin upon them. There is no contrivance to detach this fin, hence the bullets are much less perfect than those made at Woolwich.

Some considerable inquiry was made at the arsenals with regard to the Minié bullet machinery at Woolwich, as it is proposed to construct some of a similar description at the Pittsburgh arsenal.

At a private establishment near Hartford, the Committee saw some small-arm cartridge-making machinery, by means of which the operation was reduced fully one half. This apparatus resembles a small lathe; the bullet is inserted by the point into one of the spindles; another hollow spindle comes against the rear end of the bullet, and by means of a third spindle within the second, the piece of paper is wound round the bullet and secured by a string that is drawn through gum arabic. This secures the paper without any necessity for a knot. The string is cut by a knife, which forms part of the machine.

The cartridges are filled with powder by an apparatus which enables the operator to fill a great number at one time; it contains the principle of a good contrivance, but is very defective in its arrangement.

The apparatus that is employed for cutting paper and card in the United States, is much better than the knife and scissors used in the Royal Arsenal. That for cutting cardboard is a singularly simple and perfect contrivance. It consists of a stationary blade about eighteen inches long, placed with the cutting edge upwards, and upon which the board to be cut is laid. Over, but on the outside of this stationary blade, and in actual contact with it, rolls a disc of steel, which is softly pushed into contact with the fixed blade, as the operator slides it backwards and forwards with his right hand, whilst with his left he pushes the cardboard against the guide that determines the dimensions of the material that is being cut. By means of a simple application of a fixed cord, which is wound once over a groove, upon the steel disc, a greater amount of motion is given to the disc than is due to its rolling motion, and this, causing the cutting edge to slide upon the material it cuts, facilitates the operation.

In a manufactory where felt hats are made by machinery, which the Committee visited in order to see if the method employed was at all applicable to the manufacture of gun cartridges, the wool is fed into a card machine by rollers, and at the other end of the apparatus it is blown out, in detached filaments, against a cone of wire gauze, not unlike the cartridge bag, which revolves slowly. Within this cone a vacuum is kept up, which draws the flocculent wool against its meshes, and this is continued during a few minutes, during which period a considerable quantity of wool has been deposited uniformly over its surface. The gauze frame is then removed from the machine with its covering, over which a wet woollen rag is thrown, and over all, another gauze frame; and the whole is put into a vessel of boiling water for a few minutes. The substances are then removed from the gauze frames and rubbed, twisted and worked by hand within the rag; again put into boiling water, and by a series of workings and boilings, the operation is completed.

In New Jersey, the Committee saw a machine for making up packages of different powders, by means of which one lad could perform the work of twelve, and every packet made up in it is precisely identical. The material to be packed is put into a hopper, which delivers the proper quantity into a scale, and from it into a paper on an endless chain, carrying a series of sockets, and working with an intermittent motion. At each stoppage, a cross-head comes down and advances a number of the packages one stage, and thus advancing them progressively, delivers a package complete at every movement.

Sewing machines are much more extensively used in the United States than in England, for almost every kind of work where the needle is employed. A large amount of ingenuity has been brought to play upon this class of machinery, and such is the demand for it, that there are now nearly twenty different patterns, and patents. The chain stitch of the original machines is rather defective, as, if the sewing is broken at any point it is easily opened up. For some kinds of work this class of machine sews with a waxed thread, which in some degree modifies its defects by giving the thread greater adhesion to the material.

In another extensive variety of machines of this nature, a shuttle is used with a second thread; this thread being taken hold of by the principal thread and pulled by it into the hole. The two threads thus intertwine with each other at every stitch, the point of junction being in the hole in the material being operated on.

This is the description of machine that is most extensively employed; but the work performed by all of this class is inferior in strength to hand-sewing; but it has a more uniform appearance and is quite strong enough for many purposes. It is used in shoemaking, saddlery, light harness work, upholstery, and in the clothing making establishments.

A vigorous effort is being made to obtain a machine that will give the true stitch of the hand, and the Committee saw one at work in Boston which effected this by passing the needle through the cloth and taking it up through a second hole; this it did, but, as in hand-sewing, a short thread only can be used, it requires replenishing every now and then, whereas, in all the other machines, the needle draws the thread from a reel. Another objection to the last machine is its great complexity; for although the work to be done is simple when performed by hand, yet to give these same movements to an iron hand is more difficult. When this machine is improved and fully matured, it will probably supersede all others; but the stitch of either of these machines, with a good thread, would be sufficient for flannel cartridge bags.

The wooden bottoms for shot and wooden fuzes, are made at Watervliet Arsenal, in a common lathe, to which some additions have been made by one of the workmen that have increased the produce both in quality and quantity, in a manner truly surprising. The fact that an ordinary turner can turn out the sabot for a six-pounder shot in ten seconds, and keep on at the rate of six per minute, requires no comment. As this system of turning is applicable to almost everything that is made in the turnery of the Royal Laboratory, which has to be repeated, its introduction will be a great improvement.

At the front of the lathe is fixed a true straight edge, with its upper edge slightly rounded. At the back of the lathe there is a similar fixture, only its upper line is the form of the article to be produced. The instrument that holds the 'gouge' or turning tool, has a slight resemblance to a carpenter's plane, one end of the plane resting on the front straight edge, the other upon the guide at the back; it thus bridges over the material to be turned.

In commencing on a piece of rough wood, the end of the tool-holder does not rest at first upon the copy, but on a small wedge which the workman gradually withdraws with two of his fingers, until the holder touches the copy; and then passing it carefully along it produces the precise pattern without any time being wasted in gauging.

The sabot is made by cutting the plank with saws, into blocks of sufficient size to produce that article.

The lathe has an iron face plate that runs truly, having a small worm at the centre, which is the driver, to take the sabot. The face plate being perfectly true, the block of wood, even though taken from the saw, must run equally true, and hence that side of the block, which goes next the plate, is not touched further by any instrument.

The tool-holder is then held over and drawn up a short incline, which, in a few seconds, completes it.

Underneath the revolving article there is a second slide, in a line with the axis of the lathe, and which carries a vertical pivot, in which is fixed a gouge at the proper radius. The workman with his left hand, pushes the slide on to the work until it comes to a fixed stop, while with his right hand he moves the gouge in a circle a few times, which completes the operation.

The fly screw press, which is invariably used in England for making percussion caps, was not seen applied to that purpose in the United

States. The machinery usually employed differs from Mr Wright's machine, described at page 163, and consists of two spindles, one to cut out the cross and the other to form the cap, having an iron finger passing between the two for carrying the caps from the first to the second.

The machine is self-acting, one lad being able to attend several of them.

ROYAL CARRIAGE DEPARTMENT

Next to the machinery used in the manufacture of small arms, that bearing on the branch of work carried on in the Royal Carriage Department, came most under the observation of the Committee. To describe all that was seen relating to this Department would lengthen out this Report inordinately. The Committee will therefore only draw attention to such points as appear to them to be the most important.

The work that is carried on in the Royal Carriage Department being multifarious in its nature, the different things will be classed under the heads of 'wood' and 'metal.'

In those districts of the United States of America that the Committee have visited the working of wood by machinery in almost every branch of industry, is all but universal; and in large establishments the ordinary tools of the carpenter are seldom seen, except in finishing off, after the several parts of the article have been put together.

The determination to use labour-saving machinery has divided the class of work usually carried on by carpenters and the other wood trades into special manufactures, where only one kind of article is produced, but in numbers or quantity almost in many cases incredible.

As a general rule the machinery employed is of a coarse description, and the work which it produces may not in every case come up to our notions of finish; but it is produced cheaply and quickly, and the same or similar apparatus may be advantageously employed in England.

The machinery that is employed in the construction of carriages and waggons has little peculiarity, except that the sawing, planing, mortising, tenoning and boring machinery has been specially made for the particular purpose. One of the machines used in such work is a contrivance for boring square holes. The cutting instruments consist of a hollow-square chisel, with four cutting sides–the size of the hole required. Within this chisel, and combined with it, works an ordinary screw auger, which is driven by an independent motion, having the small worm at the point, and about $\frac{1}{16}$ of an inch of the auger projecting beyond the chisel.

The action of the machine consists in pushing the cutting edge of the chisel against the wood by means of a screw, which pares the round hole formed by the revolving-bit into a square one, the whole of the chips being withdrawn by the screw-motion of the auger.

The manufacture of carriage-wheels frequently forms an important branch of trade, and is carried on chiefly by the aid of machinery. The wheels are of a lighter class than is generally used in England. In the heavier class the felloe is composed of segments, as with us; but for light wheels it is made of two semi-circular pieces, which are secured to each other by a small iron plate on the inside of the joint, the whole being clamped together. These semi-circular felloes are made by bending the ashwood in a mould, under the influence of steam or boiling water, and keeping them in position until thoroughly dried, when they will permanently retain their form.

The segments of ordinary wheel-building are all cut out with saws worked by power; but the arrangement for obtaining the segment, and the mode of holding the saw, differ widely. In some the plank is laid on and secured to a sliding rest, having a fixed wooden segment to guide it correctly. In others the wood is fastened on the end of a lever, having an adjustable fulcrum, so that any segment can be obtained.

The saws are generally fixed in a frame holding two, one at each extremity, both under the same tension. This is a very simple arrangement, and seems to fulfil every condition. The frame slides on parallel bars, and may be worked above or under.

In an establishment for making wooden barrow-wheels, the segments were put under a circular knife in a powerful machine, and sliced to the finished dimensions with one thrust.

As a general rule, in the best workshops, the felloes are turned on their four sides. The outside, under the iron tire, being done after the wheel is put together. The spokes of wheels are all turned by a Blanchard's machine (see page 137), for which purpose it is peculiarly well adapted. It makes them with a perfect square or oblong at one end, which gradually passes from that into an oval, and from that into a circle.

Walnut is frequently employed for the spokes of wheels, and almost invariably for the lighter class, and seems to resist the action of the rough roads of the country in a manner truly surprising. The nave or 'hub' of the best class of carriage-wheels is frequently made of locust-wood, and the turning, boring, and mortising is all done by special machines for that purpose, but common all over the country.

After the spokes are driven into the 'hub' or nave, the wheel is fastened on a true spindle in a horizontal position, for the purpose of having the extremities of the spokes cut to the exact form, which is done in a very superior manner, by means of a circular cutter, which revolves with a high velocity, and which passes around the outside of the spoke on the sun and planet system. This is much better than the mode of doing it by the single cutter, moved end on, which is so liable to chip off the sharp corners.

After the wheel has been put together, and turned up truly on the outside, it is put into a machine for the purpose of having the hole formed for the iron bush, as the outside of the bush is an irregular figure, and difficult to fit by any ordinary system of boring. The machine is so contrived that the whole is cut out by tracing a fixed copy of the figure required in the following manner:

The wheel lies on a horizontal chuck, which has a slow circular motion. Over head is a fast running spindle, that has a cutter on its lower end, and it is arranged in such a manner that a slight amount of transverse motion may be given to it. Its position is determined by resting against the copy of the hole required. The revolving cutters chisel out the nave, which slowly turns round, and as the spindle descends it gradually shifts transversely, according to the copy.

All the iron tires, which came under the observation of the Committee in the national and private establishments, are entire. A very superior machine is employed to stretch them to perfect truth, before they are put on the wooden wheel. This machine consists of a turned circular plate, on which slides a number of segments in radial lines, the outside of them being turned to the inner diameter of the tire. The red-hot tire is then laid on the plate, the segments are slid out to the correct diameter by means of a screw and cone at the centre, and as the tire cools it is drawn into a perfect circle, thus insuring a degree of accuracy almost equal to its having been turned in a lathe. When this tire is put upon a wooden wheel that has been turned truly, the carriage necessarily travels smoothly.

Several methods of turning lance and other long poles came under the notice of the Committee, all of which produce them in a very superior manner. The poles are first sawed into square pieces, and put into a turning-lathe with the corners upon them. The turning-apparatus slides upon two parallel rods, which run along the entire length of the lathe. In turning parallel rods the sliding apparatus consists of two cutting tools with a socket between them, the first tool being set to cut the wood to the diameter of the socket in order to insure steadiness, the second tool finishing off the pole to the required diameter.

In turning taper poles, or even poles of an irregular figure, the second tool is not a fixture, but is placed in a holder, that traces a copy placed alongside, and causing the tool to rise and fall as it goes along.

In turning heavy poles, the labour of pushing the slide along is saved in a very ingenious manner, by making the steadying socket with a screwed interior, and setting the first tool so that it may leave the pole large enough to form a thread, and this thread screws the apparatus along, and is turned off by the second cutter at the further end of the socket, without having made its appearance.

The tools and machinery used for making wooden tubs and pails, have been brought to a very perfect system, which enables a few hands to turn out 1,000 per day. In making a pail the material undergoes the following operations:

1st. The wood is sawed up into plank the breadth of the stave.

2nd. The plank is cross-cut into pieces the length of the stave.

3rd. By means of a peculiarly constructed saw, the block is sliced off into segments, which completes the rough stave. This saw is the edge of a cylinder, about the diameter of the pail to be made, the stave, as it is pushed on end, being within the cylinder. This works with the same facility as a common circular saw.

4th. The rough staves are then exposed a few days, and then baked in a kiln till they are quite brown. This is done in order to thoroughly desiccate them. On being withdrawn from the kiln, they are quickly put through the following operations, so that the wood may not have time to imbibe any moisture, till it is converted into a pail.

When made in this way, the pails may be left empty under a tropical sun without being injured.

5th. The baked stave is held by a lad against the side of a plane worked by power, that gives the true bevel in about two seconds.

6th. The planed staves are thrown beside a workman, who sets the required number up into a rough pail, and slips a temporary hoop upon them to keep them together.

7th. The rough pail is slipped upon a conical chuck, having a large washer at the end to keep the staves together, and the temporary hoop is removed. A slide-rest alongside is pushed by hand quickly in both directions, which completes the outside in a few seconds. The temporary hoop is again slipped on the pail, and it is passed on to the next operation.

8th. The pail is put into a box-chuck, the shape of the outside, and by means of a slide-rest the interior is finished with a single thrust. When the slide-rest is at the bottom of the pail, the left hand of the workman acting upon a lever on a second tool turns out the circular groove for the bottom; and thirdly, by means of a tool like a carpenter's spokeshave, he rounds off the mouth. The pail is now ready to have the bottom put in.

9th. The boards for the bottom are planed in large machines on both sides, and then cut into square pieces, about an inch larger than the diameter of the pail.

10th. These squares are then gripped between two revolving plates, and cut to the diameter by means of a parting-tool, laid into a notch in the rest, the position of this notch giving the diameter. A second tool is then brought against the outer edge, so as to bevel it to fit the pail.

11th. The pail and bottom are passed to a workman, who puts them together.

12th. The pail passes to another man, who puts on the hoops.

The time required for the several operations is under six minutes.

A similar system has been applied to the manufacture of furniture. In one establishment 400 chairs are made per day, of which two in the dozen are rocking-chairs.

The material is cut out by sawing, planing, and other machinery, referred to elsewhere.

Such parts as are of irregular form, such as the arms of an easy chair, are first passed through between two planers, which finish them on the sides. The arm is then fastened in a holder, and held against a set of revolving cutters, having a templet attached, against which the holder is pressed; this determines the outline of the rounded arm, irrespective of the boy who holds it. The arm is then held against revolving buffs, to be smoothed and polished. The other parts of the chair are finished in a similar manner.

The four pieces that compose the seat of the chair are put together by tenon and mortise, the several mortises being cut at the same time by means of revolving drills, which have a reciprocating motion the length and direction of the mortise.

By passing the seat around properly shaped cutters, to which a templet is attached, its exterior form is brought out.

The various parts of the chair that are round are turned in lathes; those that are not so very ornamental, both taper and parallel, are produced on the lance-pole principle, at the rate of 30 inches per minute.

In a wooden bedstead manufactory, where 15,000 are produced annually, machinery is employed for everything, from the sawing to the sand-papering, These bedsteads are put together by having the wooden stays screwed into the posts, with a right and left hand screw. The apparatus for cutting the thread on the stay and in the post is very perfect.

The several parts of a house are got up in separate manufactories, such as stairs and staircases. Here there is every appliance for bending and twisting the wood, and working it under awkward forms. For doors, window-frames, and sashes. For this purpose special tools are employed for mortising, tenoning, and forming the mouldings.

In the manufacture of Venetian blinds and jalousies, all the cutting and planing is done by ordinary machinery; but a special machine is employed to form the pivots. It consists of a frame for holding the piece of wood, at each end of which there is a spindle, that carries a double cutter. These come end on and form the pivots, which are made at the rate of 20 per minute.

A very clever machine is employed for making wooden matches, at the rate of 15,000 per minute. An even and straight splitting wood is employed, which is cut transversely into pieces the length of the match. A piece of thin cloth is glued over one side. In this state the clumps are given to the machine, and pass in front of a revolving steel disc, which splits them into a series of thin plates, which are held together by the cloth. It is then again passed through at right angles, which converts it into matches. The cloth enables them to be all dipped together, and is readily parted from them when required.

In an extensive manufactory of wooden pegs for shoes, the tree is first cut transversely into slips the length of the pegs.

2nd. These slips are put on a machine, and scored like a ploughed field.

3rd. They are then put on another machine, and scored at right angles to the first scoring.

These two operations are to form the points of the pegs.

4th. They are laid on an iron table, which passes under a descending knife, which, at the root of the formerly cut grooves, splits the wood into slips.

5th. They are then cut at right angles, which splits the whole into pegs.

In a shoe-last manufactory, Blanchard's machine is employed, which turns them out in a surprisingly quick and perfect manner. They are then filed over by hand, and polished on a revolving sand and emery block.

Boot trees and oars for boats are all made on the same principle; indeed, this principle is one of universal application to all such forms.

Spade-handles and scythe-handles both form interesting branches of manufacture, in which specially designed tools are used to save labour; but enough has been said to show the wonderful energy that characterizes the wood manufacture of the United States.

Of the appliances that may be copied with advantage, is the system of combining a pendulum saw with a planing machine, or even a carpenter's bench. This is a circular saw, which hangs from the roof, and is used for cutting through long planks, or trimming off the ends just where they lie.

The system of smoothing and polishing wood with revolving wheels, instead of by hand, is attended with great economy, and is sufficiently nice for many purposes.

The system of bending timber is more general than with us; and a vigorous effort is being made to bend ship-knees and other timber. They are softened by steam, and the bending power is either mechanical or hydrostatic, according to the size of the mass.

A very superior kind of screwed auger is employed for boring hard wood, its construction resembling a corkscrew, having at the point a sharp gouge, which cuts from the centre to the outer extremity.

An improvement has also been made on the screwed auger for soft wood, by means of which end, wood or a flat surface at any angle may be bored with a small hole.

In this auger, that part of the screw that performs the cutting is hooked round into the form of a ram's horn with sharp cutting edges. These pare out the shavings, without bruising like the common auger, and leave a very smooth hole, adding to the capacity for holding any pin that may be driven into it.

A method of cutting hollow grooves of different circles by means of a single instrument is very ingenious. The cutting instrument consists of a wheel of any diameter filled with circular cutting tools, which revolve at a high velocity. Alongside is a table on which the work to be grooved is laid, with an apparatus that is capable of sliding in any direction. If the wood is passed in a line parallel with the spindle of the cutter, it produces a groove of the cutter's diameter, or if it is passed along the face of the cutter, that is, at right angles to the spindle, then it produces a flat surface. Any intermediate angle will produce a groove between the flat surface and that which is due to the diameter of the instrument when the wood is passed parallel with its axis.

The following is a very quick and perfect method of producing half round wooden rods.

1. The timber is cut into long square pieces.
2. These square rods are turned parallel on the lance pole principle.
3. A circular saw is set to run in the centre of a bored socket that fits the turned rod. The rods have merely to be pushed through the socket when they encounter the saw and come out in two half-round pieces.

In almost all the workshops, carriages are used to carry the material from stage to stage during the several operations.

METALWORK

The machinery and tools connected with the metal work of carriages and other articles contain some clever contrivances.

In a lathe for turning axle-trees, the principal head stock and driving gear is in the centre, and the axle is pushed through the inside of the spindle, and is there pinched with screws to run truly. A shifting head-stock is fastened at both ends of the axle, and a pair of sliding saddles are employed, each being set at the proper angle to bring out the required taper, and are self-acting.

For almost all large wheels or other heavy work that has to be chucked, vertical lathes are employed, and the face plate resembles a round table, on which the work is laid, a bracket on each side rises from a sole plate, and carries a slide rest, having some resemblance to an English planing machine, only specially adapted to boring and turning.

In producing small articles in turning lathes, the system of clamp milling referred to in page 135 is extensively employed in the manufacture of accoutrements and other articles usually made by hand-turning tools.

Much ingenuity has been displayed in the contrivance of machines for cutting plate iron with a continuous slice; one of the best is a machine about ten feet in length, having a steel edge along the corner of a bed-plate or table, and upon this the sheet or plate to be cut is laid.

In a line with this steel edge, but above it, is fixed a slide rest that carries a steel disc, which revolves as it goes along almost in contact with the steel edge. The two combined form the best shears that the Committee have yet seen anywhere.

By attaching a bracket to the slide rest, the machine can be set to cut out circles. Machines on this principle are used to cut tin plate.

The class of tools commonly used by tinmen are almost obsolete in New England States. In a well-furnished tinman's shop there are about twelve different kinds of machines employed, viz.; machines for clipping and shearing the plate; for turning over the edge to make different kinds of joint; for grooving or corrugating the plate so as to stiffen it; for grooving the edge, and inclosing a wire within it; for cutting out circles, and at the same time turning up the edge; machine for setting done [*sic*]; for closing; for turning cylinders, cones, and other forms, copies of each have been purchased.

A mixture of chloride of zinc and muriate of ammonia is also invariably used in soft soldering.

Revolving barrels for smoothing rough metal articles that are to be painted, or that have to be handled, or that are covered with grit or burrs, are employed.

As an example of their use, the best class of scythe handles, which are now made of wrought iron tubes, for the sake of lightness, are cut into the required lengths, and while straight, twenty-five of them are put into a long barrel, with a little sand, and then set to revolve for a couple of hours, which rubs off all asperities, and makes them feel quite smooth to the hand. They are then bent into the crooked shape that is required.

Very superior wood screws are used all over the United States. They are made with what is termed a gimlet point, and in the generality of wood they do not require a hole to be bored for them, as they force their own way, by reason of the taper-screwed point which is formed upon them. The machinery by which these screws are made is self-acting, and is quite different from the screw machinery used in England.

Machine-made nails are preferred to hand-made for the generality of purposes, not only because they are cheaper, but also on account of their less liability to split the wood, and their greater holding power. The only exception to this is where nails have to be clinched, the machine-made nail being more apt to fracture under this operation than the other.

There are immense nail manufactories in different parts of the country where no hand labour is employed. The machines for the smaller sizes of nails are self-acting, for the larger sizes the iron has to be supplied to the machine by a boy, who makes nails at the rate of 200 per minute.

Machine-made nails have been much improved by a change in the mode of rolling the iron, by which the fibre now runs in the line of the nail. In the operation of forming a nail by the machine, the piece of iron is first cut off; secondly, gripped in a die, which gives it the shape; and at the same instant another die comes into contact and upsets the head. In all the nail-making establishments that the Committee visited, they form part of an iron-work and use up the iron as it is made, which tends to reduce the cost.

Large spike nails, that are used to fasten the rails on railways, are also made by machinery. Here the iron is fed into the machine in the red-hot state. The bar is entered between a pair of rolls, a projection upon the roll forms the sharp point of the spike, and detaches it from the bar; a die seizes hold of the spike, and another die comes against it to form the head. This machine is placed beside the mouth of the furnace, and in this furnace, being long, the bar is heated as it is drawn through by the machine. This machine turns out the spikes at the rate of 60 per minute.

The wooden carriages used for heavy guns in the United States are constructed in a very simple but strong manner. The sides are composed of a vertical, diagonal, and horizontal beam, in the form of a triangle. In the 32-pounder carriages, and those of smaller size, the horizontal beam serves for the two cheeks, but in the larger sizes each has a separate beam, the whole being bolted to stays transversely. The American officers claim for this carriage the merit of possessing the maximum of strength with the minimum of labour and material. The trucks of the heavy gun carriages are made out of a solid block of lignum vitæ. This wood now costs $90 per ton, but has been as high as $150.

Machinery is invariably employed for cask making in the United States. There are several different patents in operation for this purpose, all of which produce the article very quickly, and of a quality sufficiently good for the use to which it is applied.

The steps on the cheeks of gun carriages are made in the form of quadrant or quarter circle, whose centre is on the inner edge of the cheek.

In the wood-planing machines of the United States it is not unusual to see them combined with a lathe headstock, placed at right angles to the planing-bed. The lathe-spindle carries a planing apparatus, so that two sides of a log are thus operated on at the same time.

Two very light and neat gun-carriages, for boat service, were shown to the Committee by Captain Dahlgren, United States Navy, at Washington Navy Yard. Drawings, with a description of them, will be found in Chapters III and IV, of Captain Dahlgren's book on 'System of Armament for Boats.'

MISCELLANEOUS

The labour-saving contrivances of a miscellaneous character that came under observation were too numerous to detail, but from the circumstance that in any newly organised manufacture the old beaten track is departed from, and that which might be expected to possess no novelty was frequently found to be most deeply interesting.

The manufacture of locks for doors and padlocks is carried on most extensively. The shells of door-locks are made of cast malleable iron, about $\frac{1}{16}$ of an inch in thickness, and have the greater part of the wards cast within them. As specimens of castings they are very first rate.

The keys are also cast, and turned by a machine on the clamp milling arrangement; those that are hollow are drilled out by self-acting machinery.

The door-handles are chiefly made of porcelain formerly obtained from England and France; but having set up machinery, which not only does all the rough work but extends even to the forming them in metal dies, and to the turning them on slide lathes on the copying principle, they are now produced of superior quality, and at a remarkably low cost.

In the making of padlocks an extensive plant of machinery is used for all the forging and finishing processes. A lock of good quality, $2\frac{3}{4}$ by $2\frac{1}{4}$, is sold wholesale at the rate of 50 cents per dozen, each of the twelve having a differently formed key, so that each will only fit its own lock.

The manufacture of clocks affords another example of the advantage of labour-saving machinery when judiciously applied. At a manufactory visited by the Committee 600 clocks are produced per day, with 250 individuals employed, many of whom are boys and girls, who have been trained to perform a single operation. The whole system pursued might be carried out in England with great advantage.

In the manufacture of the pianoforte and the melodion, a building similar to an English cotton mill is set up, with steam-engines and special machinery applied to the production of every part, both wood and metal.

Machinery has been particularly applied to the manufacture of boots and shoes; to the equalising of the leather, to the cutting out of the material, and to the lighter sewing.

In the American cities, the stones for building purposes are dressed for the builders in large manufactories by means of powerful self-acting machines.

Biscuit, both for home consumption and for exportation, is manufactured on a most extensive scale.

Whips are made in factories where all the plaiting and twisting is done by machinery.

In granaries or warehouses a water-wheel or engine is very common. A machine termed an 'Elevator' is employed for raising grain from the holds of ships into flour mills and storehouses. The elevator is constructed in various ways, but in all the grain is dealt with on a hydraulic principle, a spout descending into the hold is immersed in the grain to be drawn upwards. The elevator is a most valuable labour-saving apparatus.

It has hitherto been supposed that machinery could not be applied successfully to striking up the teeth of files. This has been accomplished in a very successful manner, one machine being capable of cutting 12 dozen files per day, one man being able to attend two of them. There is a difference of opinion with regard to the merits of hand-made and machine made files, but the great economy of the latter will ultimately determine the question in favour of the machine-made files.

The use of a small mill to grind paint is common in almost every workshop where the article made has to be painted. These mills work silently without any attendance, and as the process of grinding may be said to cost nothing, it is long continued, and hence the paint employed even on rough articles is very smooth, and gives a fine surface.

A machine called the 'Yankee Chaff Cutter' is an admirable contrivance for cutting up hay or straw. It consists of two cylinders, which work together in contact, having a series of spiral blades on their surfaces. These blades, working together, feed the machine and carry the material past stationary blades, which cut it up into chaff. Great advantage is secured by the angular position of the blades, as with the spiral form there is only one point in action at any one time. As the surfaces of the several blades, both revolving and stationary, are all ground together, the machine acts on the scissors principle, and performs its work quickly with a small expenditure of power.

The Nasmyth steam hammer, and the forging machine, are both used in the States, but neither of them so extensively as in England.

The apparatus which is most commonly employed is some modification of the trip hammer, an arrangement similar to the English being the most common.

At Mr Alger's works in Boston they use this hammer worked directly by steam. A steam cylinder at the back of the anvil raises the lever, and then allows it to drop freely. There are others constructed with the hammer placed vertically, which are lifted by cams on a revolving shaft, on a similar plan to the machines used for beetling cloth; but on this class of forging generally the Americans are considerably behind the English.

The ship building sheds of the United States' Navy Yards are fitted up in a very convenient manner, with four or five tiers of galleries running all round, having stairs of communication at convenient distances. Upon the galleries are placed the requisite wood benches, tools, and material. This arrangement saves much unnecessary travelling, and places a large proportion of the workmen under the observation of the foreman.

At Boston navy yard there is a most extensive plant of tackling and rope-spinning machinery in readiness to be used when required. A few hands are employed, who go round to work the several machines by turn, all of them being in such order that at any moment they can be set in motion.

A piece of sheepskin with the wool on it is stitched on to the inside of the cap-pouch, which secures the caps against damp and also prevents them from being brought out sticking to the soldier's fingers, and thus dropped.

India rubber as a material for pumps is extensively used in the United States.

One of the best seen by the Committee consists of a sheet or diaphragm of this material, secured to the top of a box, this box having a pipe leading down to the water, and two valves to prevent the water from returning into the well, or into the box. The action of the pump consists in raising and depressing the India rubber diaphragm. When it is raised a vacuum is formed in the box, and the water rises from the well to fill it; on depressing the diaphragm the water is expelled by the second valve.

Another way of using India rubber for raising water is to have a chain mounted with balls of that substance, which are larger than the bore of the pump. This chain is arranged like the ordinary chain-pump, and as the elastic balls enter at the bottom, they carry all the water upwards before them. These balls are connected to the chain in such a manner, that the wear from continued action may be compensated for by compressing them into oblate spheroids.

In the Navy Yard at Philadelphia, there is an admirable arrangement of floating dock. It consists of an immense platform, strong enough to hold a man-of-war, having alongside of it a series of air tight boxes. When the platform is to be sunk to receive a ship these boxes are allowed to fill with water, and the ship to be raised is floated over it, then a number of pumps (which may either be worked by men or power) are put in operation to exhaust the water from the boxes, and the platform with its burthen is thus floated out of the water.

The United States' Marines use white lead, mixed with water and a little gum arabic to clean their belts with, instead of pipeclay; they say that it does not come off so readily, and that at sea it dries much quicker than the pipeclay.

The tongs used in the United States' service for carrying large shells are peculiar, in having a link at the handle end of each of the levers, both of which are linked into a centre ring, the tendency of which is to pull them together, and being large enough to be slid on a handspike, it appears altogether to be a very convenient instrument.

The lifting jacks that are employed in the United States' Artillery consist of a vertical screw, which is raised by the rotation of a nut. In the small jacks there are four levers on this nut, but on the larger the nut is a bevel wheel, which is worked by a bevel pinion placed horizontally, thus increasing the power, and making it more convenient for acting on large masses.

The soldier's bedsteads at Watervliet Arsenal are constructed in such a way that in case of any emergency each can accommodate two men, by having one placed above the other on the top of the frame.

At Watervliet Arsenal the gun carriages, when completed, are put loosely together and thus stored. When required for use they are taken apart again and carefully fitted before being issued.

Small breast drills are much used in the United States, which in England are usually worked by a bow, but are here driven by bevel gear, in which a large wheel gives motion to a small pinion that affords a quick motion by a simple and compact arrangement. These drills seem very suitable to artillery purposes.

In the United States the 'tourbine' is invariably used in preference to the water-wheel, and is said to yield from 82 to 85 per cent of the whole power, besides possessing the advantage of being capable of being used if submerged in back water, the power being as the distance between the two water levels.

Irregular cams are much used in the United States to perform intricate operations, as an example, in fitting the iron work of a spade to the wooden handle, the latter is inserted into the former, and both are put between a pair of cams of the precise form that the crooked spade is required to be, and as they envelop the whole handle, the material cannot avoid becoming the right shape.

A very expeditious method is employed to cut key ways in the eye of iron wheels. A mandrel which fits the hole nicely, is furnished with a cutter at one side which may be set outwards by a wedge; this mandrel is pulled downwards through the wheel, carrying the cutter with it, and takes out about $\frac{1}{8}$th of an inch in depth at each pull, the whole operation taking only a few seconds.

A very efficient and strong cable nipper is used on board the American men-of-war. It consists of a sliding wedge, which is actuated by a screw and lever, by means of which one man can easily check the cable as it runs out.

There is a very compact and firm arrangement of forge anvils at Harper's Ferry Armory for forging the parts of gun locks, etc. A large mass of stone is sunk in the ground flush with the floor; on this stone are fixed two iron standards, similar to those of a heavy lathe bed. Upon these is laid a square log of wood, the length being dependent on the number of anvils required for the particular operation carried on at that forge. By this arrangement the log can be raised or lowered to suit the height of different workmen.

Major Thornton, of Governor's Island, informed the Committee that the way in which he packed flannel cartridge bags for storing effectually secured them against the moth.

Tight boxes are made and smeared internally with a heavy coating of coal tar, and lined with strong cartridge paper, the tar acting as a paste in fixing the paper lining, and by filling the cracks and interstices of the boxes, effectually prevents the entrance of insects and moisture, thereby insuring the preservation of the bags.

At Mr Kemble's foundry at Cold Spring, anthracite coal is used in forging iron by a peculiar arrangement of furnace. The furnace is constructed upon the reverberatory principle, but the requisite combustion instead of being obtained by means of a tall chimney, which is sufficient for bituminous coal, is accomplished by a series of air jets under high pressure, which is forced into the furnace, the pressure of the air being equal to three pounds to the square inch.

Mr Francis, of Williamsburgh, New York, the inventor and manufacturer of metallic life boats, etc., conceiving that the same material might be beneficially applied to military purposes, such as ammunition and store waggons, etc., which might also serve under peculiar circumstances as a substitute for boats or pontoons, accordingly produced an artillery caisson and Commissariat store waggon of galvanised corrugated iron, and similar in dimensions to those at present used in the United States' service, which he submitted for experiment to a board of officers at the arsenal at Washington on 26 June, 1854.

	Caisson	*Store Waggon*
Length	10 feet	10 feet
Breadth	4 ,,	4 ,,
Depth	2½ ,,	2½ ,,
Weight, empty	5 cwt.	6 cwt.

The first experiment was made with the caisson, mounted on a common carriage, together weighing 17½ cwt., and secured by a couple of lashings; the whole was lowered into the water by means of a crane, and eleven men embarked in it. There was no leakage, and the displacement did not exceed two-thirds of the depth of the caisson; it was first towed to some distance by a man-of-war's boat, and then brought in shore until the wheels took the ground. The lashings were cast off, two thowl pins with grommets fixed to the gunwale, and with two oars the caisson was rowed with ease and safety.

While this was going on twelve men embarked in the covered waggon, which was taken in tow by the caisson, and pulled to a considerable distance without difficulty.

These experiments were highly satisfactory, and led to the conclusion that waggons constructed on Mr Francis's principle might be employed with great advantage as pontoons or boats, in countries where economy of transport is a matter of great importance, and the usual means of crossing rivers are very scarce or not to be found.

The whole of these caissons, etc., is composed of iron, with the exception of a wooden top-rail passing outside, and another at the bottom. They are provided with a tool-box in front, are very strong, not affected by extremes of heat or cold, are perfectly watertight and easily repaired.

5

[Contracts]

In Chapter 2 of the Journal the Committee, in giving a detailed account of their proceedings have mentioned whenever they entered into contracts for machinery, at the same time giving the reasons that guided them in so doing. In this chapter they give the details of the contracts there referred to, with the dates on which they were received, and the dates on which they were accepted.

In making contracts the Committee have endeavoured to secure against confusion by having dealings with as few firms as possible, and those of the highest character; and they trust that though the machinery ordered will not be completed so soon as they could have desired, when it is delivered and set to work it will prove good and efficient, and by superseding hand-labour, be the saving of large sums to the British Government, and, at the same time, add to the uniformity and stability of the articles manufactured by its assistance.

In selecting machines for the Royal Carriage Department, the Committee have not been so much guided by the finish or solid construction of the machines ordered as by a wish to obtain those, the working details of which are most ingenious and perfect, with the view of having them as models from which, should it be necessary, more stable machines can be constructed.

Had the Committee ordered these machines to be made in iron instead of wood (the material commonly used) their price would have been very materially increased, and the number they could have purchased proportionally reduced.

The Committee think it necessary to point this out to the Honourable Board, as all the machines used for wood work in the United States are roughly constructed, and would not bear comparison in stability and appearance with the highly-finished iron machinery of England.

The first tender entered into by the Committee was one for a set of stocking machinery, submitted to them at Springfield by Mr Ames of Chicopee,

on the 17th of May, 1854, and which is referred to in Chapter 2, p. 102.
The following is a list of the machines estimated for in this tender:

No. 1 or 198.	Machine for roughing	$1,750
2 or 197.	Machine for rough turning, new pattern	1,800
	Right of patent for above	50
3 or 199.	Machine for 'spotting'	1,550
4 or 200.	Machine for sawing breech and muzzle	275
5 or 170.	Machine for bedding barrel	3,300
6 or 249.	Machine for planing sides and edges, one extra spindle	2,500
7 or 227.	Machine for bedding breech plate	2,900
8 or 116.	Machine for fitting bands	1,750
	Right of patent	50
9 or 190.	Machine for turning between bands	1,500
10 or 251.	Machine for smooth turning breech	1,350
	Right of patent	50
11 or 185.	Machine for smooth turning above the lock	1,260
	Right of patent	50
12 or 228.	Machine for bedding lock	3,350
13 or 229.	Machine for bedding guard	2,300
14 or new	Machine for boring side tang screw and pin holes	1,400
15 or 248.	Machine for grooving for ramrod	2,700
	Packing and boxing	750
	Expenses of delivery in New York	175
		$30,860

The whole to be delivered in New York within twelve months of date of acceptance for the sum of thirty thousand eight hundred and sixty dollars.

signed JAS. T. AMES, *Agent*
Chicopee, May 17, 1854.

This tender was accepted by the Committee the same day, viz., May 17th, 1854.

Whilst in New York the Committee received the following tender from Messrs Robbins and Lawrence, of Windsor, Vermont, for machines to produce different parts of the new pattern musket. This tender is referred to in Chapter 2, p. 104, and is dated New York, May 25th, 1854. The Robbins and Lawrence Company propose to manufacture for the British Government the following machines, viz.:

No. 1.	For Lock-plate.	
	4 Milling machines, $300	$1,200
	2 Drilling machines, 5 spindles, $225	450
	Tapping apparatus	300
	Apparatus for reaming tumbler hole to size	100
	1 Edging machine	360

No. 2. For Hammer.		
1 Drilling machine, 4 spindles	$225	
5 Milling machines, $300	1,500	
1 Checking machine for hammer hole and for trimming lock plate	525	
No. 3. For Tumbler.		
2 Double milling machines, $300	600	
3 Milling machines, $300	900	
1 Drilling machine, 4 spindles	225	
1 Grooving machine	200	
1 Squaring machine	300	
1 Screw (hand) machine	50	
No. 4. For Swivel or Stirrup.		
1 Drilling machine, 4 spindles	225	
No. 5. For Sear.		
1 Drilling machine, 4 spindles	225	
2 Milling machines, $300	600	
1 Double milling machine	300	
No. 6. For Main-spring.		
3 Milling machines, $300	900	
No. 7. For Bridles.		
1 Drilling machine	225	
2 Milling machines, $300	600	
No. 8. For Lock-screws.		
2 Screw milling machines, $225	450	
2 Thread-cutting (hand) machines, $75	150	
1 Slitting machine	150	
1 Pointing machine	225	
No. 9. For Cone or Nipple.		
1 Clamp milling machine	275	
1 Chasing machine	400	
1 Squaring machine	300	
1 Drilling machine, 6 spindles	285	
1 Hand machine for finishing thread	100	
No. 10. For Triggers.		
3 Milling machines, $300	900	
No. 11. For Trigger-plate.		
2 Milling machines, $300	600	
1 Edging machine	360	
1 Drilling machine, 4 spindles	225	
1 Slotting machine (for trigger)	200	
No. 12. For Trigger-guard.		
1 Drilling machine, 4 spindles	225	
2 Edging machines, $360	720	
No. 13. For Butt-plate.		
2 Milling machines, $300	600	
1 Edging machine	360	
	16,660	
Add 6 per cent for boxing and delivery in Boston	999	60
	$17,659	60

The whole of the machinery to be delivered in Boston in nine months from date of acceptance, for the sum of seventeen thousand six hundred and fifty-nine dollars and sixty cents. This tender was accepted by the Committee on the 26th of May, 1854. An additional charge of $1\frac{1}{2}$ per cent was to be made if the machinery manufactured by Messrs Robbins and Lawrence were delivered in New York instead of Boston.

The Committee having received a communication from the Honourable Board, dated 14th June, 1854, m/209, removing the limit of their expenditure for the Small Arm Factory, and authorizing them to buy or order such machines as seemed to them necessary for that establishment, visited on their route the machine shop of Messrs Robbins and Lawrence, at Windsor, Vermont, with the view of ordering further machinery from that firm, and were furnished at their request, on the 24th July, with a formal tender for the following machines for producing different parts of the new pattern musket.

This tender is referred to in Chapter 2, p. 104.

1. For Lock-plate.	
3 Milling machines $300	$900
2. For Hammer.	
1 Drilling machine, 4 sp.	225
1 Milling machine	300
3. For Tumbler.	
1 Drilling machine, 4 sp.	225
4. For Swivel or Stirrup.	
1 Milling machine	300
6. For Mainsprings.	
1 Milling machine	300
1 Drilling machine, 4 sp.	225
7. For Bridles.	
2 Milling machines $300	600
8. For screws for lock.	
1 Clipping machine	150
9. For Cone.	
2 Drilling machines, 6 sp. $285	570
1 Machine for finishing screws	100
10. For Trigger.	
2 Milling machines $300	600
1 Drilling machine, 4 sp.	225
12. For Trigger Guard.	
1 Edging machine	360
1 Milling machine	300
13. For Butt-plate.	
1 Drilling machine, 4 sp.	225
1 Edging machine	360

The machines detailed up to this are those that were struck out of Messrs Robbins and Lawrence's first tender by the Committee, as mentioned in Chapter 2, p. 107.

14.	For Wood Screws.	
	1 Clamp milling machine	$275
	1 Slotting machine	150
	1 Pointing machine	225
	1 Screwing machine	100
	1 Clipping machine	150
15.	For Bands.	
	8 Milling machines, $300	2,400
	2 Drilling machines, $225	450
16.	Set of Screw Machinery for Bands.	625
17.	For Ring or Burr of Band-screws.	
	1 Tapping machine	150
	1 Drilling machine	225
18.	For Sight.	
	1 Drilling machine	225
	2 Milling machines, $300	600
	1 Punching machine	250
	4 Milling machines for beds, $300	1,200
19.	For Sight Slide.	
	2 Milling machines, $300	600
	1 Slotting machine	150
	1 Planing machine	225
20.	6 Universal milling machines, $850	5,100
21.	12 feet lathe, with back gear and screw	475
	2 6-feet lathes, with back gear and screw, $280	560
	4 6-feet lathes, with back gear without screw, $250	1,000
22.	For Rifling.	
	2 Machines for rifling barrels, 3 ft. 3 in. long, $550	1,100
	Index for cutting mills	50
	Add for boxing and delivery in Boston	1,335
		$23,585

The whole of this machinery to be delivered, packed and boxed, in Boston, within fifteen months of this date for the sum of twenty-two thousand two hundred and fifty dollars.
signed S. E. Robbins, *President*
Windsor, *July* 24, 1854.

This tender was accepted by the Committee the same day, viz. 24th July, 1854.

Finding whilst at Windsor that Messrs Robbins and Lawrence made some very good wood machinery, the Committee requested them for a tender for the following machines for the use of the Carriage Department, viz.:

1. 2 Universal milling machines, $850	$1,700
2. 2 Rotary drills, $400	800
3. 4 Milling machines, $300	1,200
4. 2 Drilling machines, $225	450
5. 1 Tenoning machine	350
6. 1 Horizontal boring machine for wood	350
	4,850
Add for boxing and delivery in Boston	271
	$5,121

The whole of these machines to be delivered in Boston, carefully packed, within eighteen months of this date, for the sum of five thousand one hundred and twenty-one dollars.
signed S. E. Robbins, *President*
Windsor, July 24, 1854.

The Committee accepted this tender on the same day, viz., July 24th, 1854.

Having visited an establishment near Windsor, where tools for tin-work are made, and obtained a list of prices, the Committee requested Messrs Robbins and Lawrence to tender to them for these machines and procure them from Mr Whitney, the maker (as mentioned in Chapter 2, p. 115), which they did. The machines are as follows, viz.:

6 Folding machines at	$20	$120	
2 Grooving machines at	16	32	
4 Burring machines at	10	40	
2 Wiring machines at	12	24	
4 Turning machines at	10	40	
2 Setting down machines at	10	20	
2 No. 4 bending machines at	36	72	
2 Funnel formers at	36	72	
6 Hand drills at	6	36	
2 Closing machines at	16	32	
2 Lever shears at	50	100	
		588	
Add for boxing and delivery in Boston		35	28
		$623	28

These machines to be delivered carefully packed in Boston, within six months of this date, for the sum of six hundred and twenty-three dollars, twenty-eight cents.
signed S. E. Robbins, *President*
Windsor, July 26, 1854.

This tender the Committee accepted the same day, viz., July 26th, 1854.

Finding, on their second visit to Boston, that there was no regular communication by sailing vessels between that city and London, the

Report of Committee on Machinery

Committee were of opinion that it would be advisable to have all the machinery shipped from New York instead of from Boston, and accordingly wrote to the different firms with whom they had contracted, to know what additional charge there would be on delivering the goods at the former instead of the latter place.

Messrs Robbins and Lawrence agreed to deliver all their machinery in New York for an additional sum of 2½ per cent.

Whilst at Windsor the Committee ordered the following machines from Messrs Buck and Co. for the Royal Carriage Department, but their tender not reaching them till they were at Boston, they could not write their acceptance of it till then.

The following is a list of the machines tendered for by Messrs Buck and Co., viz.:

1 Large size tenoning machine with relishing cylinder	$150
1 Large size tenoning machine, double cutter and relishing cylinder	200
1 Small size tenoning machine with one cope	90
1 Small size tenoning machine with two copes	100
2 8-ft. planing machines to plane 24-in. wide, with side jointer and dead weight, and prepare rolls	570
1 16-ft. planer, dead weight, and pressure rolls, to plane 32-in. wide, and side jointer	400
1 Hut rimmer and boring machine	450
1 Power mortise and boring machine combined	110
1 Power mortise and boring machine with face	150
1 Hut mortise and boring machine combined	90
2 Horizontal boring machines	150
1 Scroll saw	65
1 Door machine	150
1 Sash-making machine	100
1 Sash-striker with feed	100
Expense of packing freight and delivery of above tools in Boston	375
	$3,250

The whole of this machinery to be delivered carefully packed in Boston, within three months of date of acceptance, for the sum of three thousand two hundred and fifty dollars.

This tender was accepted by the Committee in a letter dated August 2nd, 1854.

In answer to a communication from the Committee, Mr Buck replied that he would deliver the machinery in New York instead of Boston, at an additional cost of 2 per cent.

In Boston the Committee visited the agent of the Cutting Machine Company who, at their request, submitted a formal tender for the following machines, viz.:

1 Card-cutting machine	$25
1 Paper-cutting machine	55
2 Tin-cutting machines with circular apparatus complete, $45	90
4 Patent flexible stamps, with devices as per order $14	56
1 9-ft. machine, warranted to cut ¼-inch boiler plate	1,200
	$1,426

To be delivered on board ship, at Boston, within twenty-one days of the date of agreement, the 31st July, 1854, carefully packed, at an additional cost of 2 per cent.

These machines having been shipped before the Committee left New York for England and the bills of lading, etc., presented to them, they drew a bill for the amount, and paid the money to Mr Richmond, agent of the Cutting Machine Company, at Mr Gillespie's office, on the 16th August, 1854.

The next tender submitted to the Committee was that of Mr Wright, of Frankfort Arsenal, for machinery to produce percussion caps, cups, and friction-tubes.

The following is a list of the machines tendered for by that gentleman:

2 Percussion-cap machines, $1,300	$2,600
2 Machines for making iron cups	1,600
1 Cupping machine for friction tubes	800
1 Vertical press for further elongating and finishing tube	350
1 Machine for cutting, milling, and drilling	175
Total	$5,825

The above machinery to be delivered in the city of New York within seven months of date of order, carefully packed, for an additional sum of 5 per cent.

signed GEORGE WRIGHT
Springfield, August 3, 1854.

This tender the Committee accepted the same day, viz., August 3rd, 1854.

Shortly before they left the United States, Mr Ames submitted to the Committee tenders for the jigs and gauges, and for some additional machinery. The former articles, though very expensive, the Committee considered it absolutely necessary to order, as it is only by means of a continual and careful application of these instruments that uniformity of work to secure interchanges can be obtained.

The following is a list of the articles tendered for in the first-named contract, viz.:

For Lock-plate.	
1 Gauge and plugs for testing the drilling	$20
1 Gauge receiving for testing the edges	35
1 Gauge pattern for testing the position of bolster for main-spring	6

1 Gauge grooved for thickness of plate, bolster, etc.	$12
1 Gauge plug for testing the sizes of holes	4
2 Gauge plugs for testing tapping	6
1 Pattern for filing bolster to height	2
1 Milling tool for milling the main-spring bolster	25
1 Tool for hole drilling	45
For Hammer.	
1 Gauge pattern for testing the punching	10
1 Gauge pattern for testing milling of bolster	2·40
1 Gauge pattern for testing for straightening	20
1 Gauge pattern for testing for testing edges	12
1 Gauge grooved for testing the finished dimensions	12
1 Gauge for testing the drilling of nose	15
1 Drilling tool for nose	40
For Tumbler.	
1 Gauge receiving for testing the milling filing, etc.	35
2 Filing jigs for edges	60
1 Filing gauge for squares	18
1 Drilling tool for screw-hole	25
1 Tapping tool for screw-hole	38
1 Drilling tool for swivel axle-hole	25
For Sear.	
1 Gauge receiving for testing filing, milling, etc.	30
1 Gauge plug for testing, drilling, and milling	5
1 Drilling tool for axle tube	30
2 Filing jigs for edges	50
1 Filing jig for end of tang	5
For Bridle.	
1 Gauge receiving for testing filing	25
1 Gauge pattern and plugs for testing, drilling, etc.	20
2 Drilling-tools for holes and milling pivot	55
1 Filing jig for edges	15
1 Reaming-tool for reaming tumbler axis and holes	35
For Main-spring.	
1 Gauge receiving for testing filing, etc.	25
1 Drilling-tool for tang and milling pivot	35
1 Gauge-plate for levelling bottom edges	4
2 Filing jigs for hook and tang	25
For Sear-spring.	
1 Gauge receiving for testing filing, etc.	12
1 Gauge plate for levelling bottom edges	2·50
2 Filing jigs for edges	36
1 Filing jig for length of mill end	3
1 Clamp for drilling hole	10

For Lock-screws.	
Gauge-plate for testing dimensions, length, cutting threads, etc.	$25
For Barrel.	
Gauge for testing the counter-boring of breech	3
Gauge for testing the tapping of breech	5
Gauge plug for testing the drilling of cone seat	2·50
Gauge for screw-tapping	4
Gauge receiving for testing cone-seat drilling	25
Gauge receiving for breech	40
Gauge profile for testing the underside of tang	6
Gauge profile for top	5
Gauge plate for testing the diameter of breech tenon	12
Gauge plate for testing barrel at six points	25
Gauge plate for length and height of stud	4
Gauge for testing the position of stud from breech	10
Drilling-tool for tang screw-hole	40
Drilling-tool for vent	20
Drilling-tool for cone-seat	75
Tapping-tool for cone-seat and taps reamers	100
Tool for prick-punching vent	22
2 Filing jigs for breech	100
Gauge-nut for testing the thread of breech-screw and tap	15
For Breech-plate.	
Gauge receiving for testing the filing, etc.	25
Gauge plate for testing the exterior curves, etc.	10
2 Drilling-tools for screw-holes	75
For Guard-plate.	
Drilling-tool for screw-holes	30
Filing-jig for edges	23
Gauge receiving for testing filing	24
Gauge profile for testing the underside	10
Gauge plate for thickness, curves, etc.	12
Drilling-tool for pin-hole	12
For Trigger-plate.	
2 Drilling and milling tools for tang and trigger screw-holes, and milling tang-screw studs	30
1 Milling-tool for studs for trigger	20
Filing-jig for mortise	14
Gauges for receiving	20
Gauges grooved for testing thickness, etc.	6
2 Gauges plugs for tapping, etc.	5

For Trigger.	
Drilling tool for axis hole	$15
2 Filing jigs for edges of blade	15
Gauges for receiving, testing the thickness, etc.	20
Filing jig for edges of finger-piece	10
For Ramrod.	
Gauges plates for testing the diameter and screw	10
Gauges profile, for testing the form of head and swell	8
For Cone.	
Gauge plates for testing exterior and interior diameter	18
Gauges and nut for testing the screw and tap	10
Screws.	
2 Gauge-plates for testing lengths, diameters, screws, etc. of side, tang breech-plate, guard-plate and trigger screws	40
For Stock.	
Pattern for vertical profile of finished stock (brass)	20
Gauge pattern for testing, spotting and angle of butt end	21
Gauge barrel (solid steel) for testing grooves	25
Gauge for testing length from breech of barrel position of band shoulders	10
Gauge bands, etc.	18
Gauge for testing the position of lock-bed	30
Gauge pattern for testing the cut for the tenon of breech-screw	5
Gauge for testing the depth of guard-bed	36
Gauge pattern for testing the fitting of guard-bed	8
Pattern for testing the profile from breech of band to breech-plate	3·50
4 Gauges grooved (16 grooves) for testing the various diameters of stocks	20·00
Gauge pattern for testing margins around lock and side plate	8·00
For Bands.	
3 Gauge mandrels for testing interior dimensions	39·00

This tender for gauging implements, with the per centage for delivery in New York, packing, etc., amounted to the sum of five thousand six hundred dollars ($5,600), and the articles detailed in it were to be delivered for shipment within twelve months of the date of order. The Committee accepted it on the 18th August, 1854, in New York.

Mr Ames at the same time submitted to the Committee two other tenders, the first for

1 Rough-stocking machine	$1,850
3 Smooth-turning the butt behind the lock	4,200
1 Smooth-turning the stock in front of the lock	1,310
2 Machines for edging the lock-plate	2,400
1 Apparatus for testing power	190
	$10,370

amounting, with per centage for boxing and delivery in New York, to the sum of ten thousand three hundred and seventy dollars ($10,370), and delivery to be made within twelve months of date of acceptance.

The second tender was for a supply of auger bits already forwarded to the Committee, and received by them whilst at Springfield Massachusetts, amounting to the sum of fourteen dollars and sixty-two cents ($14 62 c.).

The Committee accepted both these tenders on the 18th of August, 1854, in New York.

A supplementary estimate for a few minor tools having been submitted by Messrs Robbins and Lawrence, viz.:

Main-spring testing apparatus	$30
6 Milling bridges	30
Apparatus for holding and stamping barrels	30
	$90

amounting in the whole to the sum of ninety dollars ($90), the Committee accepted it on the 18th of August, 1854, in New York.

Whilst in New York the Committee wrote and requested a tender for a tension and torsion machine from Mr Kemble of Coldsprings Foundry, near Westpoint, which they received and accepted on the 22nd of August, 1854.

It was to be delivered in New York, carefully packed for shipment, within three months from this date, amounting to nine hundred and fifty-five dollars ($955·00).

A tender was also submitted to the Committee by Messrs Robbins and Lawrence, of Windsor, Vermont, for a plant of machinery to produce the barrel of the Enfield rifle musket (as mentioned in Chapter 2), but as this amounted to a considerable sum of money, and the Committee did not know what arrangements the Honourable Board might have made with Mr Whitworth or others, they thought it more prudent not to accept this tender, but to wait till they had, on their return to England, an opportunity of consulting the Honourable Board's wishes on the subject. At the same time the Committee beg to submit as their opinion, that it would be much safer to get such a plant of machines from Messrs Robbins and

Lawrence, who have studied and practised the manufacture of barrels, than from those who have less experience in this branch of trade, at a time when so many arms are required for immediate use, and for future supply.

The above tender amounts to thirty-eight thousand seven hundred and forty-eight dollars thirty seven cents ($38,748, 37 c.), including 7½ per cent for boxing, delivery at New York, etc.

Conclusion

Before bringing this Report to a conclusion, the Committee venture to offer a few observations on several subjects which strictly speaking do not fall within the province of their mission, but from their importance and bearing upon the general interests of the Departments at Woolwich, it is hoped that they may not be considered as stepping beyond their legitimate bounds in so doing.

One distinguishing feature of manufacturing establishments in the United States, both public and private, is the ample provision of workshop room, in proportion to the work therein carried on, arising in some measure from the foresight and speculative character of the proprietors, who are anxious thus to secure the capabilities for future extension, and in a greater degree with a view to securing order and systematic arrangement in the manufacture.

Another striking feature is the admirable system everywhere adopted, even in those branches of trade which are not usually considered of much importance; this applies not only to the selection and adaptation of tools and machinery, and to the progress of the material through the manufactory, but also to the discipline and sobriety of the employed.

The observations contained in the Report upon American tools and machinery will best explain the nature and adaptation of special tools to minute purposes, in order to obtain the article at the smallest possible cost; for this end capital is borrowed to a great extent and sunk in establishments not only adapted to a peculiar manufacture, but where a department is set apart for the express purpose of making the special tools and contrivances required in order to obtain that end in the most economical and effectual manner. This at least applies to establishments of any importance.

The contriving and making of machinery has become so common in this country, and so many heads and hands are at work with extraordinary energy, that unless the example is followed at home, notwithstanding the difference of wages, it is to be feared that American manufacturers will before long become exporters not only to foreign countries, but even to England, and should this occur, the blame must fall on the manufacturers of England, for want of energy in improving their machinery and applying it to special purposes. The advantages in a manufacturing point of view are all on the side of our countrymen, and there is nothing made in which they ought not to be able to undersell their American competitors either in England or on the continent.

Another point, bearing on this important subject, is the dissatisfaction frequently expressed in America with regard to present attainment in the manufacture and application of labour-saving machinery, and the avidity with which any new idea is laid hold of, and improved upon, a spirit occasionally carried to excess, but upon the whole productive of more good than evil.

The care almost universally bestowed on the comfort of the work-people, particularly attracted the notice of the Committee; clean places for washing being provided, presses to contain their change of clothes, and an abundant supply of good drinking water, in many cases cooled with ice.

The Committee also remarked with satisfaction, the regular attendance and cleanliness of the workmen, and the rigid exactness with which the work is continued up to the last minute of the working hours.

A remarkable feature in the character of the native American work-men is their sobriety; water is their usual beverage, and this they use in-ordinately in hot weather, but rarely anything stronger; clear headedness results from this and gives them a powerful advantage over those who indulge in stronger potations, which will eventually produce its effect on the national manufactures, as it now does on the intelligence and char-acter of the individual workman.

In the Government and private manufactories in the United States, piecework when applicable is universally preferred to day-work, as this arrangement yields the greatest amount of work at the least cost to the employer, at the same time paying the best wages to the individual em-ployed, and the Committee, trusting the Honourable Board will not think them presumptuous, beg most respectfully to submit it as their opinion, formed after careful consideration, that the system of paying by the piece is that on which, after its machinery is in full operation, the proposed Manufactory of Small Arms could be best conducted so as to reduce as much as possible the costs of the arms made, and yet pay good wages to the workmen employed.

The following are their reasons for coming to this conclusion.

1. In a manufactory where payments are made by the piece, it is the interest of those employed to turn out as many as possible of the article they work upon, and to suggest any labour-saving device that may occur to them as likely to increase the production of the machines they attend.

2. Because where work is conducted on this plan, the supervisors may be less numerous, as any time the workman may waste by decreasing the quantity of the article produced by him affects his wages and entails no loss upon his employers.

3. When men are paid by the piece they can be held financially re-sponsible for any work they may spoil through carelessness, which cannot be done where the payments are made by the day; this is particularly important when the parts made are required to be identical, to insure interchange, and any workman who may be employed in some trifling

operation on an article that is almost finished, by carelessness in looking after his machine may spoil and render useless a large number of parts on which a great deal of careful labour has been already bestowed.

The Committee are also of opinion that in order to keep the proposed establishment in a high state of discipline, which is absolutely necessary to enable it to work well, the superintendent should have absolute power over the men employed, as it is by this means only that combinations can be effectually checked, the contagion of dissatisfaction and idleness on the part of a few individuals prevented from spreading through the mass and vitally affecting the system of the whole manufactory.

This system is pursued in all the best Government and private works the Committee have anywhere seen, in which the order and regularity of the workmen was most observable, and they consider it their duty as they hold these opinions, formed after having seen so many first rate manufacturing establishments at home and abroad, and after much careful inquiry as to how these works were conducted, to express them freely to the Honourable Board.

Owing to the difficulty of getting special machinery constructed, and the high price necessarily asked for it by first rate machine makers, who are frequently unwilling to undertake it at all on account of the trouble it entails on them, in devising and making models, the Committee beg to submit to the Honourable Board as their opinion, that a great saving would be effected to the Government, and the service would be much benefited by the establishment of a machine shop in the Royal Arsenal at Woolwich, for the manufacture of special machines only, the system of contracting for machinery in ordinary use, and which is one of the staple manufactures of the country being still adhered to.

In all the United States' armories and arsenals, etc., and most of the large private manufactories of America, this system is pursued with great advantage.

The Committee thinking it may appear strange to the Honourable Board of Ordnance, that scarcely any allusion has been made in this Report to Colonel Colt's armory at Hartford, deem it advisable to explain the reasons of this.

At the time of their first visit to Hartford, the Connecticut River was flooded to an unprecedented degree, and Colonel Colt's factory being situated on the banks of that river, the whole of the workshops on the lower story were inundated; the water standing at one time as high as eight feet in them, for which reason all work was suspended for the time. On their second visit to Hartford, they went only to the site of Colonel Colt's new establishment, now in course of erection, the magnitude of which quite astonished them.

When completed this will be the largest and finest armory in the world, consisting of two blocks of stone buildings, one 500 feet by 60, the other 500 feet by 40 feet, connected together by a third block, so that the whole is in the form of the letter H, and measures 250 feet in width over all; it is to be three stories high, and to have 58 windows on a tier in the front.

It will when finished be more than double the size of the factory originally proposed for the manufacture of small arms in England, which was to have been 600 feet by 300 and all on the ground-floor.

The Committee having submitted the two Enfield rifle muskets, pattern 1853, which they took with them to the United States, to the criticism of the officers and others at the United States' armory at Springfield, found that their opinion of it as a whole was highly favourable, but there were three points on which remarks were made which the Committee deem of sufficient importance to mention, in order that they may be submitted to the consideration of the Committee on small arms.

1. That owing to the quantity of wood in the part of the stock under the middle of the barrel, should the wood warp from damp or other causes, it would be likely to set the barrel, it was suggested that a slight reduction of the stock at this place would remove all danger of this occurring.

2. That the thick part of the bayonet-blade is too rigid, and inclined to throw too much strain upon the neck when severely tried, and that to reduce the substance at this part would add to the strength of the bayonet as a whole, at the same time that it diminished its weight.

3. That they considered the arrangement of the trigger-plate and trigger-guard of the American musket as preferable to that of the English one, both as a part of the arm, and as an article to be manufactured by machinery.

The Committee are fully aware that in the above observations, and in the Report generally, there is much room for criticism; but they respectfully request that the Master-General and Honourable Board will take into consideration the mass of matter to be collected during a tour of several months, which required to be arranged, systematized, and copied out hurriedly at its close on their return to New York, and the few days afforded them for that purpose before the departure of the steamer for England, and therefore crave their indulgence for such errors or want of arrangement as may appear in these pages.

The only other point to which the Committee beg to call the attention of the Honourable Board, is the reception accorded to them by the United States' officials and private individuals, who in every instance manifested the utmost willingness to afford or procure information on every subject, and access to every establishment, or in any way to facilitate the labours of the Committee with a liberality and courtesy not to be exceeded.

It therefore becomes the pleasing duty of the Committee to record in this Report the deep sense of obligation under which they lie to the following officers and gentlemen whose valuable assistance so materially promoted the object of this mission, and in a particular manner to Colonel Ripley commanding Springfield Armory, Major Mordecai commanding Washington Arsenal, and Lieutenant Dahlgren, United States' Army in charge of the Ordnance Department in Washington Navy Yard, whose kindness and hospitality were superadded to important services.

General Jefferson Davis, Secretary of War.
General Totten, Chief of Engineers.
Colonel Craig, Chief of Ordnance.
Colonel Ripley, Commander, Springfield Armory.
Colonel Lee, Commander, Westpoint Military Academy.
Major Mordecai, Commander, Washington Arsenal.
Major Lymington, Commander, Watervliet Arsenal.
Colonel Bell, Commander, Harper's Ferry Armory.
Lieutenant Dahlgren, Washington Navy Yard.
Mr Crampton, Her Majesty's Minister, etc.
Mr Marey, Secretary of State.
Major Wade, Fort Pitt Iron Works, Pittsburgh.
Major Hagner, Commander, Frankfort Arsenal.
Lieutenant Balch, Assistant.
Lieutenant Rodman, Commander, Pittsburgh Arsenal.
Lieutenant Brereton, Assistant.
Commodore Morris, Chief of Hydl. Bureau.
Commodore Paulding, Washington Navy Yard.
Colonel Colt.
Lieutenant Lanmann, Washington Navy Yard.
Lieutenant Mc Blair, Philadelphia Navy Yard.
Captain Maynadier, Bureau of Ordnance.
Mr Alger, South Boston.
Mr P. Wainwright, Boston.
Major Thornton, Governor's Island.
Mr Ames, Chicopee, Massachusetts.
Mr Robbins, Windsor.
Mr Abbott, Lawrence, Boston.

New York, 23*rd August*, 1854.

ROB. BURN, *Lieut.-Colonel, Royal Artillery*
THOMAS PICTON WARLOW, *Captain, Royal Artillery*
JOHN ANDERSON, *Inspector of Machinery*

NEW YORK INDUSTRIAL EXHIBITION.

SPECIAL REPORT OF MR. GEORGE WALLIS. PRESENTED TO THE HOUSE OF COMMONS BY COMMAND OF HER MAJESTY, IN PURSUANCE OF THEIR ADDRESS OF FEBRUARY 6, 1854. LONDON: PRINTED BY HARRISON AND SON.

Introduction

I have now the honour, as one of the Commissioners appointed to visit, during the summer of this year, the Industrial Exhibition at New York, and the various seats of manufacture in the United States of America, to submit my report on those departments of manufacturing industry, the examination of which was confided to me.

In thus reporting upon the growth and present development of those branches of manufactures to which my attention was specially directed, I have thought it advisable rather to treat of them in their broader features, as shown in the localities in which they are carried on, than in their individual manifestations as displayed in the Exhibition at New York. The classification there adopted has, however, been preserved, since it is identical, in all essential points, with that of the Great Exhibition of 1851, to which attention was directed in the instructions given to the Commission.

Avoiding individual criticism as much as the proper illustration of the present position of American industry would permit, I have sought rather to test the results of transatlantic skill by its own aims, and the peculiar requirements of a people whose wants it is its honourable ambition to supply, than to institute unfair comparisons, which could only lead to conclusions of no practical value. For it could not be reasonably expected that any great amount of originality would be the result of the first efforts of a people, however ingenious, whose experience in many of those branches of manufacture which in Europe have claimed the exclusive attention of skilled artizans for ages, does not date so far back as a single generation.

Manufactures, as a *result*, must, however, be carefully separated from machinery, as a *means*; since, in the latter, originality of conception, construction, and application is one of the most remarkable features in the progress of industry in the United States, but this Report deals with manufactures only and originality was not likely to result where imitation of European productions has alone been the aim of the manufacturer and the artizan. The requirements of the markets to be supplied, the competition and comparison of American products with those of England, France, and Germany, and the fact that the very workmen, by whose agency alone a commencement could be made, were Europeans, bringing with them the traditions of the workshops and the conventional types of the old world,–all tend rather to repress than to encourage any departure from the stock forms of the more ordinary and useful productions so constantly in demand, and which the absurd prejudices or vicious cupidity of the mercantile classes often compels the manufacturer not only to imitate, but absolutely to brand with the name of his foreign rival. This is equally unfair to the native as to the foreign producer, since the former is deprived of the credit and reputation due to him for his ingenuity, whilst the latter is deprived of his market, not by fair and honest competition and comparison, but by an imitation of his wares. A difficulty was thus created by those who ought rather to have encouraged than sought to repress the rising energies and aspiring ingenuity of their own countrymen, yet such is one of the processes through which American industry has had to pass in many of its branches. Happily this state of things has been much modified recently. Nor should the various disadvantages under which the skilled workman is necessarily placed be overlooked, since in a new country he becomes almost entirely dependent upon his own resources and ingenuity alone for the supply of those means and materials which, in a more advanced state of manufacturing progress, the division of labour abundantly aids him in procuring. The mutual dependence of one branch of manufacture upon another is so wide-spread and universal that, at first sight, the difficulty in commencing some of them appears so great as to convey the impression that it is insurmountable. Certainly the one thing which, more than any other, strikes the visitor to the seats of industrial skill in the United States, is the ingenuity, the indomitable energy and perseverance displayed in overcoming the early difficulties which must have stood in the way of anything like successful progress at the outset. It is not, therefore, a matter of surprise that many skilled artizans have, from time to time, returned to Europe, after an attempt to establish a manufacture, since the embarrassments arising out of almost unaided exertions and an isolated position, were too great to allow them to do justice to themselves, or to those employers whose spirit and enterprise might have induced them to embark capital in such undertakings. The pecuniary loss of the latter has frequently been inevitable, and the early history of *nine-tenths* of the various branches of manufacture now flourishing in the United States, and amply repaying their present proprietors, is that of ruin, or of enormous sacrifices on the part of those who had the hardihood to become pioneers in those arts which now promise to become, at no distant period, of vital importance to the well-being of millions of industrious men and women. Again, even to the present time, the isolation of manufactories in places at such a distance from each other that mutual aid is almost impossible, renders it imperative that each should be complete within itself, and that everything connected with its operations should be either manufactured on the premises, or kept in stock to such an extent as shall ensure a continuous supply. Thus, both the self-assistance and the laying in a stock of materials is carried to a much greater extent than the majority of English manufacturers would credit. This gives a great peculiarity to the manufacturing system of the States, so far as at present developed, and many of the manufacturers of Birmingham and Sheffield would soon close their doors if they had to furnish themselves with all the partially prepared materials for which they depend upon those whose business it is to manufacture them. Yet such is the position of large numbers of the most successful houses in the United States, and the difficulty is met with a tact and ingenuity of no ordinary kind. There can be no doubt, too, that this necessity for self-supply has been the means of originating many ingenious machines, for which the Americans have so deserved a reputation, as applicable to the manufacture of small articles, or portions of more complicated productions.

Thus the very difficulty in procuring human labour, more especially when properly skilled and disciplined, which would assuredly be the greatest drawback to success, appears to have stimulated the invention of the few workers whose energies and skill were engaged in the early development of manufactures; and to this very want of human skill, and the absolute necessity for supplying it, may be attributed the extraordinary ingenuity displayed in many of those labour-saving machines, whose automatic action so completely supplies the place of the more abundant hand labour of older manufacturing countries.

The successful application of mechanical means to one manufacture has been, as a matter of course, stimulative of their application to another, however different, and the adaptative versatility of an educated people was never more fully displayed than in the constant effort to supply their greatest want–that of skilled labour–by applications of mechanical powers to that object. Nor can the most superficial observer fail to be impressed with the advantages thus derived from the long and well-directed attention paid to the education of the whole people by the public school systems of the New England States and of the State of Pennsylvania. Here, where sound and systematic education has been longest and, in all probability, most perfectly carried out, the greatest manufacturing developments are to be found, and here it is also where the greatest portion of the skilled workmen of the United States are educated, alike in the simplest elements of knowledge, as in the most skilful application of their ingenuity to the useful arts and the manufacturing industry of their country, and from whence they are spread over the vast territories of the Union, becoming the originators, directors, and, ultimately, the proprietors of establishments which would do no discredit to the manufacturing States of Europe.

As there is no apprenticeship system, properly so called, the more useful the youth engaged in any industrial pursuit becomes to his employer, the more profitable it is for himself. Bringing a mind prepared by thorough school discipline, and educated up to a far higher standard than those of a much superior social grade in society in the Old World, the American working boy develops rapidly into the skilled artizan, and having once mastered one part of his business, he is never content until he has mastered all. Doing *one* mechanical operation well, and only that one, does not satisfy him or his employer. He is ambitious to do something more than a set task, and, therefore, he must learn all. The second part of his trade he is allowed to learn as a reward for becoming master of the first, and so on to the end, if he may be said ever to arrive at *that*. The restless activity of mind and body–the anxiety to improve his own department of industry–the facts constantly before him of ingenious men who have solved economic and mechanical problems to their own profit and elevation, are all stimulative and encouraging; and it may be said that there is not a working boy of average ability in the New England States, at least, who has not an idea of some mechanical invention or improvement in manufactures, by which, in good time, he hopes to better his position, or rise to fortune and social distinction.

At first sight, these characteristics do not appear to be conducive to that orderly, and, still less, to that economic working of a manufacturing system, which shall supply the daily increasing wants of a country like the United States. Nor are they so, but for that element of educated intelligence of which they form only one phase. On this intelligent understanding of the true position of things, and the requirements of the social system around him, the skilled workman rests his position. In an old country, with conventional arrangements, in which the minute subdivision of labour grows rather out of the ignorance of the workers, than out of their intelligence, since the great mass are only fitted to do one thing well, the great fact of such sub-division must ever be imperatively insisted upon. But, it must be remembered that in the United States, the wants and increasing demands of a new country have to be dealt with, not the least of which is, that very skilled labour which the varied occupation of each individual artizan is intended to supply. Nor does this knowledge of two or three departments of one trade, or even the pursuit of several trades by one individual, interfere so much with the systematic division of labour as may be supposed. In most instances the change of employment is only made at convenient periods, or as a relief to the workman from the monotony of always doing one thing. This is a mechanical process which an uneducated man may prefer, but which, in a majority of cases, the intelligent workman would willingly be spared if it was not compulsory through the arrangements of the system under which he labours; or if change of employment did not endanger the amount of his earnings. There is, however, one drawback to this otherwise successful violation of the economic law of sub-division. It is unfavourable to that perfect skill of hand, and marvellous accuracy, which is always to be found associated with the constant direction of the attention and practice of the workman to one thing; and this is often very apparent in most of the manufactured articles of America. They lack that perfect finish and completeness of appearance which characterize similar productions of Europe. It is true that the practised eye alone can detect the difference, but it is not the less to be noted on that account. Probably, a more extended experience and thorough practice will do much towards remedying this defect, which may be said to exist in all early efforts either of individuals or of nations.

The position of the artizan classes of the United States has been the subject of so much disquisition, and it is generally understood to be satisfactory, alike educationally and socially, that it needs no comment here. It may be requisite, however, in considering the position and development of the different branches of manufactures, to occasionally make reference to the relations in which the producing classes stand; the hours of labour and rate of remuneration, as compared with the same classes in Europe. It will be sufficient, therefore, to state here the general fact, upon which there exists some misapprehension in England, that, though the rate of remuneration is relatively higher, the number of hours constituting a day's work, especially in factory labour, is much greater; and that this excess in the hours of work, obtains generally in every industrial occupation. Employment, however, is always plentiful: labour, especially skilled labour, ever in demand. Occasionally, in some of the middle States, complaints are made of extreme home competition, and that attempts are being made to bring slave labour to bear upon manufactures, to the reduction of prices, and, consequently, of the remuneration of free labour. But, the prevailing idea is, that slave labour can never by any possibility be made to compete, or to pay in comparison with free labour. This is held equally by the opponents and advocates of the institution of slavery as it at present exists, and it is only mentioned here in order that it might not be supposed that a question of so much importance in its social, as in its economic aspect, was overlooked; since I have no inducement or inclination to meddle with so fruitful a source of misunderstanding, even in its industrial aspect.

The laws of partnership of the United States, which encourage the formation of manufacturing companies, with limited responsibility in the non-managing shareholders,* has led to a much greater development of the industrial resources and skill of the country, than, in its circumstances, could have resulted under mere private enterprise for many years to come. It must, however, be evident to every person who carefully examines the joint-stock system of manufactures, as it now exists, that it is by no means favourable to the ultimate and permanent success of manufacturing establishments generally, and, though there are many highly gratifying and honourable exceptions, those are the best managed where there are the fewest proprietors, and, consequently, the smallest number of individual interests to consult. The machinery of a board of managers, though less cumbrous than in England, is, even in its best form, too slow in its action for the proper care of interests and the execution of prompt and energetic dealings, so essential to success in manufacturing and commercial transactions. And, though the management and responsibility of these concerns is usually narrowed to the limits of safe responsible power,

* An erroneous impression appears to exist in England as to the extent of the responsibility of all shareholders or partners in the joint-stock manufacturing companies of the United States. The agent or secretary, manager, treasurer, and directors, being also shareholders, are held by the law fully responsible, to the extent of their means, for the results of the management entrusted to and undertaken by them. The limited responsibility is therefore wisely confined to the non-managing shareholders, the latter being liable only for the amount of their subscription for shares. G. W.

the promptitude of the individual mind of a single proprietor, or even of two or three, who feel that they alone can decide, as they alone are involved in the issue, is wanting.

In stating this, however, it must not be understood as any condemnation of the law of partnership, as it exists in a variety of forms in the different States of the American Union, tending, as it does, to encourage enterprise and skill in combination with limited capital, and affording means of investment of great value to a large portion of the community; but simply as placing the question of management and the organization of industry upon its true basis–that of a distinct interest, combined with full powers and complete responsibility.

As a means of developing the individual powers and natural resources of a new country, encouraging self-reliance, promoting economical habits by affording opportunities for the safe and profitable investment of small amounts of realised capital in the hands of the industrious classes, the value and importance of a limited responsibility in the law of partnership cannot be overlooked or over-rated. To it the manufactures of America owe much of their present, as, without doubt, they will of their future prosperity and success.

Before proceeding to the consideration of the various classes of manufactures which it is intended to embody in this report, it is essential most distinctly to explain, that such report is not intended as a complete exposition of the manufacturing industry of the United States of America. Neither time nor opportunity would permit of such a complete examination of the various localities in which manufactures are carried on, and though the distance travelled over in various States during the comparatively short period of eleven weeks, which my duties in England would permit me to devote to the objects of the Commission, apart from the sea-voyage, was upwards of 5,000 miles, and embraced the States of Massachusetts, Connecticut, New Hampshire, Rhode Island, New York, New Jersey, Pennsylvania, Maryland, the District of Columbia, Eastern Virginia, Kentucky, and Ohio, it must be evident that a full and minute examination of the industry of these States alone would have occupied a much longer period than I could possibly under the circumstances devote to them. The manufactories visited were those which appeared to be most likely to offer a fair type of others which could not be included, or those which I was well assured would present the most complete development of their special branches of trade.

The circumstances of the delay in the opening of the Exhibition at New York, were favourable to a more extended visitation of the leading localities than might otherwise have been undertaken. The continued incompleteness, however, of the arrangements of the Exhibition at the time of my departure for England, on the 10th of August last, prevented so complete an examination of the contributions in their respective classes as could have been wished.

In almost every instance, the basis of this report is the result of careful personal examination and inquiry. In one or two instances, however, where the distance to be travelled over was too great, or the amount of development likely to present itself was not likely to compensate for the time occupied in the journey, I deemed it advisable to communicate by letter with some responsible person, and where the information received in reply was considered of sufficient importance to be embodied in this report, my authority is given with it.

It is possible that some of the contrivances and processes described may be mere modifications, or exact imitations of processes recently introduced into use in England. If so, this must be ascribed to the fact, that no individual could possibly be acquainted with every recent improvement in even one branch of manufacture, and, as the work here undertaken involves so wide a range, it would have been an act of great presumption to have decided amongst the many industrial expedients which presented themselves, as to their originality, or their exclusive application in the United States. I have, therefore, preferred mentioning such as were new to myself, but which may be less novel to others, rather than run the risk of neglecting any really useful or ingenious process from a puerile fear of describing something already known to those interested in any special manufacture.

No fact appears more certain than that the manufacturers of the United States take especial care to be well informed on all European improvements, either in machinery or in processes; and, as traditional methods have little hold upon the American, as compared with the English artizan, processes holding out the least promise of improvement are quickly tested, and, if found worthy, are adopted either in their integrity or with such modifications as the ingenuity or wants of those applying them may suggest or require.

The extent to which European artizans, especially Englishmen, are employed in the manufactories visited, was a point I deemed it worth while to inquire into. The numbers will be found indicated under the heads of the several branches of trade, in connection with the notices of the manufactories in which they are employed.

There are, however, very few Englishmen compared with what I expected to find, or are generally supposed to be engaged in the industrial establishments of the United States. The hours of labour, as already indicated, are too long for most of those who are induced to emigrate, since they usually seek for a less demand upon their energies than they are subjected to in England, as well as a higher rate of payment, and it is no uncommon thing for the really skilled and steady artizan, who can earn a sufficient livelihood at home, to return after a brief trial in the United States. The sudden variations in the temperature, too, and the extremes of heat in summer and cold in winter, are very trying to the majority of English workmen, and if, as is too often the case, they are at all given to intemperate habits, these are aggravated by the climate, and disease and death cuts them off much more rapidly than is, I fear, generally acknowledged. It may not be uninteresting, however, to state, that some of the best and most intelligent artizans, as also some of the most successful manufacturers in the United States, are the sons of Englishmen, taken early in life to their adopted country, acclimated by habit, rendered intelligent by the education received in public schools; their ingenuity and industry stimulated and encouraged by the character of the institutions around them, they often present remarkable instances of self-elevation; and, whilst in an earnest love of their adopted country, they may be said to be more American than the Americans, their secondary pride, if I may use the term, is, that their fathers were Englishmen.

German workmen are largely employed in many departments of industry, and being generally an intelligent and orderly class of men, are highly esteemed. Hence, in some branches of the metal trades, the modes of manufacture are often more German than English, although, perhaps, originally established upon the methods of the latter. Still, taking into consideration the whole number of English, German, and French employed in the United States, I believe it would be found that eight-tenths of the artizan class are Americans, or the children of Europeans, rather than Europeans themselves.

I now proceed to the consideration of the sixteen classes of manufactures which it was my duty to examine and report upon, under their respective heads.

1

Textiles

Cotton: Class XI

The cotton manufactures of the United States appear to hold the first position in the industrial productions of that country, alike as regards extent and value; and, although their operations are chiefly centralised in New England and Pennsylvania, yet there are only seven of the thirty-one States of the Federal Union in which the spinning or manufacture of cotton is not carried on. These are Louisiana, Texas, Michigan, Illinois, Iowa, Wisconsin, and California. In all the other States, including the District of Columbia, establishments of a more or less important character are to be found. By a return printed in a report of the Superintendent of the Census of the United States for 1850, 1054 establishments for the manufacture of cotton goods are stated to be in operation, consuming 641,240 bales of cotton, and manufacturing goods to the value of $43,207,555, or about £10,000,000 sterling per annum. Of these establishments 564 are in the New England States alone; 213 of them being in Massachusetts, 158 in Rhode Island, and 128 in Connecticut: 65 only are to be found in the States of Maine, Vermont, and New Hampshire; but the 44 mills returned as being in the last named State manufacture nearly as much in quantity and value as the 286 establishments of the two States of Rhode Island and Connecticut. Of the 530 mills scattered over the remaining eighteen States, Pennsylvania has 208. Of the Southern States, Georgia and Tennessee have the largest number of establishments, those in the former being returned as 35, and in the latter 33; but, from information received in Virginia, there is no doubt the number of mills in Georgia has increased considerably since 1850.

The number of persons employed in the cotton manufactures of the whole of the States was 33,150 males, and 59,136 females.

The New England establishments are conducted upon a similar principle to the largest cotton factories of Great Britain, and spinning and manufacturing are carried on as one concern. This, however, is not the characteristic of the mills in the above States, as they are, in many instances, employed in spinning only, and in Pennsylvania, Georgia, and Tennessee, yarns are produced chiefly for the purposes of domestic manufacture by hand, which still obtains in many parts of the older States of the Union. Thus whilst in Pennsylvania the capital invested amounts but to about one-seventh of that of Massachusetts, the quantity of cotton consumed is one-fifth, the value of raw material not quite one-fourth, the number of operatives (male and female) one-fourth, the value rather more than one-fourth, the number of pounds of yarn spun and sold as yarn is above *thirty* times greater in Pennsylvania than in Massachusetts. This, to a certain extent, gives a key to the difference in the modes of manufacture in the two States, and illustrates the distinction already alluded to. There can be no doubt, however, that domestic weaving is gradually giving way, and those manufacturers, especially in Pennsylvania, who formerly did a prosperous business as spinners only, now find that the Eastern States supply the piece goods at a rate so little above the cost of the yarn, that it is not worth the while of the farmer to continue this primitive custom of weaving his own cloth. Thus the domestic loom is fast following the spinning-wheel of the early settlers, and those manufacturers who until recently have spun yarn only, are gradually introducing the power-loom as the only means of sustaining their position in the market. This was illustrated by a visit to the 'Eagle Cotton Mill,' Pittsburgh, Pennsylvania. Formerly the proprietors spun yarn alone, and did a successful trade, but, by a return with which they favoured me, I find that in six establishments under their direction they had introduced already 540 looms to the 26,000 spindles, and were manufacturing sheetings at the rate of 6,000,000 yards per annum, together with twilled cotton bags, batting, and yarns, from 5*s*. to 18*s*.; and this, in order to make the latter pay by consuming the surplus yarns themselves. In the Penn Cotton Mill, Pittsburg, the more modern system had become the rule of the establishment, and with 7,000 spindles and 207 looms, 2,730,000 yards of shirtings were produced annually, besides 240,000 lbs. weight of coloured yarns for carpet warps and cotton rope.

At two establishments at Richmond, Virginia, the consumption of the yarn, in the manufacture of piece goods, was also the rule, and this being the extent of my journey southward, it became a matter of interest to inquire as to the progress of the cotton manufacture in the cotton-growing States. Georgia, Tennessee, and North Carolina were quoted as those in which the greatest progress had been made; whilst Virginia, South Carolina, and Alabama, were the next. In Tennessee spinning would appear to be the rule, and manufacturing the exception. In Georgia and North Carolina equal attention is paid to both; whilst in Virginia, South Carolina, and Alabama the manufacture of piece goods is decidedly more extensively carried on than spinning only. Slave labour is said to be largely used, with free whites as overseers and instructors.

In the two establishments above named free white labour alone is employed. The males are heads of departments, machinists, dressers, etc., and the females are spinners and weavers. The latter are chiefly adults, though children from twelve to fifteen are employed. The average hours of work here are twelve, but vary a little with the season. *Very full time* being the rule. At least such is the statement of the manager of the James' River Company's Mill. This establishment, as also the Manchester-Cotton Company's Mill, is at Manchester, Chesterfield County, Virginia, and situated opposite to Richmond, on the James river, from the falls on which the water power used for driving the machinery is derived. The James River Mill produces a large weight of work for the extent of its machinery. The goods manufactured are coarse cottons, and average about 2½ yards to the pound; shirtings, 28 inches wide (Osnaburgs), summer pantaloons for slaves, and bagging for export to the Brazils for sugar-bags, running about 3 yards to the pound. Bagging of a lighter character, for grain, and 36 inch Osnaburgs, two yards to the pound, are also produced. The Manchester Company manufacture sheetings, shirtings, and yarns, and employ about 325 operatives, the children being of the same average age as at the James River Mill. The manager, Mr Whitehead, is an Englishman, as is also the chief mechanic. The former has just perfected a patent 'speeder,' of which the latter expressed a very high opinion. Its advantages are a greater speed, a more even roving, and a bobbin of any desirable size, which never becomes spongy in the winding.

The small development of the cotton manufacture in the States of Indiana, Mississippi, and Arkansas, or even those of Ohio and Kentucky, required no special inquiry. In Maryland, however, there were twenty-four establishments in 1850, chiefly engaged in manufacturing piece goods, such as drillings, sheetings, ducks, Osnaburgs, and bagging. The yarns produced for domestic purposes bear but a small proportion to those manufactured into cloth, and these are chiefly sold within the State for the home weaving of mixed fabrics of wool and cotton, forming coarse linseys. The wool is mostly spun by hand in the farm houses, and the fabrics when made are entirely for domestic use. In Maryland, too, bleaching is carried on to a considerable extent.

[*Cotton manufactures of the New England States.*] Having thus endeavoured to illustrate the position of the cotton manufacture in the form in which it has developed itself in the South, and, so far as circumstances would permit of an inquiry, in the Middle States, bordering on the west, the manufacturing system, as manifested in the cotton trade of New England, demands consideration. If the illustrations given show the early progress and position of this manufacture in the United States, so far as daily recurring improvements and ever increasing wants have permitted it to remain in its original form, the manufacturing towns of Lowell, Manchester, and Lawrence strikingly demonstrate the results of the energy and enterprise of the manufacturers of New England.

At LOWELL, Massachusetts, the cotton manufacture has been developed in a form which has been a theme for many writers on the economy and social bearings of the factory system; and the plans so successfully put into operation here, and carried on since 1822, have led to the erection of large establishments with their attendant boarding-houses, at Manchester, New Hampshire, and more recently at Lawrence, Massachusetts; whilst a commencement has been made at Holyoke, in the same State, by the Hadley Falls Company, which promises a result of a more extraordinary character than anything yet achieved in the United States. Each of these localities present features peculiar to themselves, and, besides the manufacture of cotton goods, other branches of production in textile fabrics are carried on.

The falls of the Pawtucket on the Merrimack River, and the Pawtucket Canal, which had previously been used only for the purposes of navigation, and connecting the river above and below the falls by means of locks, presented to the original projectors of Lowell a site for the solution of an important problem, not only in American industry, but to a great extent in that of Europe itself. This was the combination of great natural advantages with a large and well-directed capital, resulting in extensive and systematic operations for the realization of a legitimate profit; whilst the social position of the operative classes was sedulously cared for, and their moral and intellectual elevation promoted and secured. The example has not been lost, even in Europe, and the possibilities of the manufacturing system of a country being carried on without deterioration, but, on the contrary, to the consolidation and promotion of the best interests of the labouring classes, having been so unmistakeably proved, many improvements in the larger manufactories of England, not only of those engaged in manufacture of cotton, but in other branches of industry, have resulted from the enlightened and profitable system commenced barely thirty years ago by the founders of Lowell, which is now a city containing nearly 35,000 inhabitants.

In that place there are eight manufacturing corporations, exclusively employed in the manufacture of cotton goods, two of which print and dye their own fabrics, and one company (Lowell Manufacturing Company) which manufactures cotton Osnaburgs in addition to its staple production of carpets. There are thirty-five mills, besides the print works above named, belonging to these companies. (See Table, Appendix No. 1.) These produce 2,139,000 yards of piece goods per week, consisting chiefly of sheetings, shirtings, drillings, and printing cloths, varying in quality from No. 13 to No. 40. The greater portion, however, are No. 14s and No. 13s. The consumption of cotton is 745,000 lbs. per week, spun and manufactured upon 320,732 spindles and 9,954 looms. The average per spindle is 1¼ yard per day, the medium produce of a loom being, in 14's 45 yards, and 30's 33 yards per day. The number of operatives employed in the cotton manufacture by the eight corporations exclusively engaged therein is 6,920 females and 2,338 males. This, however, does not include those employed by the Lowell Manufacturing Company in their cotton mills, as the returns only show the gross number of hands engaged in the cotton, carpet, and other departments of that establishment. The average wages of females, clear of board, is $2, or about 9*s*. 6*d*. sterling, per week, whilst the wages of males show an average of $4 80 cents, or about £1 2*s*. sterling, per week. The average hours of labour per day, *exclusive of meals*, is 12, the mills commencing at 5 a.m. and closing at 7 p.m.

Of the quality of the goods produced, it will be sufficient to say that they are generally excellent of their class, and quite equal, sometimes superior, to similar manufactured in Britain. Those of Lowell may be taken as fair examples of other cotton mills in the United States, possessing the same advantages as regards power, improved machinery, and intelligent operatives. In spinning, it will be seen that the numbers are low, the finer quality of cotton goods not being produced. The No. 40s 'printers' manufactured by the Merrimack Company being the highest class. These, when printed, are of a firm and excellent fabric. The Lowell Manufacturing Company produce a very cheap and well-looking fabric for cotton trowserings at 17½ cents, or about 9½*d.* sterling, per yard. These are made up of dyed yarns in checks and stripes, and are woven on gingham looms.

In selecting the cotton manufactories of Lowell as an illustration of other manufacturing localities engaged in the same branch of industry, it must be borne in mind that each of the latter have certain features peculiarly their own; but as the statistics of Lowell, as given (see Table, Appendix, No. 1), fairly represent the rate of production, etc., in the class of goods included therein, in the best and most extended form as to economy, organisation, etc., in a group of establishments, no useful purpose can be served in quoting the relative rate of production of single manufacturing companies in groups of two or three. It will be sufficient, therefore, in glancing at the cotton mills of Lawrence, Manchester, and Holyoke, to quote the total productions from returns courteously furnished by their respective agents or managers.

As the manufacturing establishments of LAWRENCE are intended by their enlightened proprietors to comprise all the advantages derived from the experience gained at Lowell, I would beg to quote the following description of the Atlantic Cotton Mills,* from the Appendix to the

* An admirable library has been established in connection with the Atlantic Mills by the overseers and operatives. The agent, Mr H. K. Oliver, having given 100 volumes as a foundation, and loaned $50 for the purchase of books. Subsequently, the sum of $50 and 24 volumes was presented by the treasurer, Mr William Gray, and further donations from the late Mr Amos Lawrence and the Lawrence Tract Society. The library now contains 1,100 volumes; and, on an analysis of the printed catalogue of 986 of these I find that the selection of books is admirable. There are, on History, 125 volumes; Biography, 98; Religious, 60; Fiction, 187; Poetry, 56; Travels, 78; Philosophy, 16; Science, 20; Mechanics, 10; Agriculture, 10; Natural History, 7; Miscellaneous, 319.

The permanent members, who are chiefly the overseers, pay $1 per annum; those who simply become members to use the books, pay 60 cents (about 2*s.* 9*d.* sterling) per annum; these are chiefly mill hands; and about one-third of the operatives thus subscribe to the library. It must be remembered, however, that about *one-third* of the female operatives in this mill are Irish, who either cannot read or write, or have no taste for reading. One of the hands, a young man, acts as librarian, for which he receives a small

Sanatory Report presented to the Legislature of the State of Massachusetts in 1850:

'The Atlantic Cotton Mills have erected a building 600 feet in length, five and six stories in height, partly 64 and partly 106 feet in width, which is devoted to the manufacture of brown cotton goods. It is designed to contain 42,500 spindles, and 1,168 looms: 25,088 spindles and 728 looms are now in operation; and 164 male and 619 female operatives are employed. This number will be increased to about 1,200 when in full operation. The motive power is supplied by three of Boyden's improved iron turbine wheels, each 8 feet in diameter, and of 300-horse power; 12 mill-powers are devoted to these mills. The boarding-houses belonging to these mills consist of six blocks, containing 68 tenements, and are built upon a similar plan and have the same admirable arrangements for water, cleansing, sewerage, and other purposes, as those belonging to the Bay State Mills, noticed in the description of that establishment (Class XII; see Appendix, No. 3). 32 of these tenements are intended for the females; and 36, equally good, but containing fewer rooms, are intended for the overseers in the mills, and for men with families, who may also take boarders.'

By a return for the week ending 30th July, 1853, 7,783 pieces of goods were manufactured, containing 299,262 yards, and weighing 99,647 lbs. These consist of plain sheetings, shirtings, and drillings No. 14's, and plain sheetings and shirtings No. 24's, varying in average yards to the lb. from 1·47 in drillings to 5·40 in shirtings. Examples are exhibited by the Company's agent at the Exhibition of New York.

Lawrence, like Lowell, derives its water power from a remarkable fall in the Merrimack river, across which a dam, 1,011 feet in length, has been constructed. Manchester, New Hampshire, is also situated on the same river.

The Amoskeag Fall at this latter place has a fall of about 52 feet in one mile, and is formed into two dams, the head-waters of which feed canals on two levels, from which the power is used. The first or highest level has a fall of about 20 feet, – the second one of 30 feet. The power is estimated at 10,000 horses, of which less than 5,000 are at present in use; about 750 only being in the lower level or second canal. The natural construction of the fall was very favourable to an economic use of the power, – the upper dam being raised only 5 feet high; the engineering works of the dams and canals were consequently inexpensive in comparison to those of South Hadley, on the Connecticut, to be hereafter mentioned, or even the dam at Lawrence. Turbine wheels of large size are used, as at the latter place.

There are two manufacturing companies at MANCHESTER – the Amoskeag Company, and the Stark Mills. The former consist of four mills, the

---

remuneration from the library fund. The books are given out, and returned every Saturday evening, half an hour after bell-time; the requisite rooms for the safe keeping of the books being found by the proprietors of the mill, rent free. G. W.

latter of two, and it is scarcely possible to conceive anything more complete than these manufacturing establishments, with the requisite boarding-houses conveniently situated for the purposes of the operatives.* The goods produced in the Amoskeag Mills are of excellent quality, and consist of ducks, tickings, denims, drillings, sheetings, and cotton flannels of varied width and quality. The ducks and tickings are excellent fabrics, of which examples were exhibited in the Great Exhibition of 1851, and obtained a prize medal. The total productions of the Amoskeag Mills

---

* As an illustration of the relations existing between the operatives and the corporations who employ them, it may not be uninteresting to state that on the day I visited Manchester by invitation of several of the directors, the annual meetings of the Manchester Print Works Company, and the Amoskeag Manufacturing Company, were held. The operatives employed by the latter having reason to fear that in the erection of another mill, a favourite elm, one of the last of the old forest trees on the banks of the Merrimack, would have to be cut down, hastily got up a petition, which was signed by upwards of 500 of the mill hands, male and female, and presented it to the meeting of the proprietors, requesting that, if possible, the tree should not be cut down; and giving many reasons alike creditable to the good sense and feelings of the petitioners, amongst which were, 'that it was a beautiful and goodly tree,' and belonged to a time when 'the yell of the red man and the scream of the eagle were alone heard on the banks of the Merrimack, instead of the two gigantic edifices filled with the buzz of busy and well-remunerated industry,'–'a connecting link between the present and the past, and perhaps may serve as a ever-living, yet silent monitor, each autumn, as the aged and yellow leaf falls from among its fellows, to remind us of our own mortality.' 'It is, it may be therefore assumed, a *useful* tree, to say nothing of its absorbing noxious gases, and giving out healthy ones.' The petitioners repudiated any intention to interfere with the arrangements of the shareholders in the disposal of their property, acknowledging how largely they were indebted to them for 'the green enclosures and hundreds of trees which line the streets, all promptly cared for and protected at no inconsiderable expense; and though they could not but indulge a hope that the company would "spare that tree," yet,' say the petitioners, 'we shall not murmur if, upon the whole, by stern necessity, you should remove the object of our solicitude.' The company, however, decided to comply with the wishes of the operatives so properly expressed, and, as I had every reason to believe, from the position of the tree, at no little inconvenience and cost. On examining the signatures to the petition, I was struck with the admirable character of the handwriting of the majority.

There is a savings bank attached to the Amoskeag mills, in which has been deposited $137,000 (about £27,000 sterling) since 1840, on which 5 per cent. per annum is paid by the company. This does not include amounts withdrawn by operatives leaving the establishment. And it may be well to state here, that the average continuation of females devoting their attention for a time to factory-labour, is about 2½ years at Manchester and about 3 to 3½ at Lowell. At Lowell, each corporation formerly had its own savings bank for its own operatives, allowing, as in the case of the Amoskeag Company, an interest on deposits. Now, however, they are all consolidated into one general savings bank, and the deposits amount to $700,000 (about £140,000 sterling). G. W.

for the four weeks ending 21st May, 1853, was 1,597,166 yards, weighing 585,745 lbs.; and the amount paid in wages was, - males, $6,604 83 cents; females, $21,143 66 cents; being a total of $27,748 49 cents, or at the rate of upwards of £6,000 sterling per month.

Specimens of the goods manufactured are exhibited at the Exhibition of New York.

As Mr Whitworth's engagements did not permit him to visit Manchester, it may be proper to state here that the Amoskeag Company have extensive machine shops for the manufacture of locomotive engines, and the machinery for their mills. The average number of locomotives turned out this year will be five per month.

The Stark Mills Company manufacture 36 and 48 inch sheetings, and 30 inch shirtings, No. 14 yarn, and 48 inch drills, No. 12½ yarn, ducks, and seamless grain bags, and an article, a species of drilling, known as leather cloth, which is now coming into extensive use as a substitute for morocco (see Class XVI). The seamless grain bags constitute a novel and important article of trade, and are of excellent quality and make. These bags are 45 inches long, and are manufactured of various qualities and weight. The warp is a double one, and, by the construction of the loom, the 'filling,' or weft, traverses both sides, uniting the warps at the edge, instead of producing a selvage. The loom is a perfect self-actor or automaton, so to speak; it commences the bag, goes on until the requisite number of picks has been thrown in to make up the length; it then closes the bottom, throws in a given number of picks as a *tab*, and then commences another bag. All that the weaver has to do is to attend, in the usual way, to the perfect working of the machine, and cut out each bag, as from their thickness any quantity accumulated on a cloth beam would be an incumbrance to the machine. As the bags are cut out, each weaver folds and piles them by the side of each loom, and these are removed and an account taken every half-day. The bags are hemmed round the top, or mouth, by sewing machines, each machine being attended by one female operative, and the average work of each is 650 bags per machine per day.

There are 126 of these seamless bag looms at work in the Stark Mills. The average make is 47 bags per loom per day, and the speed about 130 picks per minute. This is rather high, as the general speed of power-looms, and indeed of machinery generally, is lower than in England. By this means human labour is economised, and one operative can attend to more machines. Thus, it is often found that a weaver will attend to four looms in the United States, who, in the same quality of work, would attend to only two in England. I believe, however, that the conviction is gradually forcing itself upon the mind of the British manufacturer, that machinery may be run at a speed which is not economical, either as regards the quality of goods produced, or the most profitable use of human labour.

The seamless bag loom is the invention of Mr Cyrus W. Baldwin, of the Stark Mills, and the looms are all manufactured in the machine shops of that establishment.

Pillow-cases and bed-bolsters could be as easily and as profitably produced by this loom as grain bags, but the latter are in great demand at present. Mr Baldwin has just patented an adaptation of his invention to the weaving of cotton hose for fire-engines. The experimental loom will produce 1,000 feet of hose per day, and from the perfect character of the work, there is no material escape of the fluid when the hose is filled with water, as the fabric swells, and it is as perfect as a duct as the ordinary leathern hose, and much more elastic and portable. In point of economy, too, the cotton hose must possess great advantages, its cost being but 6½ cents, or about 3½*d.* sterling, per foot, and no oiling is required to preserve its elasticity. This alone gives it a greater advantage over the leathern hose, since both labour and material are saved; the cost of the latter being 62½ cents, or about 2*s.* 9*d.* sterling, per foot, and requires renewing every three years. Thus, if the cotton hose was renewed every year, which it is not likely would be requisite except under very peculiar circumstances, an immense saving would be effected. But a double hose, which would wear longer than the leather, and still save the oiling, can be made by inserting a smaller cotton hose within a larger one; a coating of caoutchouc rendering the whole perfectly water-tight, and less susceptible of external injury than a single hose would be. It is scarcely possible to conceive a more perfect fabric than the hose produced by the experimental loom shown to myself by the inventor, who states that it is also applicable, and that he is about to adapt it, to fancy weaving. The invention is patented in England, as well as in the United States.

An improvement in the carding engine is in operation here. It consists of a series of circular saws with very fine teeth, set round a cylinder. It throws out an immense quantity of dirt which escapes the ordinary carding machines. Should flax cotton ever come into general use, this would be the most fitting machine for carding it clean, and getting rid of the woody fibre, by a mechanical instead of a chemical process, the initiatory steps towards which will be mentioned in its proper place in Class XIV.

The total production of the Stark Mills for the month ending May 28th, 1853, was in drills, sheetings, etc., 1,046,138 yards, weighing 379,132 lbs.; in seamless bags 109,544 yards, weighing 97,895 lbs. The number of operatives employed, 1,054 females and 215 males. Specimens of the bags are shown in the Exhibition at New York, but not by the company.

At Nashua, in New Hampshire, also on the Merrimack river, there is a considerable development of the cotton manufacture, but Manchester presents the highest type in that State. Manufactures of various kinds, chiefly cotton and woollen, may be said to be carried on at almost every available point in the Merrimack and Charles rivers, and indeed upon the smaller and more tributary streams of New Hampshire and Massachusetts. The largest river passing through the latter State, however, has only recently commanded sufficient attention to induce the employment of a sufficiently large capital in developing its immense natural advantages; and it would certainly appear that at HOLYOKE, on the Connecticut river, the factory system of the United States of America is destined to the fullest expansion under the most favourable circumstances as regards power, the conveyance of raw materials and manufactured goods, as also salubrity of situation for a manufacturing town.

The Hadley Falls Company have here constructed a dam across the Connecticut river, 30 feet high at the head water, and 1,017 feet long; whilst taking advantage of the peculiar conformation of the land on the banks, the application of the power derived therefrom is so arranged, that a row of mills two miles in length might be erected, and power to drive at least 1,000,000 spindles easily applied thereto. A town is laid out in a plan calculated to secure many advantages to its future inhabitants, and boarding-houses, etc. erected for the operatives employed in the two mills and machine shops already at work. Another mill intended for the manufacture of fancy fabrics is now in the course of erection. Of the two above named, one contains 18,432 spindles, for No. 14 yarn, which is manufactured into sheetings at the rate of about 5 tons weight per day. The other mill contains 30,700 spindles for No. 90 yarn, this being the finest number spun in the United States. It is manufactured into lawns, or jaconetts, chiefly for printing. There are 450 looms in this lawn mill. The goods thrown off are of a superior quality, and show that the manufacture of the finer fabrics in cotton is as likely to be successfully carried on in the future, as the coarser and more useful qualities have been in the past. There is only one other mill for the manufacture of lawns in the States. This is at the Portsmouth Steam Mills, Portsmouth, New Hampshire. Here there are 26,000 spindles driven by steam power, the yarn being No. 90's, as at Holyoke. Another mill for this class of goods, the Pacific Mills, Lawrence, Massachusetts, is in the course of erection, and will be in operation shortly.

As already indicated, there are a considerable number of mills manufacturing cotton goods, many of these, too, of large size, in detached localities in the New England States; but those visited, and now reported upon, may be considered as types of the whole.

In Pennsylvania, as already mentioned, the establishments are generally smaller, and are more isolated. There is a considerable concentration of the cotton manufacture at Manayunk, on the Schuylkill river, near to Philadelphia, but though promised the particulars as to production, class of goods, etc., it has not been furnished to me.

The manufacture of ginghams is now carried on to a large extent, particularly at South Hadley, opposite to Holyoke, and at Clinton, Massachusetts. The goods are admirable, though light in quality, of good dye, and the colours generally put in with good taste.

This, however, depends so much upon the market or class of customers for which the goods are intended, that justice to the manufacturer always demands caution in pronouncing as to the fitness or non-fitness of decoration, even in its simplest forms.

It will therefore be sufficient to say here, that the combinations of colour are often harmonious and pleasing, and sometimes even elegant, and in a material which is regularly sold at 12½ cents to 15 cents, or from

6½*d*. to 8*d*. sterling, per yard retail, and therefore within the reach of all classes.

The Glasgow Mills, South Hadley, derive the water power which drives their machinery from the head-water of the dam constructed by the Hadley Falls Company for the works at Holyoke; indeed the privilege is rented from that corporation. They are entirely devoted to the manufacture of ginghams. The yarns, 28s and 30s, are spun on 10,000 spindles, and manufactured into cloth by 300 power-looms carrying 4 shuttles each, for different coloured yarns, and acting by means of a revolving shuttle box. Each loom produces from 28 to 30 yards per day, and the total produce of the establishment is 14,000 pieces, of 30 yards each, per week, giving a total of 52,000 yards. The yarns are all dyed on the premises, and are generally clear and brilliant in colour, and the goods have a reputation for fast dyes.

The Glasgow Mills Company contribute to the Exhibition at New York a very satisfactory assortment of dress ginghams, handkerchiefs, gala plaids, and white and coloured cotton yarns.

The Lancaster Mills Company, Clinton, Massachusetts, manufacture ginghams on an extensive scale. The looms, 600 in number, together with 96 carding engines, are all in one room or loom-shed of excellent construction, lofty and well ventilated, the roof being of slate and glass, supported on iron columns.

There are 21,000 spindles and 800 operatives, chiefly females. The yarns are No. 25 warp, and No. 27 filling or weft. The goods therefore are of a little lower quality than those of the Glasgow Mills, which, as already stated, are manufactured of No. 28 and No. 30. The make and dye, as also the general selection of colours in the Lancaster Mills, are good, and the goods must be largely in demand, as the make averages 5,000,000 yards per annum, or 100,000 yards per week.

The manufacture of cotton bedquilts by a method included in Bigelow's patent for the manufacture of Brussels carpet by the power-loom, is also carried on at Clinton by the Lancaster Quilt Company. The mill is small, and was erected for other purposes. These bedquilts are good, cheap, and useful articles, in great demand. There is considerable scope for design, untrammelled by the conditions of Marseilles or quilted fabrics, from the facility with which the figured surface can be produced. The fabric, however, does not present the substance of the ordinary quiltings, but in other respects is admirably adapted for summer use or for hot climates.

Tickings are extensively manufactured, and often form a department in the large cotton establishments in the New England States. The manufacture of them, with checks, stripes, etc., also form a considerable item in the home industry of Pennsylvania, and, to a certain extent, in the factory system of that State. Some goods of this class, contributed at the New York Exhibition, manufactured by D.Lammot and Son, Lenna Mills, Delaware county, Pennsylvania, are most excellent of their kind. They are 36 inches wide, 1,100 reed, No. 30 warp, and No. 35 filling or weft, with 140 picks to the inch. It is scarcely possible to conceive a firmer or better made article, and the traditionary notion that really good tickings can only be manufactured from flax receives a severe shock when such cotton goods as these are presented for examination. The Lenna Mill runs 4,000 spindles, and works 160 looms, of which latter twenty-nine are for the manufacture of the finer qualities of ticking quoted above. It is comparatively a small establishment, situated in one of those beautiful valleys, on a stream supplying water power, with which the State of Pennsylvania abounds, and which, in presenting great natural facilities, have originated the comparatively large number of small manufacturing establishments, especially of cotton and woollen yarns and goods, to be found in that State.

The manufacture of stocking-net and of cotton hosiery of the coarser qualities is gradually developing itself in various localities, especially in the states of Connecticut and Massachusetts. As, however, hosiery comes under the head of Class XX (Clothing, etc.), this branch of industry will be noticed in its proper place.

The display of cotton goods in the Exhibition at New York, as illustrative of this branch of industry, is, on the whole, a satisfactory one; nearly every special department having one or more representatives.

It is now requisite, in order to complete as far as possible this review of the cotton manufactures of the United States, to briefly notice such branches as may not have come within the range of business of the establishments visited and reported upon, and whose contributions to the Exhibition have been indicated in due course.

The generality of contributions of cotton fabrics most generally in demand, such as sheetings, shirtings, drills, ducks, etc., are of good make and quality, and are often well finished and neatly made up. These, however, need no special remarks, and a reference to the catalogue for the names of the contributors will be sufficient.

The Silesias manufactured by the Franklin Manufacturing Company, Providence, Rhode Island, as also the same kind of goods, together with nankeens and drills of a fine quality, manufactured by Goddard Brothers, of the same place, are excellent fabrics, well dyed and finished.

The Canada plaids, adapted for vestings, manufactured at Whittenton Mills, Taunton, Massachusetts, are excellent in make and dye, and in good taste in the selection of the colouring. In cotton damasks and quilts, dyed and plain, the goods of Malcolm and Heskett, Paterson, New Jersey, are the only examples of American production of their kind. They are of fair make, and the usual character in design.

The cheapness of cotton in the United States, as a raw material, causes it to be used for many of the purposes for which flax and hemp are alone employed in Europe. The cotton sailcloth shown at the Great Exhibition of 1851 was an evidence of this. Therefore cotton is largely used in the manufacture of cordage and twine, and the examples shown in the Exhibition at New York are very excellent. The American Cordage Company, New York City, exhibit specimens of patent cordage, rope rigging, tow-lines, etc., of various sizes. These are remarkably clear in the strand, and are evidently made with great care. Cotton seine twine, and cotton lines for drift and other nets, are also exhibited, which show how largely cotton is employed in this branch of trade.

Fishing nets of cotton-twine are woven by Mr John McMullen, of Baltimore, on a loom of his own invention. These are not exhibited, but they are quite equal in every respect, probably superior, to hand-made nets.*

Cotton-wick for candles and lamps forms another item of the cotton trade in the United States, lamp-wicks being largely consumed. Mr A. Wortendyke, Godwinville, near Paterson, New Jersey, exhibits specimens of cotton-wick, counter-twist wick for patent machine mould candles, and chandler's wick. These are excellent examples of their kind.

The estimate formed of the present position of the cotton manufacture of the United States must be one of no merely apologetic character. All that has been attempted has been well done, not only in the results, but as regards the methods by which these results are attained; and if, as will undoubtedly strike the European observer, many of those developments of the cotton trade in its finer and more ornamental fabrics are as yet unattempted, it must be remembered that the useful fabrics, those absolutely necessary to the comfort of the great mass of the people, were those which would be first in demand, and most likely to remunerate the enterprise of the earlier manufacturers.

In the New York Exhibition a specimen of British calico is exhibited, which was purchased by the contributor, Mr Hagerson, at Boston, in 1813. Its width is 33½ inches, its price at that date 85 cents, about 4*s.* sterling, per yard. The same quality of cloth can now be purchased for 3½ cents, about 2*d.* sterling, per yard. An illustration like this proves how largely manufacturing skill, not only in Europe, but in America, must have progressed since that period, diffusing health and comfort, independence and happiness, by affording occupation to hundreds of thousands in both continents, whilst its influence has been felt in the social relations of life in a remarkable degree in the United States, as evidenced by the condition of the people in Lowell, Manchester, Lawrence, and other places of less concentration, but not of less skill, enterprise, and industry, or of lower tone as regards morals or social position.

Since my return to England, I have received through his Excellency, John F. Crampton, Her Majesty's Minister at Washington, two communications from Mr Henry Le Weeks, Hannahatchei, Stewart County, Georgia, making inquiry as to certain conditions in the growth and preparation of cotton, for the information of the planters of the Southern States. The questions asked appeared so important that, being desirous to obtain and transmit the best practical information, I communicated with Thomas Bazley, Esq., President of the Manchester Chamber of Commerce, and Henry Houldsworth, Esq., also of Manchester; and subsequently, at the suggestion of Mr Bazley, with Robert Hyde Greg,

---

* See Mr Whitworth's Report on this invention.

Esq., Norcliff, near Macclesfield. Each of these gentlemen furnished me with replies, to the questions transmitted, which I embodied in a reply to the communications of Mr Le Weeks.

A copy of that reply, and of the communications to which it is an answer, will be found appended. (See Appendix, No. 2.)

2

Textiles

Woollen and Worsted: Class XII

Next to cotton manufactures in extent and value, but scarcely less in importance, are the various branches of the manufacture of woollen and worsted goods now carried on in the United States. Many of these are of comparatively recent introduction, whilst others have been pursued for years under many disadvantages, both as regards the supply of raw material, and the skilled labour requisite to carry out the more perfect finish of the better class of goods. In many respects these difficulties have been in a great measure overcome, and the higher class of woollen goods produced in America will, in many respects, compare with similar fabrics manufactured in Europe. But it is more than questionable whether, as is frequently asserted, the wool of which these goods are manufactured is of American growth, unless in some special instances, or where the material has been carefully selected for the purpose. For though American wool is particularly well adapted to cassimeres and stuffs, from the length of its staple, it is not so well suited to the manufacture of broadcloths. This opinion is borne out by that of the most experienced buyers of this class of goods. There can be little doubt, however, that the position to which the manufacture of the higher quality of cloths has attained will stimulate the growth of the finer qualities of short staple wool. And when the success of the examples shown by the American growers at the Great Exhibition of 1851 is remembered, it must be quite evident that an accurate knowledge of the requirements of the woollen manufacture in its two great divisions, the one demanding a 'carding,' and the other a 'combing' staple for their perfection in fabrics, is alone needed to enable the growers of the United States to meet all the requirements of the home demand, at least. At present German, English, and Australian wools are imported for the manufacture of certain classes of goods; but, according to the report in the census of 1850, the largest proportion of imported wool comes from Buenos Ayres and the neighbouring States on the Rio de la Plata. This is of a coarse and cheap variety, costing from 6 cents to 8 cents, about 3¼*d.* to 4¼*d.* sterling, per pound. The same authority states that the imported wool amounted up to that date to about one-third of the whole raw material manufactured, but that whilst in 1850 the quantity of wool imported into the United States was 18,669,794 lbs., valued at $1,681,991 (about £400,000 sterling), a remarkable increase took place in the year 1850–51, far exceeding the importations of any previous year, being 32,548,693 lbs., and to the value of $3,800,000 (about £900,000 sterling). This would go to prove a great and sudden increase in the woollen manufactures, and this is borne out by the fact that many of the larger establishments were either commenced, or largely increased their make about this period, particularly in the coarser quality of carpets, and the manufacture of felted goods.

By the census returns of 1850, twenty-four of the thirty-one States of the Union, and the District of Columbia, had establishments engaged in some department of the woollen manufacture. The seven States in which this branch of industry had not been commenced were, South Carolina, Florida, Alabama, Mississippi, Louisiana, Arkansas, and California. The New England States had not so many establishments in operation as the two States of Pennsylvania and Virginia, and only five more than those of New York and Ohio. Thus it will be seen that, whilst the cotton manufacture is located more exclusively in the Eastern States, the woollen manufacture is extended in almost equal proportions over the whole of the Middle States, and extends itself into the western regions and towards the south. The extent of the woollen manufactures of Massachusetts, however, is seen in the fact, that whilst in the 380 mills of Pennsylvania the consumption of wool is 7,560,379 lbs., employing 3,490 males and 2,236 females, producing 10,099,234 yards of cloth and 1,941,621 lbs. of yarn, of the annual value of $5,321,866 (about £1,300,000 sterling); 119 establishments in the first-named State consumed 22,229,952 lbs. of wool, employ 6,167 males and 4,963 females, and produce 25,865,658 yards of cloth and 749,550 lbs. of yarn, of the annual value of $12,770,565 (about £3,000,000 sterling). The difference of the modes of manufacture in the two States above-named, as illustrated by the cotton trade, is here shown again in the fact, that a very large proportion of the produce of the woollen mills of Pennsylvania is yarn only, a large amount of this being consumed in home manufacture for domestic use, or in the weaving of mixed goods and carpets by hand, and this, too, in addition to the home-spun woollen yarns mentioned as being worked up with the cotton yarns produced for that purpose. The 130 establishments in Ohio, as also 121 in Virginia, 25 in Kentucky, and 33 in Indiana, would appear to manufacture the greater portion of the yarns spun therein; it is probable, therefore, that the yarns of Pennsylvania are largely used for the supply of the West in the materials for home weaving. After all, however, this department of industry is becoming daily more and more exceptional; but it is interesting as illustrating the early condition of a new country in its efforts to supply its own wants, in the absence of that larger development of manufacturing means and appliances which capital, skill, and a large and ever-increasing demand can alone establish on a firm and enduring basis.

The total number of persons employed in the various establishments for the manufacture of woollen goods in the United States in 1850 was 22,678 males and 16,574 females.

Wages, as in the cotton trade, vary very much in different localities, but the average appears to be slightly lower than in that department of industry in the New England States, whilst it is higher in Pennsylvania. In the Western States the increased average is considerable.

Many of the establishments of New England being very extensive, are conducted upon a similar plan to the larger cotton mills. At Lowell, the Middlesex Company's mills present a good type of this higher class. The number of operatives employed is, males 575, females 730. The mills run 16,340 spindles, with 75 looms for broadcloths, and 328 looms for cassimeres. The weekly production is 24,000 yards of cassimeres, and 3,000 yards of broadcloths. Cheap broadcloths are produced here, with a cotton warp, presenting a very good surface. These sell at $1½ (about 7*s.* sterling) per yard. Doeskins are the chief make, and, with beaver cloths and cassimeres, are manufactured entirely of native wool. The printed trouserings manufactured by this company are, in common with similar goods produced by others, of a very effective character. The colours are clear, and generally well selected, and the style and finish much better than would be generally supposed by those whose ideas of fancy cassimeres are set upon woven patterns only, and they appear to suit a large class of buyers, particularly for the western markets, as being at once low in price and showy in character; and the colours are said to wear better than European opinions of such modes of production would give them credit for. The general make of goods by the Middlesex Company is illustrated in the Exhibition at New York.

The Bay State Mills, Lawrence, Massachusetts, may be said to offer the best type of a large woollen establishment, manufacturing a great variety of fabrics. These mills form a model establishment, so to speak, of its class; alike as regards extent, construction, machinery, arrangement and internal economy. It has been therefore thought advisable to extract an illustrated description of it from the Report of the Sanitary Commission of Massachusetts, 1850, given in the Appendix. (See Appendix, No. 3.)

There are 1,000 males and 1,200 females employed at present in the Bay State Mills, working 98 sets (3 to a set) of carding engines, and 700 looms, together with the dye works and printing shops for flannels and carpets. In the week ending 7th May last, the production of this establishment amounted to 40,898 yards plain flannels, 3,962 yards twilled flannels, 10,159 yards fancy cassimeres, 6,770 yards satinettes, 1,030 yards broadcloths, 568 yards beaver-cloths, 1,703 yards felted carpets, 2,752 yards felted beavers, 1,540 yards felted linings, 1,464 long shawls, all wool, and 5,970 square shawls.

The flannels are all either dyed in fancy colours or printed. The latter is chiefly block work, although cylinders are used for some styles. These fancy coloured flannels are extensively used in the United States for children's clothing, and make up into elegant-looking, yet low-priced, articles of dress. The cassimeres are of good make, and the pantaloon satinettes (cotton warp) are excellent of their kind in colour and finish.

The felted fabrics manufactured are noticeable as being produced by a different method to that adopted by any other establishment.* These are excellent goods, both as regards the firmness and, to a certain extent, elasticity of their texture; and the imitation 'petershams' are superior to even the general run of common cloths for overcoats. In England there still exists, and perhaps not without cause, a strong prejudice against all felted fabrics for clothing. It would appear, however, that they are largely used in the United States in making up cheap clothing. There are some peculiarities of the people, however, as affecting consumption, which will be considered in due course.

The felted lining cloth, composed of gauze and a comparatively small quantity of wool, is a new and useful fabric. The wool is felted down upon the gauze as a back, and the substance obtained by this means is much greater than the quantity of material employed would warrant any one to expect. When the back or gauze surface is concealed by the use of the fabric as a lining, the result, as regards appearance, is very satisfactory, and it is said to wear well, as the two materials are fairly united by the felting process.

The manufacture of woollen shawls was first commenced in the United States at this establishment, and is the only one in which they are yet manufactured, to any extent, at least. They consist of gala plaids, manufactured chiefly of American wools. The general styles are well selected, and the dyes and finish well managed. The fringing machine is a useful and ingenious contrivance, by which the threads constituting the fringe of each shawl are twisted to a proper tension. It does the work usually performed by hand, and accomplishes as much as 10 females.

The printed shawls prepared at the Bay State Mills are of the usual character as regards fabric; the patterns being chiefly oriental in style. They are all printed, dyed, and finished before they leave the establishment.

There is an excellent display of the shawls manufactured by the Bay State Mills in the Exhibition at New York.

The printed felted carpets or druggets made here will be duly noticed in their proper class (XIX).

On visiting the Manhan Manufacturing Company's Works, Waterbury, Connecticut, I found the more general felting process at work for the production of felted cloths of good character, chiefly for overcoats. An excellent imitation of 'petersham' is also made here, the curled surface being well wrought and finished. Felted carpet fabrics are also manufactured, but not printed, in this establishment. At Clinton, Massachusetts, the Clinton Company, in addition to the manufacture of coach lace, to be hereafter noticed, employ 100 looms in making pantaloon

---

* See Mr Whitworth's Report.

stuffs, being mixtures of cotton and wool. The looms are similar to the gingham looms mentioned in Class XI, and the fabric produced is a well-looking and cheap article, for which there is a large sale.

In Maryland there is a considerable development of the woollen manufacture; and, in this State, the highest as well as the lowest class of goods manufactured may be said to be produced. Baltimore is the centre of this trade, and a coarse quality of fulled linseys, six quarters wide, form a species of domestic manufacture of the district 50 miles round that city. These linseys are chiefly manufactured for negro clothing, and are sold to the planters at the South for the use of their slaves. Considerable quantities, too, are sold in the West, as far as the Missouri and the Rocky Mountains, being used there for the clothing of labourers and backwoodsmen.

In the State Penitentiary of Maryland, at Baltimore, coarse garment plaids, of good make, are woven by hand by the prisoners. The cotton and woollen yarns are also spun and dyed by them. The colours are generally clear and brilliant, being selected with more discrimination, as regards taste, than is generally found in the coarse class of goods either in America or in Europe. The fabric is an useful one, and is in such demand in the Southern and Western States, that manufacturers are induced to get up imitations in competition with the article as produced in the Penitentiary.

The produce of the Penitentiary looms is about 500 pieces per week. The return promised by the Warden, of the number of looms, carding engines, and spindles, together with the number of persons employed in this department, has not, however, reached me.*

---

* The profitable employment of prisoners is a rule in all the State penitentiaries of the United States.

The labour is farmed out to a contractor, who finds the proper tools, machinery, and materials of labour. In the Maryland Penitentiary, in addition to the spinning and dyeing of yarns, and weaving the fabrics named, rag carpets are woven in looms adapted to the purpose, the weft, or filling, being composed of white and coloured rags twisted or spun into a strong cord or yarn. There are also extensive workshops, in which wooden pails and buckets are made; chiefly by machinery directed by the prisoners, and the manufacture of corn brooms, so largely used for domestic purposes in the United States, is also extensively carried on. A large nail forge, chiefly for the manufacture of railway pins by a mechanical contrivance, is another branch of industry pursued by the prisoners; and the dressing of bristles, for the use of brush makers, is about to be added to the employments already followed.

The prison authorities have nothing to do with the direction of the labour, that being in the hands of the respective contractors. They enforce the discipline of the prison, and obedience to the direction of those employed by the contractors to superintend the work.

The earnings, or returns to the State, of even the lowest class of prisoners in the Maryland Penitentiary is 62 cents (about 2*s.* 10*d.* sterling) per day. Each prisoner is tasked to earn a certain amount, according to his employment or presumed capacity. All above the amount is carried to his account,

At the establishment of Messrs Wethered and Brothers, Baltimore, the manufacture of the higher qualities of doeskins and fancy cassimeres is very successfully carried on, and the examples contributed by them to the Exhibition at New York are a most satisfactory proof that, for all practical purposes, the American cloths are equal in quality and finish to similar goods produced in Europe. This is confirmed by the cassimeres, doeskins, and satinettes manufactured by Jacob T. Seagreave and Company, Burrellsville, Rhode Island; Platner and Smith, Lee, Massachusetts; Slater and Sons, and the Vassalboro' Manufacturing Company, Maine, exhibited at New York by their respective agents. The latter company exhibit three-quarter cassimeres, made of Silician wool (4,000 warp), of an exquisite texture and beauty of finish. In fact, the contributions to this department of the Exhibition at New York, though not very numerous or extensive, go to prove very distinctly that, in the production of the finer qualities of woollen goods, great advances have been made in the United States. The efforts after novelty, however, do not always lead to successful results in European eyes, and the fancy trouserings are often extremely *outré* in pattern.

The Jacquard has been lately introduced into use in the production of figure patterns in fancy cassimeres. There are good specimens of these manufactured by the Melville Manufacturing Company, Melville, Massachusetts. The pattern being woven, the goods are slightly fulled afterwards, and the result is satisfactory when too defined a figure has not been attempted. In small patterns the fulling process breaks down the forms, and thus gives, when finished, an agreeable variety to the surface.

The satinettes manufactured by the Perkins Company, Akron, Ohio, speak well for the progress of the woollen trade in the west.

The perfection of dyeing and finish to which felted fabrics have been brought is well illustrated by the goods manufactured by the Union Manufacturing Company, and Lonnsburg, Bissell, and Company, Connecticut, and the Winipank Mills, in the same State. It is scarcely possible to believe that so simple a process as that of laying sheets of wool in succession, and, by mere vibration, working them together, could produce such fabrics as the best kind of American felted cloths certainly are.

The blankets and flannels exhibited at New York are fairly illustrative of the excellence which the manufacture of this class of goods has obtained in various parts of the United States.

---

and paid to him when he leaves the prison, or he is permitted to purchase such books as the Warden approves of. This is the mode in which many prisoners for life prefer expending their surplus earnings, and many of them have excellent collections of books.

In the New York State Penitentiary at Sing-Sing, cutlery of a very superior quality is said to be manufactured, but time did not permit of an inspection of the industrial operations in that prison. G. W.

3

Textiles

Silk: Class XIII

The manufacture of silks is comparatively exceptional in the United States, and notwithstanding many vigorous attempts, not only to establish the manufacture, but to raise the raw material on an extensive scale, little has resulted except in the production of sewing and fringe silks and twist.

The growth of silk in any considerable quantity never appears to have been fairly realised, although in the State of Connecticut, where the greatest number of establishments are still carried on for the manufacture of sewing silks, it rose from 2,430 lbs. weight in 1827 to 176,210 lbs. in 1844, being an approach to one half of the whole amount produced in the United States in that year, viz., 396,790 lbs. The decrease since that date to 1850 has been so great, that it is quite evident comparatively little is now doing in the culture of silk, since only 10,843 lbs. is given as the production of the latter year.

Taking a deep interest in this department of industry, and the probable value of silk as a collateral crop to the farmer in a country in which there appeared no difficulty whatever in the raising of the silkworm, I prosecuted my inquiries as far as time would permit, hoping to be enabled to reconcile the very contradictory opinions abroad on this question. It is acknowledged on all hands that many portions of the United States are well adapted to the growth of the mulberry tree, and the healthy development of the silkworm; but it was also quite as evident that the active habits of the people were not adapted to the work of looking after the cocooneries, or managing filatures for reeling the silk. This work, however, was attempted to be carried out in large establishments during the period intervening between 1831 and 1845. The '*Morus multicaulis*' speculations arising out of the mania for mulberry trees, and the failure of the cultivation of raw silk on a scale not at all suited to the peculiarities of its growth and culture, produced a reaction in the public mind in the latter year; and since that period the growth of silk has been comparatively neglected, even where it might be successfully carried on in the manner usual in France and Italy, if the habits of the agricultural classes were not, more or less, opposed to a somewhat sedentary occupation, requiring, for a certain period of the year, incessant attention. It was this which doubtlessly suggested large establishments for the care of the silk-worm, but when disease attacked any portion, the whole perished. It was found too that the '*Morus multicaulis*' was comparitively worthless, and that the native and best varieties of the white or Italian mulberry produced the only silk really worth growing.

The States of Virginia, Ohio, Kentucky, and Tennessee are generally acknowledged to be admirably adapted for the growth of silk, and excellent examples of the produce of each State have been, from time to time, brought forward, and there can be little doubt that, under a proper system of domestic culture, with suitable centres for receiving the cocoons and reeling the silk, an excellent material would be the result.

At Rapp's Colony, or 'Community of Economy,' on the Ohio river, Pennsylvania, the annual produce of cocoons was formerly 3,000 lbs. weight, which was reeled and partly manufactured on the spot. Mr R. L. Baker, the present director of this establishment, now states, however, that the growth of silk has been abandoned, having ceased to be at all profitable.

At Wheeling, Virginia, Mr J. W. Gill still continues to collect cocoons and reeled silk from the States above named, and this is manufactured, with the addition of foreign silk, into a variety of goods, but not to any extent.

American silk is also manufactured on a small scale, and, with the addition of the imported material, at Newport, Kentucky, on the opposite side of the Ohio to Cincinnati. Here are five or six looms, and a little throwing and winding machinery, the property of Messrs Jones and Wilson. The manufacture is carried on by the latter, an English weaver from Macclesfield. Specimens of the goods produced are displayed in the Exhibition at New York, and consist chiefly of neck-ties, vestings, and dress fabrics of very fair quality and finish, and excellent grey goods for printing. Mr Wilson, a practical weaver, understanding his business and desirous to improve it, is thoroughly impressed with the capabilities of the district around for the perfect growth of silk equal to the best Italian, but confesses that he sees little prospect of the culture being so managed as to be of much importance for many years to come. Native reeled silk costs $5 to $5·50 per pound. Imported silk can, however, be purchased for much less, notwithstanding a duty of 15 per cent. Thus this latter fiscal restriction is of no practical value, but, on the contrary, adds to the difficulty of the manufacturer, the 25 per cent duty on manufactured goods giving him little or no advantage, for the 7 per cent waste, which he cannot work up in the United States, has to be sent to England at a great loss, and no drawback is allowed on the raw material thus exported by the original importer.

There are several manufactories of sewing silks at Mansfield, Connecticut, and a few other parts of New England. In 1845 there were four mills in Norfolk County, Massachusetts, consuming 12,900 lbs. of silk per annum; in Hampshire County three mills, consuming 6,100 lbs.; and in Middlesex County one mill, consuming 3,505 lbs. Probably the largest establishment of the [silk industry] in the United States is that of Mr John Ryle, Paterson, New Jersey. Sewing silks, floss silks, and silks for fringes and gimps, constitute the production of this mill, which is a well conducted and compact concern, near the Passaic Fall, on the river of that name. The consumption of material is from 1,000 lbs to 1,200 lbs. weight of silk per week. The articles manufactured are excellent of their kind.

At Philadelphia a branch of a firm at Macclesfield, Cheshire, is carried on for the manufacture also of sewing silks. Mr B. Hooley, the proprietor, tried the production of ribbons, but did not succeed in making them at such a price as would command a market.

Ribbons, however, are manufactured by Messrs Horstmann and Sons, of Philadelphia, but rather as a means of keeping their extensive establishment for the manufacture of silk braiding and fringes in full work, at periods when the latter are not in sufficient demand, and the braiding machines are so constructed that, by the substitution of the jacquard, ribbons can be woven upon them. This concern will be alluded to under another head (Class XX).

At West Newton, Massachusetts, Messrs Plymton, Stevenson and Company, of Boston, have a small establishment for the manufacture of silk-braid and ribbons. The braids are generally tasteful in design and well made; the ribbons, however, are deficient in surface and show a want of experience in the weavers. Here, also, the jacquard machine is applied to the braid-looms when required for ribbons. There are not more than twenty of these.

The most important enterprise in the silk trade now prosecuting in the United States is that of the application of the power loom to the weaving of brocatelles at the Eagle Mills, Seymour, Connecticut. This establishment has only recently commenced work. The machinery is beautifully constructed, and works with great accuracy. The looms carry revolving shuttle-boxes, and are all made upon the premises. Ten looms are in full work, five preparing, and ten more in the course of construction. The goods manufactured are 48 inches in width, of firm fabric, but defective in the surface. This arises more from the evident inexperience of the weavers, than any defect in the mode of production. These weavers are females, selected from carpet manufactories, and, of course, lack the practised eye of the silk weaver in detecting defects. Cotton is chiefly thrown in for the backs, though linen is sometimes used. The designs are French in character and superior to those generally produced in such fabrics. The price varies from \$3 to \$3½–that is to say, from 14*s*. to 19*s*. sterling per yard; but it must be taken into account that the width is twice that of similar goods as generally manufactured in England.

A very excellent adaptation and improvement upon the French and

English 'reading off' and card-cutting keyboard machine is used in this establishment, the invention of the superintendent.

This would appear to be the first attempt to fairly introduce the manufacture of the higher class of silk goods into the United States; and there is every probability that it will be a successful one, from the admirable character of the arrangements and accuracy of the machinery, when directed by increased experience.

The goods intended for exhibition by this Company had not been arranged at the Exhibition at New York at the period of my departure, and the display in this class was very scanty.

Examples of upholstery silk damasks, brocades, and silks for ecclesiastical decorations, the production of European weavers, manufactured by the exhibitor, Mr Jacob Neustader, of New York, presented some points of excellence, and a few examples of sewing-silk were illustrative of this branch of the silk trade.

From these remarks it will be readily understood that this department of industry is comparatively in its infancy, and that, in the growth and preparation of the raw material, the United States have receded, and not advanced, since 1844.

4

Textiles

Flax and Hemp: Class XIV

The extensive use of cotton for many of those fabrics which in Europe are manufactured of flax, and even of hemp, such as sailcloth, sacking, twine and cord, has evidently not been favourable to the employment of the latter as materials. The manufactures of flax seem at present to be altogether exceptional, whilst hemp is employed to a considerable extent in most of the large cities of the United States in the production of ropes and cables, tarred and untarred, and of these, together with a few specimens of sailcloth and hemp carpeting, there are some creditable examples in the New York Exhibition. The growth of hemp in the United States, as shown by the census of 1850, is returned at 35,093 tons, but there is some doubt as to the accuracy of the return, and no means of comparison with former statements, as the gross produce of flax and hemp were given together in 1840. Manilla and Sisall hemps are imported, but the hemps of the States of Missouri and Virginia are largely used. Formerly New Jersey exported a considerable quantity of hemp.

The cables produced in the navy-yards of the several cities, where such establishments are kept up by the Government, are of a very high character.

In the manufacture of flax, improvements in the preparation of the fibre appears to engage more attention than any other point. A machine has been recently invented which seems likely to produce some satisfactory results as regards the preparation of flax-cotton, by mechanical rather than a chemical operation.

Flax is largely grown in some of the States for the seed, for oil making, and the straw is at present thrown away. In some parts of Pennsylvania and Ohio, it can be bought for from $4 to $5 (about 19*s*. to 24*s*. sterling) per ton, delivered to the consumer, and can often be obtained for the mere cost of carting away, after the separation of the seed. Flax-seed is raised at from 15 to 20 bushels per acre, usually 16 to 18 bushels. This sells at $1 per bushel, and, if more cultivated, the price would perhaps be reduced to 40 cents.

In the mechanical process above named, the straw is put through a machine and crushed by rollers. By an arrangement for a difference of speed in these rollers, the fibre is torn to pieces and reduced to a cotton wool, the woody fibre falling out in the process.

The flax, thus prepared, can be sold without further preparation to paper makers at 2 cents (about 1¼*d.* sterling) per pound. A chemical process is applied simply to bleach, after the separation. This statement as to the growth and price of flax, and the results of the mechanical operation of the machine, is given on the authority of Mr Kennedy, of the firm of Messrs Kennedy, Childs and Company, Penn Cotton Mill, Pittsburg, Pennsylvania. The samples of unbleached flax-cotton, prepared by this process, appeared, on examination, to be of sufficient length of fibre for spinning on the ordinary cotton machinery, and only to require finally freeing from the smaller particles of woody substance. For this purpose, the new carding-machine, in operation at the Stark Mills, Manchester, New Hampshire, alluded to in Class XI, seemed most especially well adapted from the thorough manner in which it clears the material submitted to its action.

Mixed Fabrics: Class XV

The mixed fabrics manufactured in America have been treated of in Class XII, alike for convenience as from the fact that they chiefly belong to that class of manufacture, though composed of other materials as well as wool.

The only fabrics calling for notice here are mouselline de laine and barrage, which are manufactured by a few firms in New England, the most extensive being the Manchester Print Works Company, Manchester, New Hampshire, and the Hamilton Woollen Company, Southbridge, Massachusetts. The produce of the Manchester Company's mill, for the week ending June 4th, 1853, was 422,389 plain de laines, 7·44 yards to the pound. The printing and dyeing of these goods will be considered in due course in Class XVIII.

The grey mousellines de laine manufactured by the companies above named are firm fabrics, of a fair class, and are woven by power. The Manchester Company employ 400 looms in the manufacture of varied qualities of de laine and barrage cloths; the coarser kinds of the former being woven of No. 37 cotton warp, and No. 40 wool weft or filling. In the finer qualities, No. 70 cotton warp, spun on the Potter mule, and No. 50 wool weft, spun on the Smith mule, are used.

The contributions in this class to the Exhibition at New York are very limited. The specimens of silk and wool flannels are good, and the fabric of the printed shawls of the usual character; the patterns of the latter being of the oriental style.

5

Leather (Exclusive of Tanneries), Furs, etc.

Class XVI

The manufacture of leather in its various forms and qualities, and its making up into articles of use, has grown into an important branch of the industry of the United States. Differing in no important point with the methods used, or the purposes to which it is applied in Europe, the manufacture of leather into saddlery, harness, portmanteaus, etc., is carried on in almost every town of any importance, and the trade is conducted on a similar principle to that usually adopted in England. Occasionally large manufactories of portmanteaus and harness are met with, but these are exceptional. At Philadelphia the manufacture of articles in leather is followed to a very considerable extent, and the harness and saddlery produced in this city is of a very superior character, alike as regards workmanship and beauty in design. Nor are the productions of New York and Boston less noticeable for the same qualities.

In portmanteaus, valises, and other articles for use in travelling, considerable ingenuity in arrangement is often shown. The workmanship and finish of the best class of goods are unexceptionable, and, even in the cheaper and lower qualities, the style in appearance is a matter of much consideration, and displays a decided advance, in point of taste, upon the unsightly character of the cheaper kind of travelling conveniences in England.

There are very few examples of portmanteaus contributed to the Exhibition at New York. In harness and saddlery, however, there are specimens showing great skill in execution, as also fitness and beauty in construction and decoration. The whips are generally of excellent workmanship, but the mountings are often in the absurd taste which at present prevails in Europe.

In furs and feathers, those shown, though of American production, are more or less of European manufacture, and in artificial hair there are some clever contrivances.

Specimens of furs from the State of Minnesota are also exhibited. These chiefly consist of marten, otter, fisher, mink, and beaver, together with buffalo robes and bear-skins, and are generally well preserved and of fine surface. Minnesota is the great fur-producing State of the Union, and the trade of 1852 is said to have reached nearly $700,000 (about £170,000 sterling).

The specimens of various kinds of leather contributed to the Exhibition at New York, show in a most favourable light the ability of the Americans as tanners and finishers of leather, in nearly all its most useful forms. As Professor Wilson reports on tanneries, it is requisite only to remark, that in fancy-varnished leathers, and in morocco, the surface, finish, and elasticity are generally of a high character.

In connection with the material of morocco leather, it may be as well to notice here an important invention already alluded to in Class XI, under the name of Leather Cloth. This is a most perfect imitation of morocco, by the application of a preparation of caoutchouc, or gutta percha, to the surface of plain woven or twilled cotton cloth. The surface is corrugated in imitation of morocco, and is coloured and varnished so as to present all the external appearance of that kind of leather. The elasticity is perfect, showing no tendency to crack, and so far as time has at present tested its durability, this appears to be satisfactory. Its cost is less than one-third that of morocco, and from the width of the cloth, it cuts to much greater advantage in the covering of articles of furniture, for which, as well as carriage linings, particularly railway carriages, it is coming largely into use. Time did not permit of a visit to the manufactory of the patentees, Messrs J. R. and C. P. Crockett, at Newark, New Jersey, but specimens of the fabric have come to hand recently, which fully confirm the opinions above given; but it is now stated, that after a trial it has not been found to answer for the lining-band of hats, to which it was at first applied, as the colouring matter is decomposed by wear, from the surface not possessing the absorbent quality of leather.

As the trade in raw hides and the manufacture of leather must, from the extent of the latter, form a most important item in the productions of the United States, it is to be regretted that, up to the present time, the census returns for 1850, in this department of industry, have not been made public.

6

Paper, Printing Types, Bookbinding, etc.

Class XVII

The manufacture of paper is carried on to a great extent in most of the Atlantic States, but unfortunately the census returns, so far as published in a collective form, do not contain the statistics of this important branch of trade, which must have increased to a great extent within the last few years, as the quantity of printing papers alone consumed in the United States must be immense.

The whole production is by machinery, not more than one or two houses making hand-made papers. The machines are all adaptations and improved applications of Fourdrinier's invention, modified to suit the wants or ideas of the manufacturer, who would appear to be the stationer also, as far as the making up of the paper into reams, quires, and packets is concerned, and its distribution to the retail trader, even in the ornamental form in which it reaches the public. In this respect the general commercial dealings in paper are very different to those of England, and the function of the wholesale stationer is an exceptional one in the United States.

The State of Massachusetts would appear to be largely engaged in the manufacture of paper. At Lee, Berkshire County, in that State, there are 19 paper mills, employing a capital of about $200,000 (about £50,000 sterling). In Norfolk County, Massachusetts, there are 17 mills, and in Worcester County, 15 mills, employing a capital of £100,000 sterling in this manufacture. In 1845, up to which date the last general statistical information on the State of Massachusetts is published, there were 89 paper mills, consuming 12,886 tons of materials, and making 4,765 tons, giving 607,175 reams of paper per annum, the value of which was $1,750,373 (about £430,000 sterling), and employing 1,369 operatives; and this certainly gives no exaggerated view of the general position of the paper trade in nearly all the New England States, New York, New Jersey, and Pennsylvania, at the present date.

The materials used are chiefly raw cotton and mill waste. Linen rags are imported from Europe, but the principal consumption would appear to be cotton, either as above named or in rags. The general character of the printing paper is of a low quality, with a very small amount of dressing or size. In writing papers the make is quite equal to the general run of European papers, but the finish is not always so perfect. It is stated, however, that whilst the Americans try to imitate the English finish, the latter are trying to imitate that of makers of the United States.

A considerable number of paper mills are engaged in the manufacture of wall papers only, for printing as hangings. These vary in quality, but are chiefly of the coarser and cheaper kinds; and as there is a constantly increasing demand in this branch of the paper trade, the manufacture extends rapidly.

The Ivanhoe Mills, Paterson, New Jersey, has the reputation of being the most complete establishment of its kind in the States. This manufactory is erected on the Canal, by which the head water of the Passaic Fall, on the Passaic river, is converted into mill power.

The buildings are admirably arranged, and whilst every department, from that for the reception and picking of the rags and cotton waste, to the making up of the paper, is complete within itself, the whole is so connected as to be perfectly progressive; thus economising labour, and preserving the orderly arrangement and cleanliness of the whole establishment.

Book papers only are manufactured, and two sets of operatives are employed, one for night and the other for day; so that the works always go on, except on the Sabbath.

There are ten vats or engines for grinding the materials for making the paper. These are larger than those generally used in England, and this appears to be the rule throughout the United States. Each vat contains pulp sufficient for making 180 lbs. weight of paper. In England they usually contain 120 lbs. only. The bleaching of the material is effected in a number of airtight chambers of large size, built of stone, with iron doors, the whole so constructed that no chlorine can escape.

The paper machine at work in the Ivanhoe Mills is one of great accuracy and beauty, wood framing being avoided as much as possible in the construction, and bright brass substituted. The whole has a clean and elegant appearance, with perfect freedom from rust; whilst the action is firm, smooth, and steady.

If the inspection of this establishment was satisfactory as to the mechanical and economic arrangements, and the excellence of the book-papers produced, a visit to the Carew Manufacturing Company's Paper Mill, South Hadley Falls, on the Connecticut river, Massachusetts, showed that the older mills were quite equal to the production of the first qualities of writing papers, though the internal arrangements might be less consecutive and complete than those of the Ivanhoe Mills. The Carew Company's papers are firm, well-finished examples of the best class of writing papers, pure in colour, and well glazed. The pressing is chiefly effected with steam-heated cylinders covered with paper, through which the sheets pass in quick succession, and a very good surface is obtained.

The cutting of the paper into sheets is effected by a very simple but ingenious contrivance. A knife is set in the wheel round which the paper revolves as it leaves the machine; a flat board, swinging on a pivot, comes in contact with the edge of the knife at stated intervals, thus measuring off the sheet and cutting it at the same operation. The sheets fall into a receptacle below the revolving cutter, and are removed at stated intervals as they accumulate.

The water used in the Carew Mill is not from the Connecticut river, from which the power is obtained, but from an artesian well sunk in the rock on which the mill is built. The water is of great purity, and the supply is most abundant.

It has been already stated that the paper manufactures of the United States fulfil the functions of the wholesale stationers of England. Of this the Carew Company are an example. In the above mill the paper is made, cut into sheets, pressed, the edges cut and gilt, ruled by machinery when required, and made up into reams, quires, and packets, in such covers as the retail stationer to whom it is supplied may require; the styles of making up being more or less imitative of those adopted in Europe.

The wrapping papers of the United States are of great variety, both as regards quality and colour.

Envelopes are usually made of a light buff-coloured paper; but the chief portion of the fancy stationery at present used would appear to be imported from Europe.

The printing operations are extensive and well conducted, particularly in bookwork. The printing of newspapers alone forms a large item in the industry of the country. In the New England States, according to the Abstract of the Census of 1850, there were 424 newspapers; in the Middle States, 876; in the Southern States, 716; and in the Western States, 784; and the following table shows the daily, weekly, and monthly issues and aggregate circulation, as given by the above authority:

Number of newspapers published in the United States in 1850

	Number	*Circulation*	*Number of copies printed annually*
Dailies	350	750,000	235,000,000
Tri-weeklies	150	75,000	11,700,000
Semi-weeklies	125	80,000	8,320,000
Weeklies	2,000	2,875,000	149,500,000
Semi-monthlies	50	300,000	7,200,000
Monthlies	100	900,000	10,800,000
Quarterlies	25	29,000	80,000
	2,800	5,009,000	422,600,000

With an educated people, taking a vital interest in all public questions, the newspaper press is likely to increase even in a greater ratio than it has done during the past decade. The number of German emigrants has caused the establishment of newspapers for their use, and at Cincinnati alone there are four daily newspapers published in the German language.

Type founding is carried on to a great extent at Boston, New York, and Philadelphia, and there are single establishments in several other of the large cities. The whole of the type used in the United States, besides a large quantity exported to the British provinces and the various States of South America, is produced in these foundries, and the statements made in the Report of the Jury for Class XVII in the Great Exhibition of 1851, appear to be so far correct, that a reference thereto will be sufficient for all practical purposes here;* and the same may be said with reference to the sketch of the progress of the paper manufacture, given in the same Report.†

At the type foundry of John K. Rogers and Company, Boston, Massachusetts, machines for casting the smaller bodies of type are in regular use. A pump is used for forcing the melted metal into the mould, and a workman turns out 90 brevier type per minute, and smaller kinds at a more rapid speed. This machine, or an adaptation of it, is in general use in the type foundries of the United States, and some few have been lately exported to England.

The larger kinds of type are still cast by hand in the usual way.

Ornamental type is not manufactured to any great extent; the forms, however, of the ordinary bodies are neat and elegant, with a clearly-cut edge. The metal used is an alloy of lead, tin, and antimony, each preponderating according to the type required; 75 per cent of lead, however, is stated as the average of that metal.

The use of gutta percha for the purpose of stereotype appears to have been extensively experimented upon in the United States; and at the Smithsonian Institute, Washington, a method invented by Mr Josiah Warner, of Indiana, has been adopted for the purpose of realizing a plan suggested by Professor Jewett for stereotyping catalogues by means of separate titles. The titles of the books to form the catalogue being set up in any convenient number, a matrix is made therefrom, and a stereotype plate cast in gutta percha. This is sawn into the number of titles of which it is composed, and the alphabetising is accomplished by the simple assortment and arrangement of those titles, which are fixed together in the requisite pages. By this means the books added to any library during the year may be inserted in their proper places, and an annual catalogue published at a comparatively small cost, containing all the recent additions to the library.

It must be evident that the gutta percha plates, thus divided into titles, might be used for the formation of matrices from which to cast metal titles, if such are desired. All the practical difficulties in the way of the realization of this plan have been, it is now believed, overcome, and the Smithsonian Catalogue may be shortly expected to illustrate the practical

* Jury Reports, 8vo. edition, page 410. † Idem, page 443.

benefits to be derived by public libraries in the adoption of the plan above described.

The few examples of type sent to the Exhibition are good, being generally of pure and tasteful forms, and remarkably clean in the casting.

There are specimens of combination type, exhibited by the manufacturer, Mr John H. Tobitt, of New York. There is generally amongst practical printers a great antipathy to change the form and arrangement of existing 'cases;' and, however plausible a plan of combination types, in which a word or portion of a word can be 'set up' at a single operation, may be, the single 'body' has still the advantage in one important point, -that if damaged, a letter only is lost, instead of a whole word, if but a single letter of which it is composed is injured.

The specimens of bookbinding exhibited do not call for any special remarks. They are of the usual character, so far as the external appearance is concerned, and the style of gilding and the taste displayed in the decorations are altogether based on European modes: a redundancy of unnecessary ornamentation being the leading feature, as in England.

In commercial account books the workmanship and finish are generally excellent.

Of stationery and paper the display is very limited, and the show cards and labels, exhibited as specimens of ornamental printing, are of average merit.

7

Printing and Dyeing

Class XVIII

In the operations of printing and dyeing the fabrics manufactured of cotton, wool, and a combination of the two, the manufacturers of the United States have been largely aided by the emigration of artizans from Europe, who, taking with them the experience gained by the almost unremitting attention which has been paid to these important departments of industry during the past half century in England, France, and Germany, have been enabled to establish on a firm foundation the commencement, at least, of an invaluable adjunct to the manufacturing processes in woollen and cotton goods, now so successfully carried on for the supply of the demands of the American population.

The introduction of calico printing has been one of no ordinary difficulty, consequently the results have been of slow growth, and accompanied by many failures. It would appear now, however, to be fairly established; and although dependent in a greater degree probably than in almost any other branch of industry upon the imported skill of British workmen, either as designers, engravers, or printers, yet the tact of the American is nowhere seen to greater advantage than in the steady and skilful devotion of his energies, both manufacturing and commercial, to the development and progression of a trade requiring more than ordinary skill to conduct with pecuniary profit. The fluctuations of fashions, the uncertainty of the style likely to prevail in the market, from the influence exercised by European goods, - the adaptation in colour and quality to an almost endless variety of market, - all contribute to place the beginner at no slight disadvantage; and this applies to a nation as well as to an individual.

No means present themselves for obtaining a statement of the quantity of cotton cloth printed in the United States. The two largest concerns of this class, however, are the Merrimack Print Works, belonging to the Merrimack Manufacturing Company, Lowell, Massachusetts, and that of Messrs Jacob Dunnell and Company, Pawtucket, near Providence, Rhode Island. The latter was the first establishment of the kind in the United States, and the production of the two concerns amounts to 30,000,000 yards per annum.

Messrs Dunnell's Print Works are erected on the banks of the Pawtucket River, and the internal arrangements are excellent as regards the economy of human labour, and the scientific application of the best known and thoroughly established modes of calico printing.

In cotton both madder and steam prints are produced; jaconettes and cambrics, under the generic name of lawns, form a considerable portion of the madder work. Mousellines de laine are also printed in considerable quantities when they are in demand; but latterly this firm has found that steam prints or all wool (de laine) fabrics have been preferred by the class of buyers whom they supply. Steam prints, which have been somewhat out of favour, were again in demand in certain markets, chiefly in the Southern States, and these, though all within the boundary of the Federal Union, require special styles to suit them. Showy patterns, large in form, and strong in contrasts of colour, are those chiefly sold in the South and West. In the Middle States, especially in New England, smaller patterns, more subdued in colour, and approaching to the best character of goods used in England, are required. In fact, in these latter States, the neat madder prints find their chief market.

The print shop at Messrs Dunnell's is very lofty and well arranged, and contains ten machines, all of which are usually at work. In steam prints as many as ten colours or tints have been produced in one pattern: the average is four, and the result is, as usual, more satisfactory. The machines are run at much greater speed than in England, in order to make up, as far as possible, for the higher rate of wages paid in the States. Thus, the workman having more hours to work, and a greater quantity of work to get through and be responsible for, has his attention ever on the stretch to avoid the production of bad work. The average amount got through is 12,000 pieces per week.

The dye-house arrangements are within a very small space, considering the work got through.

An ingenious contrivance for untwisting the cloth as it comes from the dye-house is in successful operation, by which one boy does the ordinary work of six men. This untwisting machine receives the cloth through a wooden funnel, similar in shape to the hopper of a mill, into a trough of the width of the fabric, forming an inclined plane. The boy attending to the machine assists the opening of the cloth, which is chiefly effected by an undulatory motion, applied as it descends from the funnel; this partial spreading out is rendered more complete by the passage of the fabric over the curved edge of a piece of wood, across which it is drawn on the ascent up the inclined plane or trough, which conducts it to a series of wooden rollers with spiral edges of brass fixed in their surface. These open the cloth completely, in a similar manner to the iron spreading roller attached to every calico printing machine.

An adaptation of a French method of using the chloride is adopted.

This is effected by passing and wringing the cloth in its open state through the chloride, or 'chemick,' as it is called in England, and then washing out by passing over three large wooden cylinders placed above each other, over which an abundant stream of water flows as they revolve. The wringing out of the water by other cylinders leaves the cloth perfectly free from all excess of dye and chloride, and it is then dried and folded from the machine.

The drying process is conducted in a very economic manner, by a contrivance somewhat similar to the ordinary washing machine used after bleaching.

The steam is also economised to a very considerable extent. It is first worked at high pressure, then taken off at 10 lbs. to the square foot for bleaching purposes, and then again at 1 lb. to the square foot for heating the decoctions of dyestuffs. By this means little or no steam is lost, and the saving is said to be about 33 per cent in fuel over the English method. No steam was visible in the bleach-house, whilst in bleach-houses generally, the upper part is full; and, of course, this is all waste.

The earthenware bleach-rings mentioned in Class XXV, page 292, are used in both the bleach and dye house.

The goods printed by Messrs Dunnell bear comparison with most of the same class printed in Europe. The steams are clear and bright in colour; the madders, especially the lawn fabrics, clean and well finished; the prints being generally pure and of excellent tint.

The designers and printers employed by this firm are nearly all Englishmen. French designs on paper, together with the current styles in cloth, are obtained from Paris; but it often happens that if any of these are produced as soon as obtained, they are too early–in fact, too novel for the home market. Thus a pattern printed one season and unsuccessful from its novelty, has sold well in a subsequent season requiring the same class of goods, the importations from Europe having made the buyers acquainted with the style. For it is a singular fact, that however much these buyers are supposed to give the fashion, and thus rule the market in America as in Europe, their principles of judgment or taste, as it is called, are so low as rarely to enable them to pronounce whether a pattern is good or bad until it begins to sell, or they find it hanging upon their hands. It has thus happened that a style pronounced totally unsaleable, has in a subsequent, or possibly a more advanced period of the same season, become quite the rage. These are points upon which the calico printer has to calculate, if he can do so, alike on both sides of the Atlantic.

The goods printed by Messrs Dunnell are brought into the market by a great sale at auction in New York, at the commencement of each commercial season. All are cleared out at a blow, and this plan appears to have answered their purpose most admirably. The buyers at once take the responsibility of making the goods sell; the printers are left free to attend to their productions for another season, without the necessity of selling out a dead stock at the end of the current year. This appears to be a bold and original course, suited more to the purposes of the American market than consistent with the commercial notions and usages of European manufacturers.

At the Merrimack Print Works, Lowell, Massachusetts, madder colours only are printed. There are twelve machines employed upon these, printing in from one to four colours, and the daily produce is 40,000 yards, or about 8,000 pieces per week. The goods printed are manufactured in the cotton mills of the same Company (see Class XI). These are of excellent quality, and the character of the printing and dyeing places them on a par with similar fabrics manufactured elsewhere.

The patterns are generally well selected, being of a neat character and similar, in many respects, to the goods known in England as 'Hoyles.' The smaller kind for children's wear, are especially noticeable for the same qualities, which render the prints of such houses as Hoyles and Liddiards so much in request in England; and when it is considered that this class of calico print is largely used in the United States for the summer clothing of males, especially for boys, the demand would appear to be far beyond the present means of supplying it.

The Merrimack Company, as also Messrs Dunnell and Company, contribute largely to the Exhibition at New York, and the goods there displayed, though simply selected from their ordinary productions, are characterised by the qualities already indicated.

The water of the Merrimack river appears to be better adapted to the dyeing of cotton than of woollen. This point claimed particular attention at Lowell, from the comparison of the colours dyed for carpets by the Lowell Manufacturing Company, and the dyes at the Merrimack and Hamilton Print Works; and the impression was confirmed on examining the dyes of the cottons and de laines printed at the Manchester Print Works, Manchester, New Hampshire. On mentioning this to the manager of the works at the latter place, he confirmed the opinion I had formed on this point. This question of the selection of a water site for dyeing and printing is a most important one in the United States, since it is quite certain that in no country is there a greater variation in this respect.

The Hamilton Manufacturing Company, Lowell, dye and print a large quantity of the goods manufactured by them. The dyed goods are chiefly for the China market, and the home market of the West. These consist of drills, and are usually blue or brown. The average weekly produce in this department amounts to 36,000 yards.

The prints are all madders in from one to six colours, and the weekly work of the five machines employed is returned at 110,000 yards.

The styles of the patterns are generally darker than those of the Merrimack Company, but are good examples of madder work.

The internal arrangements of both the Merrimack and Hamilton Print Works are very good as regards economy and facility for work, but do not call for any special notice.

At the Manchester Print Works, Manchester, New Hampshire, both cottons and mousellines de laine are printed; the goods being manufactured by the same Company (see Class XV).

Ten machines are employed, and these print from single coloured madders up to eight colours in steams, six colours being the average in mousellines de laine.

The cottons printed here, however, are nearly all madders, the designs being of the usual character; about 500 pieces (15,000 yards) being the average weekly production.

In mousellines de laine, 15,000 pieces (45,000 yards) are printed weekly, and the patterns are mostly of a season later than those of Europe, for reasons already given as affecting the sales. The designs are adapted to the markets, which are chiefly in the Northern and Western States, and the styles are generally showy, rather than elegant.

The printing department is well arranged, and the scouring of the grey de laines is now attended to more than it formerly appears to have been, from the quality of some of the colours. Seventeen machines are now used, where formerly eight were considered enough; and the absolute cleanliness of the fabric before printing seems to be recognised in its full force, – a fact that the scientific printers of France discovered years ago, but which British as well as American printers paid in costly failures for finding out.

A folding machine, for grey cloth, is attached to the dyeing machine of the scouring room. It consists of a swing frame with rollers, through which the cloth passes. As the frame swings on a pivot, by which it is attached to the upper part of the frame, it lays out the cloth as it descends from the rollers in smooth and even folds, ready for carrying to the printing machine, instead of allowing it to be crushed into an unsightly mass as it passes from being dried in the grey.

In printing, the pieces are run up from the machine to drying-rooms, descending again to the machine for examination by the printer. The de laines are folded and carried to the steam, but the cottons are run up to 'ageing' rooms, from whence they are in due time run off again to the dunging and dye vats. Carrying about by human labour is thus saved to a very great extent, and a cotton piece is scarcely touched by hand, from the time it enters the machine until it is dyed and has to be untwisted from the wringing after dyeing.

The steam here is economized as much as possible. A small steam-engine is attached to pump the condensed water back into the boilers. Thus water at 120° to 140° temperature is used instead of cold water, and fuel is saved by not allowing the already heated water to run away in waste.

Time did not allow of a visit to the Hamilton Woollen Company's Print Works, Southbridge, Massachusetts, but the agent of this Company, in common with those of the Merrimack and Manchester Company, and Messrs Jacob Dunnell and Company, furnished me with duplicate patterns of the goods contributed by them to the Exhibition at New York. These are good specimens of mousellines de laine. The colours are clear, brilliant, generally well selected and harmonised; the patterns partaking of the usual character. A specimen of cashmere furniture, printed with copper rollers, in the chintz style, in ten colours, is admirable, in the clearness and distinctness of the large masses of colour. The design is of the floral type usually adopted for furnitures.

The Hamilton Company's Works print about 20,000 yards per week, having four machines at work, and another in the course of erection. They retain four designers, three Scotch and one English, and these produce nearly all the patterns they use.

A department for the engraving of copper rollers or 'shells,' as they are generally called in the United States, is attached to each print works, in which the requisite patterns are engraved, and the workmen employed are chiefly Englishmen.

As an improvement in the use of the ordinary rollers and the mandril, an invention has been lately patented, as applicable to the use of copper shells, by which a great weight of copper may be saved by making the shell itself very much thinner, and inserting therein another roller with a groove or slit running longitudinally, into which a mandril is fitted with a metal notch running down the slit, thus expanding the internal roller to the full circumference of the inside of the copper shell, and rendering the whole sufficiently firm and solid for use in the machine. The value of this mode of using copper cylinders in printing has, however, yet to be tested. The saving in the weight of copper at present employed will be considerable if the invention succeeds, and the necessity for 'turning off' a pattern will be also superseded, and a standard circumference may be adopted for the cylinders, with a gain in the comparatively small amount of metal required to constitute the thickness of the 'shell.'

Felted table covers of a good quality are printed at the Bay State Mills, Lawrence, Massachusetts. These are entirely blockwork.

Silk handkerchiefs, of India fabric, imported in the grey, are printed at the Falls of the Schuylkill Print Works, near Philadelphia. Specimens of these are displayed in the Exhibition at New York, and show good work.

The specimens of printed goods exhibited are, on the whole, fairly representative of the present state of this branch of industry in the United States, and evince progress of no ordinary kind.

Printed and 'extracted' cassimeres and satinettes for trouserings and vestings are a novelty, and the specimens exhibited, as also those produced by the Middlesex Company, Lowell, are calculated to give a favourable impression of the character of this mode of dyeing woollen goods for general use. European prejudices would undoubtedly be against such modes of finishing the surface of such materials; but in the United States a novelty of this kind, commended by cheapness and fair appearance, readily finds appreciation, or, at least, obtains a trial of its merits.

8

Carpets, etc.

Class XIX

The manufacture of carpets appears to have been of steady growth. Commencing with the cheaper and more useful kinds, or, at least, those most in demand, it has progressed until it now forms a very important item in the industry of the United States.

In the New England States, where it is carried on to the greatest extent, the weaving may be said to be altogether by power. Hand-loom weaving, however, is still the chief means of production in the States of Pennsylvania, Delaware, and Maryland, and the manufacture consists of two and three ply Kidderminster, Venetian, and a little Brussels. The out-door or domestic system, as it may be called, prevails to a considerable extent, although there are factories in which looms are set up. One of these at Philadelphia has generally from 200 to 300 at work, and the proprietors employ out-door weavers in addition. This system has already been alluded to in Class XI, as applied to the manufacture of checks, tickings, etc. The weaver takes out work from the manufacturer, and the master weaver, or 'boss,' as he is called, employs other weavers to work, probably three or four looms, in a shed attached to his own house. He is responsible to the manufacturer from whom he receives the materials, and to whom he delivers the work when completed, and receives payment, giving such wages to his workmen as leaves him a small profit for superintendence, use of looms, etc., supposing the latter are his own property, which is frequently the case. In fact the system pursued in the States above named is similar to that which still exists in some of the manufacturing districts of England, except that the American 'boss'* is a nearer approach to a small manufacturer making up the materials of another.

---

* The term 'boss' is also generally used in the United States instead of 'master' or 'employer.'

The wages of the working weavers vary from $8 to $12 per week, the average being about $10–say £2 1*s*. 8*d*. sterling. This rate is given on the authority of a carpet manufacturer at Philadelphia, and one of his master weavers.

The rag carpets manufactured by the prisoners of the Maryland Penitentiary, Baltimore, have been already mentioned when reporting upon the other industrial productions of that prison.

The Lowell Manufacturing Company, Lowell, Massachusetts, manufacture carpets very extensively, employing 200 power looms, producing 25,000 yards per week. The goods consist of two and three ply ingrain, and are good fabrics of their class, being firm in make and excellent in dye, so far as permanence of colour is concerned. In some colours, too, there is much clearness and brilliancy, but, as already stated (Class XVIII), the water of the Merrimack appears better adapted to the dyeing of cotton than of wool. The designs are varied, and of the usual character of such goods.

The designers employed are French and Scotch. Two female students of the Boston School of Design are engaged, who, after a fair trial, and the requisite technical instruction, have been successfully employed in 'drafting.' They have also made one or two very creditable attempts in design.

Tufted and chenelle rugs also form a part of the productions of the Lowell Company's Carpet Works, about 50 per week being manufactured.

A very complete establishment for the manufacture of carpets of various kinds is situated at Thompsonville, Massachusetts, erected and worked for a considerable period by the Thompsonville Manufacturing Company. The works, however, had not been in operation for several months at the date of my visit.

The specimens of goods shown were of a fair quality; the arrangements of the engine-house, mill, weaving-rooms, dye-house, etc., being very complete, and of a most substantial character. When in work, 27 carding engines are employed. There are 127 power looms for ingrain carpets, turning out, when at work, from 25 to 28 yards each per day; and 7 power looms for Venetians, usually making 125 yards each per day. The hand looms comprise 42 for Brussels, 10 tufted rug looms, 16 setting Axminster, and 4 filling looms. This establishment being about to pass into other hands, would, in a few weeks from the period at which the above information as to its capabilities was obtained, be again in operation; but great difficulty was anticipated in getting together the requisite number of skilled operatives, as those formerly employed have sought and obtained employment in other districts.

In this respect the American manufacturer is peculiarly situated, and he must often carry on his business at little or no profit, perhaps at a considerable loss, in order to keep together the agents by which, when the demand comes, he can alone supply it. Hence the anxiety to settle down the operatives around the mills; to render their condition and social position such as shall absolutely attach them to their employers and the locality. The latter, perhaps, being the most difficult, from the migratory tendencies of a people so restless, and always so alive to any new contingency which promises to better their condition, however distant the field of operation may be.

The most interesting carpet manufactory in the United States is, without doubt, that of the Bigelow Carpet Company, Clinton, Massachusetts. In this establishment the manufacture of Brussels carpets by power is fully and completely carried out, and a fabric manufactured, which, for evenness of surface, fineness and strength of make, is of a most unexceptionable character.

There are 30 power-looms always in full work, weaving 5-frame Brussels carpet. The production of each is from 20 to 24 yards per day; in special cases 32½ yards are manufactured. Pieces of 60 yards each, of the widths ¾, ⅞ and 1 yard constitute the usual make. The 'fit' or 'register' of the patterns is of the most accurate character; thus giving a great advantage in the making up.

The weavers are all females, one attending each loom.

The dyes are excellent, the water being obtained from wells sunk for the purpose. Its quality is evidently favourable to the more brilliant colours, the reds and blues being especially clear and bright.

The designs are of the usual character, chiefly floral, though a few good geometric diapers indicated a tendency to a more healthy style. The patterns are generally well drawn and coloured, and are of European production.

An experiment was going on in a new loom, or rather a new application of the present loom, in the manufacture of velvet pile. By an arrangement of the wire, a knife-blade, with the edge uppermost, forms one end, and when this is drawn out by the action of the machine it cuts the pile. In order to secure the face-threads, two picks had to be thrown into the back for one at the face of the fabric, and the problem in the course of solution was the relative proportion of these threads. The results of the experiment were so far satisfactory as to show that success was certain, and that when once the splitting or division of the pile, by the tension occasioned by the back threads being too thick, was accomplished, a beautiful fabric would be the result. The produce of the experimental loom was about 15 yards per day, and it was anticipated that this would be the average in the other looms in course of construction.

The printed felt carpets manufactured and printed at the Bay State Mills, Lawrence, Massachusetts, are remarkable examples of their class. The patterns are generally well selected, and suitable to the fabric, the colours good, clear, and brilliant; whilst the printing is more accurately, fit than is usually found in this class of goods. These felted carpets sell at 90 cents (say 3*s.* 9*d.* sterling) per yard of 2 yards wide.

The carpets exhibited in the New York Exhibition are of a very miscellaneous character, many being imported. Those of American manufacture, however, are generally such as above stated; the best being 'ingrain.'

The manufacture of floor-cloth appears to be one in which, ultimate in the Americans will succeed, so far, at least, as the use of the materials are concerned; the dry character of the atmosphere being very favourable to the rapid drying of the work.

There are a few admirably printed specimens in the Exhibition, but, with the exception of one pattern, in imitation of oak *parquettage*, they are as thoroughly wrong in design, and as antagonistic to everything like the true principles of floor decoration, as the generality of such things are in England. For instance, one specimen has its surface ornamented with a portrait of Washington, and a view of Mount Vernon alternating in panels, surrounded by a wreath of flowers and the American Eagle! Yet this is intended for a floor-covering, and of course to be walked upon!

Coach lace is manufactured in considerable quantities in some of the New England States.

The Clinton Manufacturing Company, Clinton, Massachusetts, have 100 looms employed in the production of coach lace, the materials being varied according to quality, and consisting of, more or less, worsted, cotton and silk. The designs of some of the best qualities, in which silk is freely used, are very good. The looms are of the same construction as the Brussels power loom; in fact, the latter is an extension of the principle of the coach-lace loom, since the invention was first applied to the manufacture of coach lace.

The manufacture of fringes, tassels, etc., appears to be exceptional (see Class XX), but the specimens contributed to the New York Exhibition are, generally, of fair make.

Embroidery appears to form but a very small item in the industrial productions of the United States, although in the Exhibition there are numerous contributions of ladies' work, chiefly in Berlin wool and embroidered quilts.

The specimens of commercial embroidery are generally good, and in New York, as in other large cities, persons are employed by the various millinery and upholstery establishments in needlework decorations for articles of dress and furniture.

9

Wearing Apparel, etc.

Class XX

As might be expected, the manufacture of ready-made clothing forms an important item in the industry of the people of the large cities of the States, such as New York, Boston, and Philadelphia, and that, in a country where male labour is in such demand, females are more largely engaged, and, on the whole, better paid for their work than they are in Europe. Still, during the last summer, attention has been called to the condition of the needlewomen,–especially the shirt-makers of New York, and facts brought to light of a parallel character to those which a few years ago formed such prominent points in an inquiry into the industrial and social position of the same class of females in London.

The general statistics of the clothing trades, spread, as the various occupations connected therewith are, over the whole country, presented so many difficulties that, after several attempts to obtain accurate information on this point, I was compelled to abandon it. There can be no doubt, however, that could accurate data be obtained, the result would be of a very extraordinary character; as in a country where all classes of the people may be said to be well dressed, and where the cast-off clothes of one class are never worn by another, the manufacture of the cheaper kinds of clothing must be carried on to an enormous extent.

Cincinnati, Ohio, appears to be a great central depôt of ready-made clothing, and its manufacture for the Western markets may be said to be one of the great trades of that city. The system pursued is that chiefly of out-workers. The articles of dress being cut out in large quantities in the warehouses of the dealers or manufacturers, and distributed to those who undertake to make them up at their own houses, or to master tailors or 'bosses,' who, paying the hands they engage either by the piece or by the day, undertake the making up of large quantities of clothing. The various sewing operations are carried on by the latter in workshops, in most instances well adapted to the peculiarities of the trade, and latterly sewing machines, of varied construction, have been largely employed. In one of these establishments, the manager, or 'boss,' stated that a person skilled in the use of the machines in operation there, would do as much work as ten ordinary needlewomen, and that by those in which the needles acted vertically, any character of seam required for ordinary clothing, either in right lines or curves, could be sewn. The successful action of this machine is further exemplified in the manufacture of boots and shoes, so far as the stitching of the upper leathers is concerned.

At Louisville, Kentucky, and St Louis, Missouri, the manufacture of clothing is also extensively carried on; but Cincinnati may be considered as the great mart of ready-made clothing for the Western States, and, in a measure, for those of the South also. In 1851 there were in the latter city 108 establishments, employing 950 hands in their own workshops, and upwards of 9,000 females, either at their own homes or under 'bosses.' The proprietors are chiefly German Jews, and most of the operatives are Germans.

Under-clothing, as made up from the piece, such as shirts, etc., is manufactured upon the same system, and mostly by the same firms. This applies generally throughout the larger cities.

The manufacture of stocking-net, and of hosiery of the coarser qualities, is gradually developing itself in various localities, especially in the States of Connecticut and Massachusetts.

The Waterbury Knitting Company, Waterbury, Connecticut, manufacture cotton drawers, under-shirts, and merino wool knitted articles. By a regulation of the directors, however, strangers are not allowed to see the machinery; but judging from the character of the goods, the machines are, in all probability, either of French construction, or upon the same principle as the circular looms of Jouve or Gillet. The former has a patent right for the United States, which is held by the Enfield Manufacturing Company, Thompsonville, Massachusetts. This establishment is under the direction of an Englishman, Mr W. G. Medlicott, and 400 operatives are employed. There are 3,000 spindles for cotton yarns, and 13 sets of carding engines for wool. The company has at work 58 circular knitting machines of various dimensions, after the invention of Jouve, of Belgium, for which, as above stated, it holds the patent right, and 23 machines made by François Gillet, of Troyes, France. The articles produced are knitted cotton drawers, under-shirts in cotton, cotton and wool, and all wool. All these are of fair average quality and make.

Elastic webbings used in the making up of clothing, suspenders, gaiters, etc., are manufactured in the State of Connecticut, and the establishment of Messrs Hotchkiss and Merriman, Waterbury, affords an illustration of the extent and mode of operation. There are several other concerns of a similar character, but this appears to be the largest. They spin the cotton yarn in one factory where 200 persons are employed, manufacture the buckles and metallic mountings in another, and at a third factory the elastic webbing is produced; the threads of caoutchouc for intermingling with the warp being prepared on the premises, the weaving being effected by power. About 150 persons are employed in the latter factory, where the webbing is afterwards cut to the requisite sizes for the various articles into which it is to be made up, the leather trimmings punched out to the proper forms, and then assorted into convenient quantities with the necessary metallic appendages. The materials thus prepared and assorted are distributed in the villages and farm-houses around, to be made up by females, in many instances at their hours of leisure from domestic employment, and by others as a means of obtaining a livelihood. Even little children, under the age at which the law of the State allows of their employment in manufactories, can be usefully engaged in some portions of the work thus undertaken at home. A waggon is used for sending round the materials and collecting the finished work. This visits a given district at stated periods, taking out fresh work and bringing back that distributed on the former journey. About 600 persons are thus employed at their own houses by Messrs Hotchkiss and Merriman.

The goods are made up into dozens and half-dozens, and usually packed in ornamental boxes for distribution to the retail dealer. They consist of the commoner kind of such articles as used in England, but are of excellent make and quality.

It will be seen that the method pursued in the making up of these articles is similar to that adopted in Scotland and the north of Ireland by the embroidery houses of Glasgow and Belfast.

[*Boots and shoes.*] The boot and shoe trade of the United States is of a very extensive character, and the systematic manner in which it is carried on worthy of being understood and adopted elsewhere. A scale or series of sizes is adopted, say in women's and children's shoes from 1 to 6, and even higher numbers, the half constituting a size between each. The various portions of the boots and shoes are cut out to these sizes and half sizes. These are put up with all the requisite trimmings necessary to complete the articles, in sets of 60 pairs for the common kinds, and 24 pairs for the finer qualities.

Being cut out and made up into sets, they are sent to be 'fitted' for the maker – that is, the various parts of the upper leathers are stitched together. Much of this is now done by one of the various kinds of sewing machines in which the needle acts vertically, as the force required to pierce the leather is more directly applied than in those in which the needle acts horizontally or in a segmental curve. The neatness, accuracy, and strength of stitch is superior to hand work. The upper leathers thus 'fitted' are then sent to the 'binder,' who finally prepares them for the 'maker,' by whom they are soled and heeled. Being complete in make they then go to the 'trimmer,' whose work consists in punching the string holes, stringing and putting on buttons, and in ladies' shoes, bows, rosettes, etc.

Soles are cut out by machinery. A knife with a curvilinear edge is set in a frame and worked with a treadle after the manner of a lathe. By a lateral motion in the machine it can be adapted to the cutting of any requisite width of sole, and being once fixed to a given width, the process of cutting is very rapid, and material is saved by the leather being cut at right angles to the surface, instead of diagonally as by the ordinary knife.

When finished the goods are made up in boxes containing 1 dozen of assorted sizes. They are then sent in cases to the wholesale dealer who supplies the retailer. A case contains 5 boxes, making up the 60 pairs of assorted sizes of which a set of the commoner kind consists as manufactured.

Boots and shoes are manufactured and the workmen reside in all parts of the New England States, but chiefly in the States of Massachusetts, Maine, Vermont, and New Hampshire.

The finer quality of boots for gentlemen are chiefly made at Randolph and Abington, Massachusetts; the heavier kind of shoes, and the coarsest kind, usually called 'brogans,' at Danvers, in the same State. These 'brogans' are chiefly manufactured for the Southern markets, for the use of slaves, and are similar to the shoes worn by the miners of South Staffordshire.

Shoes for females are chiefly made at Lynn, Reading, Woburn and Havrehill, all in the State of Massachusetts.

The following table, compiled from the 'Statistics of the Condition and Products of certain branches of Industry in Massachusetts, for the year ending April 1st, 1845,' will show the extent of the boot and shoe trade in the six above-named towns at that date:

Amount of production in boots and shoes, 1845

Towns	*Kinds*	*Number of Pairs Made*	*Males employed*	*Females employed*
Randolph	Boots	227,131	815	649
	Shoes	332,281		
Danvers	Both	1,150,300	1,586	980
Lynn	Boots	2,000	2,719	3,209
	Shoes	2,404,722		
Reading	Shoes	274,000	358	385
Woburn	Boots	909	425	484
	Shoes	350,920		
Havrehill	Shoes	1,860,915	2,042	1,680

At the last-named town 129 machines were employed in cutting sole leather.

Lynn is still the town most extensively engaged in this trade, and, as regards the methods of business and routine of manufacture in the article of ladies' shoes, an extract from the report of the Sanitary Commission of Massachusetts, 1850, is here quoted.

'It requires, on the average, about sixty days to convert stock into articles ready for final sale; and, in the process several classes of persons are employed. The head manufacturers furnish the capital, and superintend the whole operation. The warehouses or stores of the better class generally contain, 1. A counting room, where the general business is transacted; 2. A leather room for keeping the soling leather; 3. An upper stock room; 4. Two clicker's or cutter's rooms, one for the upper leathers, and one for the "stuffs," or soles; 5. The bound shoe room; 6. The trimming room; 7. The sales and packing room. A "last" room, and others of less importance, are also sometimes provided. The stock is put into the hands of the "clickers," who cut it into "sets" of shoes, varying somewhat in number and size, from 28 to 50, No. 2 to 7, according to the size and quality of the stock. These sets are numbered, recorded, and packed in boxes, to be sent to the operatives or workmen, to whom they are charged. When returned, they are credited, and go into the trimming room, where they are finally prepared, by females, for market. They are afterwards taken to the sales-room and packed. Generally the sets, as they go from the cutter, are kept together until their final sale. These shoes are sold at the warehouses of the manufacturers, and are seldom or never consigned.

'The "closing" and binding of the shoes is done by females or "binders," and the other parts, or "bottoming" by males, or "workmen," "jours," or "journeymen." These operatives do not live in Lynn exclusively, but many of them reside in other parts of the State, and in Maine, New Hampshire, and Vermont. The binders receive from 2 to 5 cents per pair for children's shoes, 3 to 5 cents for misses', 3 to 9 cents for ladies' shoes, and 6 to 12 cents for gaiter boots. The workmen receive for bottoming, or making shoes, 5 to 17 cents for children's, 10 to 20 cents for misses', 10 to 25 cents for ladies, and 15 to 33 cents for gaiter boots. All the labour is paid for by the piece. Idle time here receives no compensation, and none need be spent. Full employment can always be obtained by competent workmen. The binders earn from $3 to $4, and the workmen from $3 to $9 per week, according to inclination, ability, and time employed (the latter averaging about $5), out of which they pay their board, which in Lynn is $2 to $2½ for males, and $1¼ to $1¾ for females. The net earnings of the females are about half as great as those of the males. The females seldom bottom the shoes, though they might very properly do it.

'Some of the workmen manufacture small lots of shoes on their own account, which they sell to other manufacturers, or have shops, in which they let to other workmen or journeymen "berths," or the right to place and use their "kits." These shops are commonly small buildings in a yard near the dwelling-house, or rooms in the barns. The price paid for a berth is $1 50 cents, $3, and sometimes $5 annually, according to accommodations, the inmates agreeing to share equally in the expenses of warming the rooms. The chips of their work are saved when fuel is not needed, and burned with wood in box stoves, open at top, when it is needed; and, in many shops, these chips supply nearly half the fuel. Very little coal is burned.'

A return, of which the following is an abstract, was kindly furnished by Mr Alonzo Lewis, of Lynn, who has devoted special attention to the statistics of the manufactures of that place. This shows the present state of the boot and shoe trade in that locality.

Shoe manufacturers–men who manage the business and employ workmen	167
Cutters–men who cut the shoes and shape from the stock	321

Cordwainers–workmen who make the shoes	4,132
Binders–females who bind the shoes	7,170
Number of pairs of shoes of all kinds made 1852–53, including women's and children's shoes, boots and gaiters	4,952,300
Value $3,706,000, say £900,000.	
Capital employed $1,500,000, say £375,000.	

The boot and shoe trade, as carried on in the United States, is altogether of a domestic character, although, from the methods adopted, it partakes of the essential features of a well-regulated factory system. The character of the operatives stands very high for intelligence and probity, and such is the confidence of the manufacturers that they never ask for a reference, still less for a security, when a new hand applies for work; and instances of the loss of materials by the dishonesty of the workmen are very rare indeed. Sets of shoes have been lost occasionally through intrusting them to travelling shoe-makers, who afterwards proved to be professed vagabonds; but this has never yet affected the integrity of the workers generally.

For the convenience of the operatives residing in distant localities, the materials, in their prepared state, are collected from the manufacturers by express-men or carriers. These deliver them to the workman for whom they are intended, and on receiving the work made up, delivers that to the manufacturer, and then receives the payment due to the former for their labour. The remuneration of these carriers is generally a small per centage on the amount.

The 'brogans' or 'negro shoes,' are often made by small farmers, who fill up their leisure time with shoemaking, especially in the winter, when outdoor labour cannot be attended to. The ready-money thus obtained contributes very materially to the comfort of this class of persons, and they pride themselves upon paying the State taxes when they are proprietors, and a great portion, if not the whole, of their rent when tenants, with the proceeds of their handicraft.

The examples of boots and shoes contributed to the Exhibition at New York are very numerous, and illustrate in a satisfactory manner the skill and ingenuity employed in this department of industry. The taste displayed in the ladies' shoes is very considerable, whilst the excellence of workmanship, especially in gentlemen's boots, show how thoroughly the division of labour which prevails in this manufacture is favourable to satisfactory results.

The manufacture of hats, in all the varieties of wool, fur, silk, and cotton felt, is largely carried on in the States of New York, New Jersey, and Connecticut. Time did not permit of any examination into the methods and processes of manufacture, but in all probability these differ very little from those of England and France. No statistics as to the extent of the hat trade could be obtained without much personal inquiry, but there is no doubt this department of industry has increased enormously during the past eight or nine years, and that great improvements have taken place in the quality and style of the hats manufactured. There is a highly satisfactory display of specimens of almost every kind of hat produced in the United States in the Exhibition at New York, and all, or nearly all, of these present unexceptionable evidence of skill in manufacture, and considerable taste in the style and finish.

The straw bonnet trade gives employment to a large number of persons in New York City as sewers, as also in several distinct localities in the New England States.

The straw braid or plait is chiefly imported from Italy, France, Germany, England, Switzerland, and China, and embraces nearly every quality; the latter (Canton straw) being largely used for the Southern markets, which are chiefly supplied by the New York manufacturers. 'Swiss lace' has been latterly in the greatest demand for bonnets, and the importations of the other kinds have been proportionably less. A coarse straw is obtained from Canada, which is chiefly used for the manufacture of men's hats.

The domestic manufacture of straw braids or plaits is almost exclusively confined to the New England States. These consist principally of split straw, narrow and fine, and almost equal to the English. Also, of a small proportion of Devon, a wide straw braid of an inferior quality, and of single 11, and other braids in smaller quantities and of inferior qualities. The English 'patent' and 'whole' straw have been imitated in New England, but with little success.

In the manufacture of bonnets, besides the New England States and New York City, a little is necessarily done in nearly all the large cities. Palm-leaf hats, for men and boys, are made up at Worcester, Massachusetts, and other parts of New England, in large quantities. The 'Shakers,' in several of their communities, manufacture a close cottage bonnet of the same material.

The number of straw hats made up in New York has been estimated at upwards of 1,000,000 per annum, and there are probably from 2,000 to 3,000 sewers engaged in making up straw hats during the season, the average wages of which are probably $3½ (about 15*s.* sterling) per week. The work is done either at the homes of the workers or in workrooms provided by the employers, according to the system adopted by the latter; in some instances, both 'in' and 'out' workers are employed. The same complaints of unremunerated labour, as already quoted in the case of the shirt-makers, are made by the straw-sewers, and the relation of this class of female operatives to their employers is very similar to that which exists in London and other large cities.

The bonnets exported from England would find a much readier and more extensive sale in the United States, if style of make was more attended to, and more of the elegance found in the French thrown into their manufacture. As it is, the best houses procure their styles and patterns from Paris, and these are imitated, as far as possible, in the making up of the imported braids by the New York sewers.

In the Southern markets the demand for straw goods is continuous throughout the year, owing to the warmth of the climate; in the Northern States, however, it is chiefly confined to the spring and summer.

The value of the importations in straw goods, that is, in the braid or plait, and made up, has been estimated to amount to $4,500,000, or about £1,000,000 sterling, and the value of those of American manufacture to about $500,000, or about £110,000 sterling.

The demand in the United States for military clothing and accoutrements is very great, from the extent to which the formation of volunteer corps is carried, and the trade thus created is an important one.

The military costumes exhibited at New York are of a very varied character, the contributions of Messrs Horstmann and Sons, of Philadelphia, being the most important. The varied manufactures of this house ought, strictly speaking, to be classified under the heads of Class XIII, Silk, in which the ribbons, etc., have been already alluded to; Class XIX, Lace, and Class XX, Clothing (that now under consideration), as also Class XXII, Ornamental Metalwork. For convenience, however, it is placed here, because the exhibits are under this head in the Exhibition at New York, and the chief productions are military clothing and accoutrements.

The manufactory of Messrs Horstmann is the most complete of its kind in the United States. In the weaving department, gold lace, silk fringes, bindings, etc., are woven by power in looms to which, as required, the jacquard machine can be applied, as already stated (Class XIII), for the manufacture of ribbons. Hand looms are employed in the manufacture of braiding. Machines of beautiful construction are used for covering cords with silk and gold thread, and by a proper adjustment these produce a pattern on the surface of the cord. As the work done is of a very high character, as regards quality and make, the speed of the machinery is much slower than usual, a low-pressure engine supplying the power.

The manufacture of military ornaments forms another department, Electro-gilt brass ornaments, sword-handles, scabbards, and the usual bullion decorations, being produced on a very extensive scale. The sword-blades are all imported, chiefly from Germany.

Messrs Horstmann are the recognised manufacturers of military equipments to the War Department at Washington, and the military clothing, etc., contributed to the Exhibition at New York are of a very high character, alike as regards design and execution. Having recently erected a very large and well-arranged factory within the city of Philadelphia, the whole establishment presents an example of system and neatness rarely to be found in manufactories in which handicrafts so varied are carried on. Female labour is, of course, largely employed in the weaving and making-up departments, and formerly in the cutting of fringes. This, however, is now performed by a machine with a circular knife, so arranged as to cut the thread on the diagonal. The double fringe as it leaves the loom, being either run off the beam or placed upon a roller for that purpose, is divided much more exactly than it could be by hand, and at so rapid a speed as scarcely to admit of a comparison with hand labour. Any width of fringe can be thus cut, the machine being so constructed as to be easily adapted thereto.

Iron and Other Metal Manufactures

Introduction

The present extent and future prospects of those manufactures already fairly established in the United States in which iron is the principal material used, required a much more extended and detailed examination than time and the distance to be travelled over would permit of my devoting to them, when taken in connection with other departments of industry requiring equal attention. The progress of the past ten or twelve years would appear, on all hands, to have been very great; and many establishments which were scarcely commenced at the beginning of that period, are now in a position to stand a fair comparison with similar manufactories in England.

Pennsylvania is the largest iron-producing State in the Union, although by the census of 1850, twenty-one States are returned as producing pig iron, and only two, Florida and Arkansas, as not having establishments for the manufacture of iron castings; whilst in nineteen States wrought iron is made.

In the production of pig iron 377 establishments were in operation in 1850, and of these 180 were in Pennsylvania, 35 in Ohio, and 29 in Virginia; the remaining 18 States having a much smaller number each.

The capital invested amounted to $17,346,425 (about £4,500,000 sterling); the produce being 564,755 tons per annum, employing 20,298 males and 150 females.

In the manufacture of iron castings, 1,391 establishments were engaged. Of these 643 were in the States of New York and Pennsylvania, 323 in the former and 330 in the latter; 183 others being in the State of Ohio. The capital invested amounted to $17,416,361, or about the same amount sterling as in the manufacture of pig iron. 322,745 tons of castings are produced per annum, giving employment to 23,541 males and 48 females. The value of the castings, and other products, being estimated at $25,108,155, or about £6,250,000 sterling.

Wrought iron is manufactured at 422 establishments in 19 States. Pennsylvania has 131, New York 60, New Jersey 53, Tennessee 42, and Virginia 39; the remaining 97 being situated in 14 other States. The capital invested was $14,495,220 or about £3,500,000 sterling; 13,178 males and 79 females being employed. The quantity manufactured amounted to

278,044 tons, the value of which, with other products, was $16,747,074, or about £4,100,000 sterling.

There can be no doubt that a very considerable increase has taken place in the make and manufacture of iron since the returns, from which the above facts were taken, were made in 1850; and from the energy, enterprise, skill, and industry of all concerned in this manufacture, and the importance attached to it as a permanent source of national wealth and prosperity, its future progress cannot fail to be more than commensurate with that of the last few years.

The manufacture of articles of utility in metal, especially iron and brass, is chiefly carried on in the States of Connecticut and the cities of Philadelphia, Pittsburg, and Cincinnati, as also to a very considerable extent in Boston, New York, and Baltimore, in all the departments connected with ship building and heavy machinery.

The manufacture of the lighter articles in metal appears to be chiefly located in the State of Connecticut, in the valleys of the Naugatuck and Housatonic; the mills and manufactories being built on the banks of those rivers and the smaller streams running into them, from which the requisite power is derived to drive the machinery employed. Thus, those natural advantages which presented themselves for the promotion of manufacturing enterprise in the cotton and woollen trades in the larger streams of the New England States, such as the Connecticut and Merrimack rivers, are equally obvious in the smaller streams of the State of Connecticut, and have been as readily seized upon for the establishment of a variety of metal trades in which, in addition to skilled handicraft, a large amount of highly ingenious machinery is constantly and most successfully employed. These various trades will be considered under their respective heads.

10

Metal Manufactures

Cutlery and Edge Tools: Class XXI

The present position of this department of manufacture can only be illustrated, with one exception, by consideration of the contributions to the Exhibition at New York, since time only permitted of a visit to a single manufactory–that of the Waterville Manufacturing Company, near Waterbury, Connecticut. At this establishment pocket cutlery only is made.

About 100 workmen are employed, and amongst them a few Sheffield workmen. Nearly all kinds of the pocket cutlery, of the best qualities, are made, and the taste displayed in the forms and general getting up of the goods is excellent. The materials, more especially the steel, used being of the best quality, the goods, of course, correspond; and from their excellence in temper as well as in finish, are in constant demand.

The factory is very systematically arranged, affording every facility for the economic production of the articles manufactured. The stock of materials of all kinds constantly kept on hand is necessarily large, as there is no possibility of obtaining a supply on any sudden emergency, as no similar establishment is to be found within a considerable distance. Everything is, therefore, as far as possible, made upon the premises, and portions of pocket-knives, the making of which, in England, almost constitute trades of themselves, are here produced in the ordinary routine of business, owing to the isolated position alike of the manufacture as of the manufactory.

The pocket cutlery displayed at the Exhibition at New York is generally of good quality and tastefully got up, and to the specimens of table cutlery the same remarks apply.

In the larger kinds of edge tools there are some admirable illustrations of the extent to which the manufacture of these useful articles is carried on, and of the attainment of a high degree of excellence, both in material, workmanship, and finish. Screw augurs and augur-bits of graduated sizes, patent expansion-bits which set to any size, and sundry carpenters' and other tools, all displaying either improvement in construction or excellence of finish and make.

A button-hole cutter of ingenious construction, with a gauge to measure the size of the button-hole, and two or three assortments of tailors' shears, are admirably adapted to the uses for which they are intended, the finish being unexceptionable.

The largest and most extensive display of edge tools is made by the Collins Manufacturing Company, Hartford, Connecticut. These consist of axes, adzes, etc., together with hammers of various sizes, pickaxes, and the ordinary run of heavy tools. Other houses exhibit hammer-heads of good workmanship and material.

The imperfect manner in which, from the cause already assigned, this department of industry in the United States was examined, is to be regretted, as from the examples shown at the Exhibition, its successful prosecution is apparent, so far at least as the more useful articles in constant demand in such a country are concerned. That a considerable number of European workmen are employed in this manufacture would appear to be more than probable, as the finish of the articles often shows an amount of skill only attainable through long practice. In the manufacture of files, for example, the best workmen are invariably Englishmen.

11

Metal Manufactures

Iron, Brass, and General Hardware, including Lamps, Chandeliers, and Kitchen Furniture: Class XXII

In nearly all the large cities, iron foundries of greater or less extent are to be found, cast-iron being largely employed in the construction of buildings both of wood and brick; and in Philadelphia, as also to some extent in other cities, whole elevations of houses, used as retail shops in the principal streets, are of cast-iron. In these cases the construction of the building is usually modified to suit the material of the front, and, in some instances, an approximation is made towards adapting the decorative part of the elevation to the material and the construction. In general, however, the ordinary architectonic forms, as used in stone and wood, are followed, and the whole painted and sanded in imitation of Connecticut red sandstone, a material now much used in building. The construction of some of these elevations is at once simple and effective, alike for strength as architectural effect, and there appears to be very little difficulty in taking out an old front and substituting a new one, as the whole is well braced together by tyes and screws – the side walls sustaining the structure in all essential points. In Philadelphia there are some admirable examples of this adaptation of cast-iron to architectural purposes, and others were in the course of construction, in which more or less of originality in the matter of design and decoration is attempted. For retail shops of several stories where light is an object, and heavy pieces of masonry tend to lessen the size of the windows, these cast-iron elevations appear to be peculiarly well adapted. The dryness of the atmosphere, however, presents a great advantage, from there being less tendency to oxydization than in a more humid climate. It would appear probable that this use of cast-iron will eventually produce a style of street architecture, as applied to retail shops, of a different character to that which now prevails, and which is in imitation of European modes alike of construction and decoration.

Ornamental castings for architectural purposes, such as balustrades, railings, etc., are produced in large quantities in New York, Boston, Philadelphia, and other large cities, and these are usually copies or adaptations of similar work made in England. A few manufacturers aim at originality of design, with a greater or less degree of success.

At Philadelphia, the garden decorations, ornamental cast-iron work for cemeteries, monuments, etc., manufactured at the foundry of Mr Robert Wood, are good examples of their class. One or two verandas and garden fountains were very superior in design to the ordinary run of such things, being well adapted both to the material and the purpose for which they were intended. Mr Wood was engaged in the production of a cast-iron statue of the late Henry Clay, 15 feet high, to be placed upon a Doric column of the same material, about to be erected by the citizens of Pottsville, Schuylkill county, Pennsylvania, – a work requiring no ordinary amount of skill in moulding, etc.

The ornamental and decorative iron castings of various kinds produced at Boston, Massachusetts, may be illustrated by those manufactured by Messrs Chase, Brothers, and Company, as these are fair types of the best class of work, both as regards execution, skill, and the application of art to the embellishment of the useful, to which this firm has paid more than ordinary attention, and that too with some success, when the difficulties in obtaining good and workable designs are taken into consideration. The iron railings manufactured by them are of a very superior character, both as regards the construction and decorative arrangements of the parts, a few being more severe as regards geometric quantities and architectonic adjuncts than would be considered commercially prudent by many English manufacturers; and the severity of style would in all probability really militate against their sale, where exuberance of ornamentation is considered an essential. Some of the articles of garden furniture and decoration are also good, but, in general, the ultra-natural types, so much used in Europe, are followed in all their absurdity. This is observable too in the umbrella stands, and hat and coat trees, not only of this house, but of all the others of which time permitted an examination.

The iron bedsteads, manufactured by Messrs Chase, are fair average examples, and are of a similar kind to those manufactured in England. The peculiar talent of the inventors of the United States does not appear to have been much directed to those useful articles of furniture, in which so wide a field still lies open for an improvement, alike in construction as in decorative adjuncts. Brass is scarcely used as a mounting, and the method of japanning is susceptible of a more than ordinary amount of improvement.

The pernicious system of covering ornamental iron castings with dark green paint, and rubbing the projecting parts with metallic powder, in imitation of bronze, is, of course, as rife in the cast-iron articles of the United States as in those of Birmingham and Sheffield. In toilet mirror-frames, too, made of cast-iron, woods and metals of all kinds are imitated. Occasionally a manufacturer sets himself against these 'shams,' and insists upon the sensible course of either black-leading or blacking with black bituminous varnish, gilding a moulding or rosette here and there, but never suggesting any other material than that of which the article is really composed.

The manufacture of cast-iron mantels for fireplaces, in addition to stoves and grates, is largely carried on in various localities. The mantels produced by the rival Marbleized-iron companies at New York are certainly very remarkable and exceedingly useful articles of their class. These are covered with a preparation of enamel, the report on the application of which, however, belongs to the class of Mineral Manufactures (Class XXVII). The decorative effect of these substitutes for the more costly material of marble is very good. Some of these cast-iron mantels, as produced by the manufacturers of grates and stoves, are merely japanned black, and being carefully got up are admirable in point of workmanship, and the distribution of the material to the points requiring the greatest amount of strength; and, except when elaborate ornamentation is attempted, the designs are pure and architectonic, though perhaps the latter might be objected to as tending to conceal the real nature of the material.

It is probable that, after the castings for millwork and machinery, the greatest weight of metal would be found to be consumed in the manufacture of the ordinary stoves used for domestic purposes. These stoves are peculiar in their construction, being chiefly adapted to burn wood as fuel, though many of them will also consume coal. They are at once the warming apparatus and kitchen ranges of the great mass of the people, and the varieties are almost innumerable, any improvement in construction being usually the subject of a patent. The manufacture of these articles may be said to be distributed all over the United States; but there are certain localities in which they form an immense staple trade.

The New England States and the State of New York may be said to supply their own wants, as also some of the North Western States, and the British provinces. Philadelphia and Baltimore manufacture stoves for the Southern and South Eastern portion of the Union; Pittsburgh, Pennsylvania, and Cincinnati, Ohio, supplying the demands of the Western and South Western States. These latter affording a wide field for inter-emigration, the wants of the settlers create a constant demand in the two last-named cities, both of which, from their favourable position as regards the supply of raw material and fuel, together with ready means of transport throughout the whole of the Mississippi and Missouri districts, have the supply of the markets of the West. One house at Pittsburgh now manufactures 10,000 stoves per annum, having given up every other branch of the iron casting trade to devote attention to the production of stoves and tea-kettles only.

It is estimated that at least 20,000 stoves per annum are manufactured in Pittsburgh, and 50,000 in Cincinnati.

The patterns or designs of the various houses differ very much. Some of the enterprising obtain designs from, and even have patterns modelled in Europe; and they register, or rather take out a patent right for the design, which can be done apart from the question of construction. Piracy is, of course, complained of, as also invasions of patent rights in construction. Much ingenuity is often displayed in the arrangement of the parts of a stove, and the adaptation of the decoration to strengthen and sustain those portions requiring the greatest amount of metal. Great efforts are made after novelty, alike in construction, ornamentation, and in name, for every stove has a distinct title by which it is known in the market; and the euphony of some of these is often more amusing than appropriate. Except for office stoves and those required for public buildings, European designs are not generally so well adapted to the use of the ordinary article as the less ornate but more common-sense designs of the Americans. There is, however, a wide field for a better style than as yet prevails, and as an absolute necessity exists for a certain amount of decoration of surface alike to strengthen the panels, sustain the angles, and hide defects in casting, which would be too apparent on a mere plane surface, the ornamentation adopted often partakes of the character of an excrescence rather than of a decorative adjunct.

Generally the designs of cast-iron articles manufactured at Cincinnati, are superior in point of fitness to those of Pittsburgh, since the latter are of a more conventional and antiquated character; and it is more than probable that the large number of Germans settled in the former city, influence in a large degree the character of its productions in this respect. One of the best and most original productions in cast iron which came under my notice in the United States was the side of a newspaper-stand in the public reading-room of the Young Men's Mercantile Library Association at Cincinnati. It is constructed of a well-arranged perforated ornament, admirably adapted to the support of a double desk or 'lectern,' upon which the newspapers are placed for perusal. It was manufactured by Messrs Horton and Macy, of Cincinnati. The general character of the productions of this firm in stoves, grates, etc., is superior in point of taste to much of the same work done in England, except in very costly articles. Unfortunately no opportunity presented itself for an interview with the principals of this house, owing to the limited period devoted to the manufactures of Cincinnati as compared with the time required to do them full justice, or the number of European, as compared with American workmen in their employ, would have been ascertained.

The general character of the cast-iron work of the United States is admirable, alike for the purity of surface in the material, and the skill shown in the moulding. The iron being in many instances smelted with charcoal, is of a firm quality and closer grain, so to speak, than that used for similar work in England. Hence the castings produced are sharp in detail and even in surface, and require a very small amount of dressing or filing to complete them. The character of the charcoal-made iron is shown in a remarkable degree in the quality of the wrought-iron nails manufactured at Pittsburgh. These are exceedingly tough, bend like wire, and are very different from the brittle articles of a similar class usually produced in England. There are 14 or 15 establishments for the manufacture of nails in the above city, producing from 8,000 to 10,000 kegs of

100 lbs. weight each, giving about 1,600 tons of nails per week. The manufacture of bar, hoop, and sheet-iron, and iron-wire, is also carried on in four or five of these manufactories. About 2,500 workmen are employed, and the value of the produce is upwards of $4,000,000, or about £1,000,000 sterling.

There are about 30 large foundries and many smaller ones at Pittsburg, employing 2,500 operatives, and consuming 20,000 tons of pig iron annually in the manufacture of various castings. The produce of these foundries is estimated at about $2,000,000.

Tacks of copper, zinc, and iron, hob-nails, and rivets, are also manufactured at Pittsburgh by Messrs Campbell, Chess, and Company. These are produced by machinery, and the articles are well and clearly made with good points and well-formed heads. All the ordinary sizes are made in iron and copper; but only the larger sizes in zinc, as the friability of this metal appears to be unfavourable to the manufacture of the smaller kind.

The manufacture of locks, latches, door furniture, etc., is also largely carried on at Pittsburgh; and the Novelty Works of Messrs Livingston, Roggen, and Company, and the Excelsior Works of Messrs Edwards, Morris, and Company, illustrate this department of industry.

In the Novelty Works 450 workmen are employed, of these 150 are moulders and casters. The character of the small castings produced in this establishment is excellent, and very little filing is required in the fitting of the various parts of the articles together. Locks, latches, and bolts are manufactured in great variety, chiefly of a comparatively cheap kind for the supply of builders in the West. A cheap, simple, and ingenious lock, chiefly for bedroom doors of hotels, is manufactured here. It has two bolts and two keyholes, but there is no connection between them. Thus, the lock cannot be opened except from the side on which it is locked.

Scale beams, platform scales (Fairbank's patent), for counters, railway offices, etc., are also manufactured in large quantities and of good quality.

Paint and coffee mills of various kinds and of a cheap character, the latter being fitted into wooden boxes, form a very important item of business; 450 dozens of coffee mills being made every week.

At the Excelsior Works similar articles are produced, but not to the same extent. About 150 workmen are employed, and the general run of work is the same as at the Novelty Works, except that each has contrivances in locks, latches, and fasteners, peculiar to itself.

A lock with a wrought-iron extension spindle, to adapt it to the varied thickness of doors upon which it may have to be fitted, is a clever and useful invention as made at the Excelsior Works. It consists of a perforated plate acting upon angular notches cut into the spindle; the plate being easily screwed at the requisite distance indicated by the thickness of the door upon which the lock has to be fixed.

Probably the most extensive, and certainly the best conducted and most systematically arranged establishment, for the production of miscellaneous hardware articles in the United States, is that of Messrs Miles Greenwood, and Company, Cincinnati, Ohio. In addition to foundries for large castings common to nearly all establishments of this class, the manufacture of the smaller cast-iron articles is fairly and successfully established, and most of those coming under the denomination of 'builders' hardware,' which a few years ago were almost entirely supplied from England, are now produced here in immense quantities, to supply the constantly necessary requirements of the Western States.

In the important item of butt hinges there can be no doubt of the great superiority of those manufactured by Messrs. Greenwood, alike as regards the general quality of the metal as in the adaptability of strength or weight of material to size. In the finish of the joints great accuracy is obtained, whilst the labour of filing is saved by grinding the joints of the hinges on stones adapted to the purpose, and driven by steam-power. About $15,000 or $20,000 worth of these butt hinges are produced yearly.

Most of the hardware articles for domestic use, usually manufactured by a large class of the hardware establishments of England, are also made here; and the general finish of the articles is certainly of a superior character, except where more than an ordinary attempt is made at ornamentation, and then the results are by no means satisfactory.

Malleable cast iron is also manufactured by Messrs Greenwood into a great variety of articles usually made of wrought iron. These consist of braces or bit-stocks, screw-wrenches, bed-keys, chest-handles, gun-mountings, saddlers' ironmongery and coachware, kettle-ears, thumb-screws, nuts, etc., and have the reputation of being very excellent substitutes for the more costly wrought-iron articles.

Works for the smelting and rolling of copper for use in manufactures have been established on the banks of the Monongahela river, near Pittsburg, by Messrs G. H. Hussey and Company. The ore is from Lake Superior, and is conveyed from thence by way of Cleveland on Lake Erie to Pittsburgh. It consists of large solid masses of native copper, cut with a cold chisel, for convenience of transport, into pieces weighing from 3 to 4 cwt., and is often of great thickness. It is also found in 'pearls' or washings, of very good quality. The copper, when smelted, is cast into bars and ingots, or rolled into sheets, and in these forms is sent to the various seats of manufacture in the United States. Messrs Hussey have one furnace constantly employed in reburning the 'slagg' of former refinings, from which they obtain 2½ per cent. of copper.

In addition to preparing the copper in sheets and bars, copper tubes for steam boilers and cooking dishes, adapted to the stoves so extensively used for domestic purposes, are also manufactured. The latter being tinned inside after the manner of ordinary hollow ware, form durable and useful utensils.

The great abundance of coal in the neighbourhood of Pittsburgh gives it great advantages in the smelting of ores, and in those manufactories in which mineral fuel is an essential. Its cost is $1½ (about 7*s.* sterling) per ton.

The rolling of metals into sheets, and drawing into wire, is carried on to a very considerable extent in New England, and silver, copper, brass, and German silver are thus prepared for use in various manufacturing localities in those States. At Waterbury, in the State of Connecticut, which may be said to be the centre of the various manufacturing towns and villages of the valley of the Naugatuck, the final preparation of metals for manufacturing purposes forms an important branch of industry. Connecticut, as a State, is considered as the chief seat of the metal toy and and tin-ware trades; and, although the manufacturing villages and establishments are spread at irregular distances along the banks of the Connecticut, and in the valleys of the Naugatuck and Housatonic Rivers, the aggregate of industry is very considerable, and is largely on the increase from year to year.

Waterbury and its trades may be taken as a fair specimen of the manufacturing towns and villages alluded to, and is said to have been the seat of the first button manufactory established in the United States. From this has arisen other analogous trades which could be readily carried on in the same establishment, often by the same or similar machinery, or by the same operatives.

The Scovil Manufacturing Company now carry on the manufacture of metal buttons as fairly established, some forty years ago, by the Messrs Scovil. The buttons of this house are good examples of the kind of goods in demand in America, and are chiefly manufactured by machines invented and constructed upon the premises. One machine for punching out the blanks for spherical buttons and raising them to the required convexity in one operation, does its work at the rate of 280 buttons per minute; whilst another machine, fitted with six punches, strikes out 1,800 plain blanks in the same space of time. An ingenious machine is also employed for 'milling' the edges of the ordinary plain gilt buttons, and does the work of 10 or 12 girls.

The military buttons manufactured in this establishment, as also the general range of ornamental articles of the same class, are in better taste than the generality of similar goods produced in England. Less effect is aimed at, and extravagant subjects in high relief are avoided; most probably from a want of dexterity in sinking the dies. The results, however, are much more satisfactory than those arising from the misdirected ingenuity which too frequently characterises the ornamentation of the same kind of articles as produced by the English manufacturer.

Like many other branches of industry, the button trade of the United States has had great difficulties to contend with at the outset, from clashing with certain supposed commercial interests; and even now, unless the American manufacturer can afford to fix a price which will allow a large margin of profit, as compared with European products, the merchant prefers obtaining his supplies from England, France, or Germany.

A beautiful automaton machine for shanking buttons has been lately introduced by the Benedict and Burnham Manufacturing Company, Waterbury, and is in operation in their wire-drawing establishment, being the invention of a mechanic in their employ. The blanks being cut in thin brass, are put into a curved feed-pipe, and descend by their own gravity to the level of the machine. Each blank is carried by the machine under a punch which stamps out the centre hole. The shank is made below by another portion of the machine, from a continuous wire carried along horizontally. From this, the wire to make the shank is cut off and bent, being pushed up at the instant the blank, with the centre hole stamped in it, comes in a vertical line therewith. Another punch descends with a hole in the centre, to allow of the doubled wire forming the shank to go into it, and this gives the blank the requisite concavity and forces the brass tightly round the wire, after which another punch, with a wedge-like edge, descends and opens the wire, spreading it within the concavity, and thus the back of the button is completed. The machine does this in the most perfect manner, at the rate of 180 or 200 per minute. All that is required of the attendant is, to feed the tube with blanks, and when one coil of shank-wire is exhausted to supply another. It is impossible to conceive anything more complete in its way than the machine at work at the period of my visit. Two or three others were in the course of construction, and in a more or less complete state.

Florentine buttons, and tufts for upholsterers, and covered nail buttons, in silk and Utrecht velvet, are also manufactured at Waterbury.

The extensive use of the Daguerreotype in the United States has created a great demand for the plates, metal mountings, or 'trimmings,' as they are technically called, connected with the cases in which the pictures are fitted. The plates for the pictures are invariably manufactured of British copper, as being morc perfectly refined and possessing a more even and sounder fibre than the native copper, therefore better adapted to this purpose, where evenness of surface is indispensable. The Scovil Manufacturing Company, whose productions in buttons have been already quoted, carry on a large trade in the manufacture and supply of these daguerreotype plates, together with the 'mats,' cases, and hinges for the latter. They also manufacture brass, German silver, and silver-plated hinges for cabinet furniture, pianofortes, etc. These are mostly very plain, or, when ornamented, are simply so in outline; and they are well-finished and tasteful-looking articles of their class.

The daguerreotype plates are prepared by the rolling process, the silver being plated upon the copper in the course of the operation, and this is afterwards cut into the various sizes required. The 'mats,' or ornamental mountings, usually fitted into the cases for keeping in the picture, are engraved and chased for the most part by machinery, and by an ingenious adaptation of eccentric movements, the 'sight' of the metal frame is corrected to the true circle, oval, or oblong with turned corners, in the same operation by which the chasing is performed.

So successful has this branch of the metal toy trade of Connecticut become, that it is confidently asserted that the German manufacturers, who formerly supplied these articles, are nearly driven out of the markets of the United States.

Candlesticks and candle-lamps also form a portion of the trade of Waterbury, and the Scovil Company are successfully engaged therein;

but, from the fact that oil is extensively used for domestic illumination, and the lamps most generally used are of a peculiar and somewhat primitive character, the manufacture of candlesticks or candle-lamps is not likely to become of much importance.

Whilst upon this point, it may be well to remark, that the lamps alluded to, and which supply the place of the common candle in domestic use, are articles of great consumption, and are manufactured in immense quantities, chiefly of tin and pewter, and even of common glass and earthenware. They are of various sizes, and are adapted to one or more wicks, according to the purpose for which they are required. These are almost universally used, except when wax tapers or gas are adopted. Attempts are sometimes made to render them more or less decorative, especially the larger kind, with two or wicks for table use; those manufactured of glass, either cut or pressed, being the most successful in this respect. From the peculiar construction required, however, there can be little doubt that a pleasingly ornamental article might be produced by the combination of metal tubes and glass,–the former being used as a column, and the latter material manufactured as the oil-holder and base, or the column and base might be manufactured entirely of metal, and the oil-holder alone of glass. At present the article is unnecessarily clumsy and inelegant. It is, however, in its integrity, and has not yet been overlaid or disguised by false or crude ornamentation, in which use has been bidden defiance to for the sake of decorative effect.

Wire-drawing is an important branch of industry in many parts of Connecticut, chiefly for the manufacture of pins. The Burnham and Benedict Manufacturing Company, Waterbury, and the Waterbury Brass Company, are both largely engaged in wire-drawing and metal rolling; but the chief manufacture of the latter company is that of brass kettles, or pans, of a novel and excellent character and make. Instead of casting them, as is frequently done in England, the article is 'spun' up from a flat plate, by powerful machinery, constructed for the purpose. Nor is this process confined to the smaller sizes, since they range from 1 to 20 or 30 gallons and upwards. There is great quality of strength throughout, with a less weight of metal than usual. The brass, from the rolling and spinning processes, is more consolidated and tougher than when cast, and therefore is not so easily fractured; yet the whole is effected without the process of annealing. The vessel is strengthened with an iron wire worked into the rim in the process of manufacture. A most useful, and even elegant-looking, though plain utensil is thus produced at a moderate price, and is especially well suited by its lightness, capacity, and durability, to the purpose of the emigrant to the Western States or to new countries, for which markets it is chiefly manufactured.

The manufacture of pins and hooks-and-eyes by machinery is an important feature in the trade of Waterbury, as also of other places in the State of Connecticut. The American Pin Company, whose works are at the above named place, manufacture these articles largely, turning out 1,200 packs of solid-headed pins per day, each pack containing 255 dozens of pins. The machinery for manufacturing hooks-and-eyes is very complete, and is somewhat similar to that employed for the same purpose at Birmingham. A machine employed in the American Pin Company's Works for 'papering' the pins is most ingenious, and accomplishes the object with great certainty and rapidity. By a regulation of the company, strangers are not allowed to see the pin-making machines.

The American Pin Company contribute specimens of their manufacture, both in pins and hooks-and-eyes, to the Exhibition at New York.

In stamped brass some progress has been made in several parts of the States. The Waterbury Hook-and-Eye Company, besides manufacturing the articles thus indicated by their title, have just commenced the manufacture of stamped brass articles, such as window cornices, curtain bands, etc.; for the establishment of which trade German operatives have been engaged, but the results could not be fairly judged of, as these workmen had only recently commenced operations. In this branch of the metal trades, however, the Burnham and Benedict Company contribute some admirable specimens to the New York Exhibition, as also excellent examples of rolled brass in large plates.

The manufacture of ladies' hair-pins by automatic machinery has just been commenced by Messrs Blake and Johnson, manufacturers of case-hardened steel rollers and machines for the use of working jewellers, Waterbury. This machine is of their own invention and construction, and is remarkably effective. A quantity of wire is coiled upon a drum or cylinder, and turns round upon its axis as suspended from the ceiling of the workshop. The point of the wire being inserted into the machine, and the power applied, the wire is cut off to the requisite length, carried forward, and bent to the proper angle, and then pointed with the necessary blunt points, and finally dropped into a receiver, quite finished all but lacquering or japanning. The pins are thus made at the rate of 180 per minute, and the machine goes on without any immediate superintendence being required until the whole coil of wire is exhausted.

The manufacture of ornamental brasswork, as applied to the purposes of lighting, forms a progressive and important branch of industry in several large cities, but more especially in Philadelphia. In the establishments of Messrs Cornelius, Baker, and Company, and Messrs Archer and Warner, of that city, as also in that of Messrs H. N. Hooper and Company of Boston, Massachusetts, the rapid progress made of late years in the manufacture of ornamental gas-fittings and table lamps, etc., is fully exemplified.

The chief portion of the work is cast, little or no ornamental stamping being attempted. It is scarcely possible, however, to conceive better work than the generality of these ornamental brass castings. At Philadelphia especially, the greatest attention has been paid to this point, and a peculiar advantage is derived here from the fact that the sand obtained in the vicinity of that place is of so fine a character as to require no sifting for use, and the finest castings are easily made, so far at least as material goes. The pattern is simply modelled in wax, and from this a brass pattern is cast direct, no white metal being used. The brass pattern is carefully and thoroughly chased, and from this all future work is produced. Thus, the shrinkage and variation of size between the white metal pattern and the brass casting, often found to exist in castings made from the former, is avoided, and the register of the two sides of a branch, or other portion of a chandelier or gas bracket requiring to be fitted together, is more perfect than it otherwise would be. The brass pattern, too, takes a sharper and more decisive chasing than white metal, and, as the castings are never chased, as from the fineness of the sand they are sufficiently sharp and effective without it, the accuracy of the pattern is of the first importance, and all that is required to be done after the castings leave the foundry is to file off the very small amount of superfluous metal retained in the casting, and fit the parts together.

The bodies of chandeliers, whether vases or dishes, are invariably spun up from the flat metal plate, instead of being stamped, as is usually the case in England. This is the old method of producing these portions of lamps and similar articles, and appears to have been introduced into practice in America by German workmen. It is not confined to small bodies; but is used for the production of larger sizes than are usually considered practicable. Very large bodies, however, are generally hammered up.

Discs of plate metal, for the purpose of spinning up, are cut by a machine with two wheels, having the sharp edges working against each other, after the manner of a pair of shears. Those circular cutters work with great ease and rapidity, giving great facilities to the workman, and presenting an elegant method of doing laborious work with the greatest possible ease and certainty.

In annealing the spun work, after the first process of raising from the flat plate, it was formerly found that the metal cracked, more particularly in the bottom angle. As the first form from the plane is a simple truncated cone, the second process of spinning, after annealing, gives the requisite curves to the sides. To prevent this cracking during the annealing process, it has been found that the simple bending or squeezing in of the sides of the cone, until the circle becomes a somewhat elongated ellipse, and then placing a quantity within each other, has the desired effect, and cracking rarely if ever takes place. This is stated on the authority of Mr Cornelius,*

* If any additional proof was required as to the value of accurate scientific knowledge as applied to a manufacture in which mechanical invention and great skill in chemistry and metallurgy is so essential to complete success, it will be found in the fact that the gentleman above-named received an education specially adapted to the requirements of the business which his father, the late Mr Cornelius, was about to establish in Philadelphia; and that his studies in abstract science and the various discoveries and expedients, both mechanical and chemical, as resulting therefrom, have given the house, of which he is now the head, an immense advantage over both foreign and domestic competitors. The system, order, and accuracy which prevails throughout this establishment is full evidence of the influence of a mind reaching as far beyond the ordinary traditions of the workshop and foundry, in a scientific sense, as in the practical result it goes beyond the mere dilettantism of speculative science *sans* application. G. W.

of the firm of Messrs Cornelius, Baker, and Company, by whom it has been successfully adopted in practice.

In this establishment, consisting, as it does, of three distinct factories, one for casting and soldering, another for machine-work, and a third for filing and fitting, 700 workmen are employed. Commenced about 30 years ago, it has gone on increasing its sphere of action to the present time, a constant attention to the scientific principles of metallurgy and mechanism having tended to its prosperity from the beginning; when, in the establishment of a business, at that period so novel in its character, and so doubtful as an enterprise, it was more than probable that everything, except traditionary modes of action, would be rather repudiated than encouraged. In point of economy, however, science has been found to be the cheapest as well as the best assistant, as indeed it ever will be when combined with a thorough knowledge of the work to be done by its aid.

In the dipping process, as pursued in these works, great modifications are made in the character and strength of the acids used. It was found that from the variation of temperature at Philadelphia, ranging, as it does, from below zero in the winter to 96° and 98° in the shade in the summer, nitric acid became unmanageable during the hot season, as its fumes were given off so rapidly as to injure the health of the workmen. The accurate scientific knowledge, however, brought to bear upon this point, –one, too, involving the very existence of the trade, except at a frightful destruction to human health and life,–has obviated every difficulty, adapted the acids to the temperature, and the dipping department is comparatively free from noxious fumes, even under the highest of the above temperatures. On the day, June 20th, on which I visited the works, the thermometer stood at 98° in the shade.

The result is equally satisfactory as regards the colour of the work when dipped, some novel effects being produced, and a singular purity of colour obtained.

In lacquering, considerable improvements have also been made. It was found that the lacquers made after the English formula lost colour very quickly, from the extremes of temperature already noted, and that during the months of July and August, when the due point of the barometer is reached in Philadelphia, the red lacquered work always streaked in the direction of the marks of the spinning tool on the broad surface of metal. After a series of experiments, carried through several months, Mr Cornelius succeeded in making a lacquer, which he states to be quite permanent under any variation of temperature.

The usual methods of decorating burnished surfaces by varnish pencillings and dippings, for coloured effects, are adopted here, as in Europe. The lacquering is all done by men, no females being employed.

The columns of table lamps are made of sheet metal formed into tubing, and fluted upon a mandril, by means of a wheel acting in the direction of the axis. This is said to have been adopted by Messrs Cornelius, Baker, and Company before it was attempted in England, and they were enabled by these means to go into the market with a great advantage in price over the cast column lamps imported from Europe, as the cost of that portion of each article was two-thirds less in Philadelphia than in Birmingham.

The manufacture of lamps for the consumption of lard still forms a considerable item of trade, especially for the Western markets; but these are not so much in demand as formerly, owing to the more extensive use of gas. At least 150 patents have been taken out, from time to time, in the United States for contrivances for effecting the consumption of lard for the purposes of lighting. The lard lamp manufactured by this firm, however, appears to have been one of the most successful; the principle of the candle being kept in view, and the heat applied in the direction of the point of illumination.

Great attention is paid both by Messrs Cornelius, Baker, and Company, as also Messrs Archer and Warner, to the perfect accuracy of all their gas-fittings. The gas-works of the city of Philadelphia are celebrated for the perfection to which the manufacture of gas is carried, and the thorough scientific principles upon which every detail of the establishment is carried out, not the least important of which is the uniform gauge of all fittings; so that any part becoming defective is at once repaired without trouble, and, of course, at a less expense than when a constant variation in the gauge of the fittings is permitted.

At Messrs Cornelius, Baker, and Company's manufactory, the sawing of the slit of the gas-burner is executed with the greatest accuracy, and, although not done with such rapidity as by the process usually adopted in England, the greatest exactitude is obtained as to the quantity of gas the burner is capacitated to consume; and whenever a new saw is substituted for an old one, the slit is carefully tested as to its capacity for consumption, by means of a gas-meter placed by the side of the workman. In short, guessing is avoided in everything connected with this establishment.

In the fitting of the pipes the same accuracy and care is manifested. The screw is turned in a lathe to prevent the possibility of splitting the pipe, which is more or less invariably done by the ordinary screw-plate; but the crack being very minute, is not discovered until after having been some time in use, when the gas begins to escape, and continues until a permanent leakage is established.

In order to secure the joints completely, a composition of wax, resin, and venetian red is applied to the tap, which is sufficiently hot to melt it; the pipe is then screwed in, and the joint is at once sounder and cleaner than when cemented by the application of white lead, the usual material employed for this purpose in England.

Amongst the workmen employed by Messrs Cornelius, Baker, and Company, are eight or nine Englishmen,* the rest being Germans,

* One of the modellers and designers employed in this manufactory at the date of my visit, was a student of the Birmingham School of Design at the period I quitted my duties there, *pro tem.*, to proceed to the United States in the business of this Commission.

Another modeller from Birmingham had been so employed for some years. G. W.

French, and Americans; but the majority are undoubtedly native workmen.

Messrs Archer and Warner employ about 225 workmen, and are engaged in precisely the same trade as Messrs Cornelius, Baker, and Company. The remarks as to the character of the work produced by the last-named firm, especially gas-fittings, and the perfect division of labour, which is not so general a feature in American as in European manufactories, applies with equal force to both establishments, and though that of Messrs Archer and Warner is not so extensive, its operations are carried on in a systematic and efficient manner, the results being shown in the articles produced, which are excellent of their class. The designs are adapted to the markets to be supplied, and as these are generally of a third or fourth rate character, according to the English standard, a redundance of ornament, rather than purity of style, is the chief point aimed at.

In no instance can the chandeliers, gaseliers, or lamps manufactured in the United States be said to come into competition with the better class of productions as manufactured in Birmingham. The demand for the larger kinds being chiefly for use in hotels, public saloons, etc., a showy and attractive article, at a comparatively cheap rate, is that most in demand; and though some of the details of those produced for these purposes are admirably modelled, there is often a great lack of that congruity in the parts by which a perfect *ensemble* can alone be realised.

The smaller articles, for domestic use, are often more effective and far more tasteful than the more ambitious examples, less ornamentation being aimed at; for, as lightness in weight of metal, and cheapness in cost, are the chief points, redundancy in decoration is not possible. At the same time as a certain amount of effect is essential, and this is attained by a repetition of the same details in varied forms of application; the chains used in the cheaper kinds of chandelier are formed of links, which, though very ornamental and effective, are scarcely touched by the file, being sufficiently accurate, when put together, to answer the purpose of effect and give the appearance of a large amount of work where very little really exists.

A peculiar kind of girandole, chiefly purchased by the artizan class as a chimney ornament, is largely manufactured. The base and vertical support is of brass with glass prisms, and often with two branches for candles. The choice of subject for the vertical portion is often ludicrously inappropriate, yet the sale of these things is said to be immense, from their showy character and low price.

Comparing the prices of the low-priced articles, it may be calculated that an English chandelier which would cost $18 (say £4 5*s*. sterling), is very fairly imitated, or more correctly speaking, its place in the market is supplied by the American manufacturer at $11½ (about £2 15*s*. sterling).

The manufacture of decorative brass work for ships, steam-vessels, etc., is largely carried on by Messrs H. H. Hooper and Company, Boston, Massachusetts, in addition to the manufacture of chandeliers. The cast work is generally in good taste, and there is more artistic breadth of effect in the ornamental portion of the articles manufactured by Messrs Hooper, than in those produced at Philadelphia; the latter being florid and showy, whilst the former are more massive and simple.

Spun work is even more largely used than in the manufactories at Philadelphia, and the skill and dexterity of the workmen in this department is very great.

Parabolic reflectors, for ships' lamps, are manufactured by Messrs Hooper by a stamping process. These reflectors are generally of a much larger size than usually attempted by this process; but it is effected with complete success, both as regards accuracy of form and lowness of price, as compared with hammered work.

Bells of an excellent character are also cast in this foundry. It may be remarked, however, that the most extensive bell foundry in the United States is that of Messrs Meneley, Troy, State of New York.

The contributions in ornamental brass and bronze work to the Exhibition at New York are generally good, the most extensive display being made by Messrs Cornelius, Baker, and Company, Philadelphia, and the remarks already made as to the character of their productions, apply to their contributions to the Exhibition, as those fully bear out the opinions expressed as to the general excellence of the work, and its adaptation to the markets it is intended to supply.*

In ornamental bronzes, the specimens of native manufacture are so mingled with those of European production, as to prevent a proper estimate of the position of this branch of industry. There are, however, two or three establishments in New York, the proprietors of which employ French and German workmen in the production of many articles of mere ornament or exceptional use, but which can scarcely be considered as coming under the denomination of works of art in metal.

In lamps there are numerous contrivances for the consumption of oil, lard, camphine, etc., These are all more or less of a decorative character, but do not call for special notice after the remarks already made as to the manufacture of these articles.

The specimens of brass joints for gas and steam tubing, manufactured by Mr Samuel Griffiths, New York, and contributed to the Exhibition, are sufficient proof of the admirable manner in which those useful articles are manufactured.

In locks the contributions are pretty extensive; but want of time and the difficulty in obtaining access to the goods, from the lack of authorised attendants, and the unarranged state of some portions of this department, prevented that close examination into details which was requisite in order to form an accurate opinion. The general finish of the articles, however, appeared to be good.

Iron safes for banks, counting-houses, etc., are manufactured to a large extent by many hardware houses. These are as varied in construction

---

* A prize medal was awarded to this firm by the jury (Class XXII) of the Great Exhibition of 1851, for chandeliers.

as the many contrivances of a similar kind for the safe custody of cash and books are in England, and are the subjects of as many patents as the authorities at Washington can be induced to grant.* The workmanship of nearly all the safes exhibited at New York is of a very excellent character, and the style and finish generally tasteful and appropriate.

In hollow ware, Messrs Cresson and Company, of Philadelphia, to whose foundry other more pressing calls prevented a visit, exhibit various culinary and household articles of a good and useful character, the produce of their manufactory.

The specimens of tin-ware show a good character of work, and, as already stated, this branch of the hardware trade is largely carried on in various parts of the State of Connecticut, whence most of the examples exhibited at New York come. A useful and well-made series of machines and tools for the use of tin-plate works, manufactured by Messrs Roys and Wilcox, Mattabesset Works, East Berlin, Connecticut, and exhibited by them, show several novel contrivances for economising labour in the manufacture of this ware.

Gimlet screws, in which the point of the screw supersedes the use of a gimlet in making the requisite hole, are largely manufactured by the New England Screw Company, and now extensively used in the United States; and the variety of sizes exhibited at New York show the applicability of this simple but useful contrivance to be more extended than is generally supposed.

Carriage springs would appear to form an important item of manufacture, and the examples shown at the Exhibition, at New York, are of good make and finish.

Miscellaneous hardware, chiefly of Connecticut manufacture, articles of wire-work, bed-springs, and sundry contrivances more or less useful and ingenious, show the extent to which the manufacture of all kinds of articles in every day use is now carried on in the United States; and that whilst the fitness, strength, and proper construction is carefully attended to, there is no lack of executive skill in substantial objects of utility.

Manufactures in Britannia metal will be reported upon with those departments of industry coming under the head 'Works in Gold and Silver, and their imitations,' as these usually form a branch of those establishments in which gold, silver, and electro-plated wares are carried on.

As the application of cast iron to architectural purposes has been already alluded to, more especially in its application to external construction and decoration, it may be desirable to notice here that in the Library of the Congress of the United States, now in the course of construction in the Capitol, at Washington, the whole of the interior fittings are of iron. The piers supporting the book-shelves, and the balustrade of the gallery, which is carried all round the room, are of this material, cast in

---

* By the patent laws of the United States, the Commissioner of Patents decides whether or not any contrivance or invention is a fit subject for an exclusive right.

ornamental forms, with medallions, also in iron, of Washington, Franklin, and other eminent American statesmen, These latter are in high relief, and are most admirably modelled and cast. The book-shelves are also plates of iron; and whilst the whole is thus rendered fireproof, it is also highly ornamental.

12

Metal Manufactures

Works in Precious Metals, and their Imitations, Jewellery and other Personal Ornaments: Class XXIII

A love of display, which is certainly one of the characteristics of the people of the United States, affords ample encouragement to those branches of industry in which gold, silver, and precious stones are employed; and, as a matter of course, the imitations of works of this class are equally in demand amongst those whose means will not permit of an indulgence in the more genuine articles. At present this tendency towards the more luxurious forms in which the decorative arts are employed is undirected, so far as the great mass of the people is concerned, by any principles of taste, and is, therefore, completely at the mercy of the unreasoning judgment of the manufacturer, whether domestic or foreign, who caters to what he supposes to be the prevailing taste of his customers. The American manufacturer, however, is in some respects wiser than his foreign competitor, and in many instances leaves the ultra-ornate to be supplied from Birmingham and Sheffield, and directs his energies to the development of a better and less exuberant style, which he finds is demanded by the more refined amongst his countrymen. Thus works are occasionally produced which would not discredit any European house. Still the manufacturer has to consider the demands of the market, and, as far as possible, regulate the character of his staple productions thereto; and, as this market gets its tone in a large measure from the abominations which the European manufacturer chooses to believe are best suited to the wants and wishes of his Transatlantic customers, the American producer has to follow the lead thus given. The shopkeeper, however, is perhaps more to be blamed than any other person. He assumes that his customers want an enormous display of unmeaning decoration, and therefore orders little else from the manufacturers. The public have, consequently, little choice; and constantly seeing ugliness and exuberance in ornamentation, its taste and judgment becomes vitiated and depraved, and the simplicity and fitness of pure forms appear tame rather than beautiful. A chaste style of decoration, therefore, appears to be plainness itself amidst the glittering nonsense usually presented for inspection. Thus the incessant effort to meet supposititious demands on the part of the public results in creating the very want of taste and judgment complained of in the abstract, but practically catered to and encouraged. The more enlightened of the manufacturers of the United States, however, are strongly impressed with the conviction, that they and their customers of the shopkeeping class are the real instructors or vitiators of the taste of the people; and though at present borne down by the style and character of the goods most in demand, as arising from the causes above named, yet with an educated people they believe that this cannot last for any great length of time. It would be well, therefore, for those foreign manufacturers who supply these articles to the American markets to consider how far a purer taste will affect the demand for their productions, as the time may quickly come when the doctrine 'that anything will do, so that there is enough of it,' will no longer be true of the kind of goods it is now their province to supply so largely.

It has been thought desirable to make these remarks preliminary to reporting upon the present state of the class of industry under consideration, as there exists, in England at least, very erroneous views as to the present and growing tastes of the people of the United States. For whatever the past might have been, or the present may be, the future is likely to be very different. The daily increasing intercourse with Europe is pregnant with great changes. Persons of even moderate means now make a point of visiting, not only England, but those continental States in which the arts have flourished for ages, to search out and carry back with them new ideas of art, and of correct principles as applied thereto, which their education and the eminently practical turn of their minds will not allow to slumber on their return home. Looking with contempt upon those articles of luxury which formerly pleased a merely puerile taste, these individuals influence their countrymen more largely than may be thought possible, since the educated intelligence of the latter renders them equally alive to the value of sound principles; and a few excellent examples in the way of illustration have an influence, and convey an amount of instruction, which the uneducated could never feel or understand. English manufacturers have it especially in their power to either assist in the formation of a purer taste in the American people, or, by neglecting their growing judgment, to oppose and retard, but not crush it. By assisting its development they are likely to secure to themselves, or, at least, share very largely in a market for a long time almost entirely their own, but which is gradually being occupied by their Transatlantic competitors. By neglecting the progress of a more enlightened and purer taste, they will as inevitably lose it altogether.

The manufacture of gold and silver plate is more or less carried on in nearly all the larger cities, especially New York, Boston, and Philadelphia. In the last named it partakes of the character of a settled trade, there being some twelve or fourteen establishments in which a considerable number of persons are employed, and the productions of which are of a varied, but for the most part of a useful as well as an ornamental character. Table services, and all the articles of utility comprised in suites of plate for domestic use, form the staple articles; and these are manufactured in large quantities.

The workmanship is usually sound, but it often happens that on close examination a deficiency in that nicety of finish, especially in the chasing, which characterises the best English work, is observable. Still it is rarely found that the equally, or perhaps more objectionable practice of over chasing, to the destruction of the artistic effect of the details, is committed. The fault is evidently that of timidity of handling; but there is a wisdom in leaving off at the right time, which the elaborate chasings of English works rarely display.

In most of the manufactories a few European workmen are to be found; but the Americans engaged in this department of industry are usually of a superior class, and it is remarkable how soon they get into the system of those amongst whom they are thrown as mere learners. In this as in other branches of industry, their minds being thoroughly prepared by education, they seem to seize upon and master even very difficult points in manipulation and construction, as it were by mere instinct. It is, however, more than probable that this rapidity of conception is unfavourable to that perfect and complete execution which is often characteristic of the more plodding and painstaking workman; and on the whole there is, as already stated, less finish than is to be found in kindred works as produced in England. This opinion applies to the general run of productions in the class under consideration, whether in gold and silver, or the less costly, but often equally meritorious productions in German silver, electro-plated wares, and Britannia metal. At Philadelphia Messrs Bailey and Company produce excellent articles in silver, many of which are in good taste. The gold filigree and pearl work manufactured in this establishment is, when the French taste is not followed, both superior in design and workmanship. There is, however, in these trinkets, as in other things, too great a tendency to follow the mere conceits of the European works of the same kind, such as close imitations of natural forms, etc., rather than rely upon the good sense which suggests constructions adapted to the nature of the material used. When this is done it is almost invariably with success.

Messrs Conrad, Bard, and Son, also of Philadelphia, manufacture gold and silver plate, but a considerable trade is carried on by them in the manufacture of spoons and forks. For this purpose, they use a machine invented and manufactured upon their own premises, by which the production of these articles is very much facilitated.* Two circular dies or

---

* A similar machine to this is said to be in operation at Paris. The principle is the same as that invented by Krupp, the exclusive right to use which, in England, has been secured by Messrs Elkington, Mason, and Company, of Birmingham, and now so successfully worked by them for the manufacture of spoons and forks. The machine of Messrs Conrad, Bard, and Son, is suitable only to very thin work; and for the want of a self-acting guide, as

rollers are sunk with forms of the articles to be rolled out. These are usually spoons of two sizes, a fork, and the side of a knife-handle. As the intaglio of one die is accurately adjusted to, and agrees with that of the other, one forming the *obverse* and the other the *reverse* of the pattern, both sides of the article are perfected at once, and the rollers accomplish the work of a stamp press in a much more effective and economical manner. The rollers are about 5 inches in diameter, and are of course manufactured of the best steel and case hardened. Being set to the thickness of the articles required, and that of the sheet metal from which they are to be made, the perfect passage of the pieces through the rollers is secured by a series of notches sunk in the margin of each figure upon the die, the impression of these being in the superfluous metal surrounding the work when delivered from the machine. This has to be cut off, and is effected by a circular saw, the prongs of the forks being cut in a similar manner. It is an ingenious and useful invention as applied to light articles in silver, taking up a small amount of space in a workshop and doing a much larger amount of work than a stamp press.

The Ames Manufacturing Company, Chicopee, Massachusetts, are largely engaged in the manufacture of silver and electro-plated wares, and the castings in that branch of their establishment were of a very excellent character. Attention is chiefly confined here to articles most constantly in demand, castor-frames, dish-covers, breakfast and tea services, and the most decided requisites of table suites, being the staple. There was much taste and great excellence of workmanship in some of the examples. In others, the usual mistakes in following European patterns were visible.

This extensive manufactory is curious from the great variety of articles and the multiplicity of operations carried on. Brass cannon are cast with great success, and a colossal statue of De Witt Clinton, as also a work of some excellence, the Angel of the Resurrection, both by Mr H. K. Brown, the sculptor, the latter being a contribution to the Exhibition at New York, have been lately produced, and in a manner highly creditable to the skill of all parties concerned. This is a novel branch of the metallic arts in the United States, and one requiring no small amount of technical knowledge to get through with success.

The artistic castings in iron produced here are equal to anything by the best houses in Europe, and is abundantly proved by the medallions in altorelievo of Washington, Franklin, and other eminent men, executed for the library of the Congress in the Capitol at Washington, as a portion of the decorations of the interior fittings, which, as stated before, are entirely of iron.

Swords also form an important branch of manufacture by the Ames Company, the blades, hilts, and scabbards being all produced upon the

---

applied to Krupp's machine, the work is liable to come out in a curve instead of a right line, and the amount of scrap metal and dust arising from the margin surrounding the article, as delivered from the machine, must be a serious drawback to its profitable use as compared with the more perfect machine in use at Birmingham. G. W.

premises. There is also a manufactory of turbine wheels, cotton machinery, planing machines and lathes.

A few English workmen are employed and one or two Germans as modellers; the rest of the workmen are all Americans.

At Dorchester, Massachusetts, Messrs R. Gleason and Company manufacture plated wares of fair average quality, employing about 100 workmen. These are white metal with electro-plated surface. Some of the castors made here are very tasteful and original, and the forms of some of the tea and breakfast sets are also good. The dish-covers are mostly after English models. The general style of the article, as also the workmanship, especially as regards strength and durability, is highly to be commended considering the influence of the ordinary demands of the market for the light, cheap, and showy; but there is certainly a lack of perfect finish in the details.

The Britannia wares of this house are also good.

Lamps in brass form a department of Messrs Gleason's trade. Die-work is chiefly used here, with some spun work in those portions of the articles to which it is most applicable.

Messrs Reid and Barton, Taunton, Massachussetts, produce Britannia metal goods on a considerable scale, it being the chief branch of their trade. The electro-plated articles of this house, however, are of excellent character.

In Britannia metal originality of design is aimed at, and with some degree of success; but they copy any successful patterns from Europe, more especially those produced at Sheffield.

They employ about 100 persons, but only one Englishman. The remarks already made as to the lack of accuracy in finish applies to Messrs Reid and Barton's productions, yet the effect of the work is generally good, and their contributions to the New York Exhibition are at once highly creditable and pleasing specimens of a useful class of articles.

The contributions in this class displayed in the Exhibition at New York are varied in character, but presenting certain points of excellence, and the ordinary over-ornamented articles are the exception.

Messrs Bell, Black, and Company, of New York, as also Messrs Tiffany, Young, and Ellis, of the same city, besides being extensive importers of European plate, are also manufacturers, and their contributions are of a good character. A tea service of California gold, with a *plateau* of silver, is remarkable for the elegance of the smaller pieces and the general excellence of the workmanship, and amongst the variety of articles exhibited by this house are some of elegant and tasteful form and ornamentation. The engraved work, too, is good, and of better design than usual. Two ewers exhibited by Messrs Tiffany are also of good form and excellent in decoration, and a dinner service also contributed by them is neat and elegant, being the very reverse of the usual mode.

Messrs Jones, Ball, and Company, Boston, Massachusetts, exhibit a very satisfactory series of the ordinary articles in silver for domestic use.

Forks and spoons of good make are contributed by Messrs Rogers, Brothers, Hartford, Connecticut.

The exhibits in Britannia metal goods are of a useful character; the 'turned' ware, manufactured at Troy, New York, by Mr John H. Whitlock, being chiefly noticeable for their simplicity and utility; those of Messrs Hattersly and Dickenson, of Newark, New Jersey, being more ornate, but of good forms.

The contributions in jewellery do not call for any special remarks; but the manufacture of gilt toys, and imitations of gold and silver ornaments for personal decoration, forms a very considerable item in the industrial productions of some parts of New England.

At Waterbury, Connecticut, the Waterbury Jewellery Company manufacture the ordinary gilt articles set with coloured glass and artificial stones. The designs are more like the Paris productions of the same class than those of Birmingham, but both French and English articles are imitated, and the work is of average quality. Machinery is extensively used in this manufacture, and with success as regards the result as seen in the quality of the work.

Attleborough, Massachusetts, is the chief seat of the gilt toy and jewellery trade of the United States, but an intended visit was obliged to be abandoned for want of time.

In 1850 there were nineteen establishments for the manufacture of jewellery, employing 381 operatives, the value of the articles produced annually being $478,200, or about £100,000 sterling. In this place the first manufactory of gilt buttons, as also the first manufactory of gold and gilt jewellery and silver ware, were established in the United States.

The manufacture of bronzes, etc., so far as at present carried on, has been alluded to in Class XXII.

13

Glass, Porcelain, Furniture

Glass Manufactures: Class XXIV

The glass manufactures of the United States are fairly established, and there can be no doubt of the present healthy position and future progress of this branch of industry. With superior materials in abundance, and skilled labour increasing with the development of the trade in which it is employed, the progress from the earlier productions, in which articles of common use were the staple, to those rich and decorative works which belong to luxury rather than to utility, has been steady and continuous; and though from the position of the people the more useful articles are those to which the manufacturer directs most attention, yet there is a laudable spirit abroad which compels him to attend to those branches of his trade wherein his European rival and instructor excels in the artistic elements of production.

It must be remembered, too, that the method of manufacturing glass by means of metal moulds, in imitation of cutting, now generally known as 'pressed glass,' is an American invention, and was first introduced into England in 1834, by Messrs Richardson of Stourbridge.

Pittsburgh, Pennsylvania; Boston, Massachusetts; and Jersey City, New Jersey; are respectively seats of the glass trade, although there are many manufactories spread over different portions of the Federal Union, especially in the New England States.

Pittsburg is chiefly the seat of the glass trade for the supply of the South and West, and its productions are confined exclusively to the more useful articles for domestic use, of which there are eight establishments, employing 500 operatives, and the manufacture of window glass, of which there are eleven establishments, employing about 600 hands, and seven furnaces for the manufacture of phials. Cylinder glass, of a common character, is the chief make in window glass; and with materials in abundance either close at hand, or within the range of the water carriage on the Ohio, Alleghany, or Monongahela rivers, which unite at Pittsburg as a centre, every facility exists for the prosecution of the glass manufacture to an enormous extent. Coal, too, is cheap and in great abundance.

The Fort Pitt Glass Works are now carried on by Messrs Curling, Robertson, and Company. The managing partner is a native of Birmingham, but has been at Pittsburg many years. His father was a working glass-maker, who emigrated in 1812 as a gardener, to evade the laws then in force against the emigration of skilled mechanics, etc. About 100 workmen are employed here, and there are three furnaces of five pots each. The make is mostly in pressed glass, and all the more useful articles are manufactured; these being of good form and clear colour. The peculiar wants of the consumer are consulted, and the goods are sent to all parts of the United States. The chief markets are those of the South and West.

All the materials are prepared on the premises. Missouri sand is used for the finer articles, and Alleghany sandstone, when found free from iron, is used for the commoner purposes after being pounded and burnt. These materials are obtained in great abundance in Pittsburg.

Messrs Bakewell, Pears, and Company carry on a similar trade, and produce an excellent assortment of useful goods which are entirely of pressed work. The average price of tumbler glasses is 2*s*. per dozen. Pressed glass decanters are manufactured by both the above named firms very successfully. These are considered as novelties in American manufacture. They are generally first made as tumblers, and then worked up in the neck, etc., afterwards. The result is a cheap presentable article, very much in demand.

A new design for a 'pressed' decanter just completed by Messrs Curling, Robertson, and Company, appeared, from its form, to have been made by the 'plunger' going in at the bottom, thus making the body and neck at the first operation, the bottom being closed afterwards.

The metal moulds for the manufacture of pressed glass are made upon the premises of each manufacturer; and as the production of a novelty is a point aimed at by each, as an important element of success, the getting up of new designs demands a considerable amount of care and attention.

The Boston and Sandwich Glass Company, Sandwich, Massachusetts, have an extensive establishment in which a good style of work is carefully carried out. There are four furnaces of ten pots each; the make of glass is about 60 tons per week, and 500 persons are employed. The general run of useful articles in cut and pressed glass are made by this firm. The forms are generally tasteful and well adapted to the material and mode of manufacture, especially in pressed glass, of which very large examples are produced. Costly cut articles, however, are not manufactured.

The Boston and Sandwich Glass Company make large quantities of the lamps peculiar to the country, already described in Class XXII as being used instead of candlesticks. The exceedingly low price at which these articles are sold creates a great dmand feor them, as the glass offers facilities for cleaning which no other material presents, the oil being so easily wiped off the surface; whilst a decorative effect is produced at a cheap rate by the adaptation of the forms to the pressed methods of manufacture.

Probably the most complete establishment in the United States for the manufacture of all kinds of glassware is that of the New England Glass Company, East Cambridge, Boston, Massachusetts.

This is an extensive concern, conducted with admirable system, the productions comprising every kind of flint and ornamental cut and coloured glass, many of the specimens of the latter approaching the Bohemian standard in quality of colour and gilding. A lightness of form, too, has been obtained in some articles which indicate considerable skill and accuracy in manufacture and finish.

About 550 operatives are employed. There are two 12-pot and three 10-pot furnaces, of from 1,500 to 1,800 lbs. of metal each.

The silex used is a sand of beautiful quality, the produce of Berkshire County, Massachusetts. The company manufacture their own minium, the material in pig lead being brought from the State of Missouri, in which lead has been of late years worked to a considerable extent.

The glass when manufactured is very clear and pure in colour, the pressed work being remarkably smooth, even, and free from *stria*. The designs and forms are generally well adapted to the material and mode of production. Amongst the large examples, a centre-piece of a tazza form, with the patera or dish from 2 feet to 2 feet 6 inches in diameter, and the stem and base in proportion, afforded abundant evidence of the extent to which this branch of the business has been carried, and the facility with which large articles are now produced by the pressed method.

The cut glass showed care and skill; the bottles are generally of excellent form and make, clean in the neck, and not over-ornamented with cutting. A satisfactory result, too, is obtained in some articles by the combination of cutting and pressing. The smaller details are pressed, and the larger and broader parts cut.

The opal glass manufactured at this establishment is better than the great portion of that produced in Europe. It is clear in body and pure in colour. Bone is used instead of arsenic for the finer qualities. For common articles the latter material is employed as usual.

Gilding is executed very successfully, and it is stated that other houses in the United States have attempted to gild their own work and have failed, this being the only establishment in which it is now carried on.

The imitations of Bohemian glass are good, and some of the designs are of a superior character, whilst others are of the usual inadaptive types, and are over-ornamented. The Venetian paper weights manufactured here are very excellent examples of their class.

Articles in glass, silvered and engraved by Kidd's process, are extensively manufactured, and are in great demand in the United States from its effective and showy appearance.

The New England Glass Company have paid a large share of attention to the production of a good quality of glass for chemical and philosophical purposes; and this now forms another important branch of business.

The contributions of this firm to the Exhibition at New York sustain its well-earned reputation. A complete service, pressed, in a neat diamond pattern, and another of cut glass, are very elegant in character, and beautiful in workmanship.

The other contributions in glass do not call for special remark.

Time did not permit of an inspection of the several glass works established in Jersey City, and Brooklyn, Long Island, both opposite to New York City, some of which have a reputation for producing good articles of their respective kinds.

14

Glass, Porcelain, Furniture

Porcelain, and other Ceramic Manufactures: Class XXV

The manufacture of porcelain is quite in its infancy in the States, but there can be little doubt that its progress will be, at no distant period, both rapid and effective, as materials in abundance are to be found in many portions of the vast territory of the Union. When the value and properties of these materials are more thoroughly understood than at present, by those most interested, it would be contrary to the genius of the people of the United States if efforts are not made to realize every advantage which their working can give.

At the Queen's Ware Pottery Works, Birmingham, Pittsburgh, Pennsylvania, a ware similar to that known in England as 'queen's ware' is manufactured. The forms of some of the articles are good, and the syrup jugs, mounted with Britannia metal, an article much used in the Western States, are very fair specimens of their kind. All the attempts at decoration were in relief only, and though a blue (cobalt) glaze is used, still the colour of the clay makes the result anything but satisfactory. The oxyde of cobalt, by which this blue glaze is produced, is manufactured in large quantities by Messrs Coffin, Hay, and Company, Philadelphia, who export it to Europe. The ore is found in great quantities near both the Missouri and Mississippi Rivers. The fine clay of the State of Missouri is obtainable in great abundance, and that, with the porcelain clays of the State of Delaware, is said to have been already exported in small quantities, at least, to England.

An experimental specimen of the former clay, tried with a blue glaze, was very pure in colour and clear in the fracture.

The proprietor of the Queen's Ware Pottery Works is an Englishman named Bennett, and it is a subject of some regret to report that he was the only person throughout the whole of this inquiry who refused to myself the information asked for. Personal and merely private questions were always carefully avoided, as calculated to prove embarrassing to those who, in a spirit of candour and liberality, were willing to furnish all needful information on public grounds only. Mr Bennett, however, declined to give any information as to the progress and present position of the trade in which he has been successfully engaged for some years past. It is a source of infinite satisfaction that a course at once so puerile and so obnoxious to common sense, was taken by a foreigner rather than by a native of the United States, and that an intelligent American gentleman, deeply interested in the porcelain trade, did all in his power to compensate for the absurd course pursued by Mr Bennett, who, in endeavouring to keep secret the resources and power of a country to whose hospitality and encouragement he owes his present position, simply excited a greater amount of curiosity and inquiry.

The manufacture of the common kind of earthenware is established and carried on at Liverpool, at some distance from Pittsburg, on the Ohio River. Here there are several manufactories, but want of time did not permit of a visit. The chief markets for the produce of these potteries are found in the Southern and Western States, and from the information in which subsequent inquiries resulted, it would appear that Pittsburg and Liverpool are likely to become the great seats of the porcelain and earthenware manufacture, from the fact that fuel is cheap, materials of various kinds abundant, whilst the demand must be constantly on the increase as the Western States become more populated. It is scarcely possible, therefore, to imagine a wider field for enterprise and skilled labour than is here presented, or one in which there would be a greater certainty of ultimate success.

Messrs O.A. Gager and Company, Bennington, in the State of Vermont, manufacture a ware known as 'Fenton's patent flint enamelled ware.' This is produced from a very white clay found near Charleston, South Carolina, and the beds of which this firm has secured the right of working for twenty-one years. It takes a beautiful glaze, and presents a remarkably clear and transparent appearance. It is chiefly used in the manufacture of the mottled-brown ware, but presents a white fracture, as the mottled surface is simply in the glaze. Of this kind of ware, even at the present rate of production, $1,500 worth per week is manufactured, for which a ready market is found. This ware is got up in imitation of a much commoner article usually manufactured of yellow clay. The present trade, however, appears to be merely a beginning, and a large development, both in the kind and quality of production, may be easily foreseen; as the facilities are great, and an enormous demand awaits a corresponding supply.

The useful article of bleaching-rings is manufactured of this porcelain clay, and finished with the brown glaze. These are largely used by bleachers, dyers, and calico printers, for running pieces of cloth through from one part of the dye or bleach-house to another in the various processes. They are preferable to the glass rings usually employed, as they are less expensive, and are not so liable to break under the expansion and contraction of the iron stays in which they are fixed for use. The demand for this article alone is greater than the manufactory can at present supply.

A parian ware, of good colour and surface, is also manufactured by this firm, from materials obtained near the manufactory, in the State of Vermont. It is composed of the flint, quartz, felspar, and clay obtained from the Green Mountain district and the adjacent rivers. This ware is unglazed, and will not stand the test of hot water. It presents, however, a valuable and useful material for a large class of ceramic productions not as yet manufactured in the United States, especially the more ornamental articles for which parian is so largely used in England.

In the Exhibition at New York is a very full exposition of the various kinds of articles and wares manufactured by this house.

In the higher branches of the porcelain trade, the productions of Messrs Haughwout and Dailey, New York, take a fair position, though the articles contributed by them to the New York Exhibition are by no means equal to the best examples shown in their own establishment. Without being manufacturers of porcelain, they employ about 100 persons, male and female, in the decorations of the various articles they import in the *bisque* state. Many of the workmen are Englishmen, the females being chiefly employed in the gilding and burnishing.

The articles consist of the usual stock pieces, such as flower-vases, dessert and toilet services, decorated with landscapes, figures, fruits, flowers, etc., and profusely gilded. These are equal, sometimes superior, to the average productions of the same class made and decorated in Europe. Exuberantly ornate, they are supposed to be adapted to the market, according to the prevailing idea of the requirements of the purchaser, as already alluded to in the consideration of thc influcnce of a suppositious public taste in design in works in gold and silver. The worst feature, howcver, is, that here, as in Europe, if the purchaser's taste is above the average of that which the manufacturer or shopkeeper fixes as his standard, the intended customer must come down to the latter, or give up his intention to purchase, since the less ornamental, but really more tasteful article, is not to be had. The shopkeeper, believing that a redundancy of decoration is the thing most in demand, and which he thinks best, has not provided for those whose taste may be in advance of his own, and, worse still, discourages any attempt in the manufacturer or the workman to produce something better, because, in his opinion, it will not *sell*.

There are some useful porcelain articles, such as sign letters and door furniture, amongst the few exhibits of this class contributed to the Exhibition at New York, in which there is both elegance of design and skill in execution, both of painting and gilding.

One branch of the ceramic manufacture, as connected with door furniture, is peculiar to the United States, and this is, the very extensive use of earthenware knobs for the fittings of common locks. These are usually made of a striated mixture of yellow and red clay, and finished with a strong brown glaze. Arrangements are made in the moulding for the insertion of the spindle, and they are exceedingly useful and serviceable articles. In testing the strength of one of these knobs, it required a very heavy blow to fracture it. They are extensively made at Pittsburg and Liverpool, Pennsylvania, as large quantities are consumed in the former place by the miscellaneous hardware houses, in fitting locks and latches.

15

Glass, Porcelain, Furniture

Decorative Furniture, Upholstery, etc.: Class XXVI

The manufacture of furniture is carried on to a great extent in nearly all the large cities and towns of the United States. In the former, extensive establishments exist for the production of decorative furniture, the constructive portions being for the most part prepared by machinery,* by which labour is greatly economised, as well as perfect accuracy of fit in the same article of furniture.

No department of industry is more largely represented in the New York Exhibition than that of furniture. This may arise from the fact that there is a large number of manufactories in New York itself, thus affording facilities for the better display of this class of productions.

At Cincinnati, Philadelphia, and Boston, there are also a number of very extensive establishments. In the former city, German workmen are largely employed, and the carvers, generally, are either Frenchmen or Germans, with a few Englishmen.

Elaborate carving is the rule rather than the exception, and this is fully borne out by the examples shown in the Exhibition, the best of which are evidently the work of Europeans, and the designs partake more of French and German than of English characteristics.

The American woods are not so much used as might be expected, walnut and oak being those which are generally selected.

Owing to the imperfect state of the arrangements, one contribution of furniture, said to be of very novel construction, could not be examined. In the space of an ordinary sideboard, it was stated that a bedstead and complete chamber suite was contained. As it was locked up, and no key to be found, its merits were unascertained, nor could any one connected with the Exhibition state who it belonged to, or where it came from.

The ordinary class of useful and cheap furniture, so largely manufactured for the Western States, was not represented in the Exhibition.

---

* See Mr Whitworth's Report.

This is to be regretted, as much ingenuity and economy of material is often shown in the production of this class of goods.

In papier mâché, the productions are all below mediocrity, being chiefly imitations of the worst style of pearl inlaying and japan work executed in England. Everything approaching to purity of design, or in the inlaid work, to geometric arrangement of the parts, is as carefully avoided as it generally is in Europe.

Paper hangings are manufactured to a large extent, and are printed both by block and by cylinder. They are chiefly, however, of the cheaper kind and of the usual character as regards design. The better class of papers approach the English and French papers in the character, though not in the elaboration of the workmanship. Messrs Perkins and Smith, New Bedford, Massachusetts, contribute some good specimens in dining, drawing, and bed room papers. They are all blockwork, and are as clear and as artistic as the style of design would permit.

A peculiar article in paper-hangings is largely manufactured for the Western States. This is about 35 inches wide, and is known as 'curtain paper.' An ornament, within a panel, is printed, extending to the length of about 1½ yard, and those are cut off and used as substitutes for roller blinds, by a large class of people in the West.

Painted wall decorations are evidently growing in public favour, and the extent to which this style of ornamentation is carried in some of the hotels and saloons of the larger cities is calculated, if directed by a purer taste, to do much for art in its higher forms. Unfortunately the upholsterer, rather than the artist, is the director, and the arabesque or pictorial panel is consequently a very secondary matter as compared with the hangings, the carving and gilding, and the carpet. In New York, several of the public buildings are very fairly decorated, mostly by French or Italian artists. The Metropolitan Hall, an admirable room, used for public meetings, concerts, and lectures, is effectively decorated, in the Italian style, with panels, arabesques, and a coffered ceiling, the general tone of the whole being pleasing and satisfactory. The lighting of the room by gas is admirably and tastefully arranged by running a series of gas-burners along the cornice of the gallery, these corresponding with the decorative members. The result is an equal diffusion of light all round the sides of the room, the centre being illuminated by a chandelier.

Carving and gilding, as applied to mirrors, pictures, and window cornices, is another important item of industry in this class, and in point of excellence of workmanship, is on a par with the usual English productions. The style of ornament, however, is at once exuberant and heavy.

On the whole, the decorative arts as applied to furniture are vitiated for any present influence on public taste, by an overwhelming tendency to display; fostered to a large extent by a class of persons whose object it is to crowd as large an amount of work into as small an amount of space as possible, and who prefer charging for labour rather than skill and taste. This reacts on the public mind, and habituates it to redundancy and over-ornamentation rather than to purity and simplicity resulting in the really beautiful.

16

Copyright of Designs

The law for the protection of copyright in designs forms part of the patent laws of the United States, and the office for granting copyrights, or rather patents, forms a department of the Patent Office at Washington, with a special officer to attend to applications of this class. The protection is granted for configuration and arrangement of the ornaments. None but citizens of the United States are entitled to the privileges, which are granted after due examination as to originality, etc., for seven years; or half the time granted in a patent for invention, etc. The fee paid is one-half the patent fee, or $15 (say £4 sterling) instead of $30, but no portion of this is returned if the patent or copyright is not granted, or the patent is withdrawn, as is the case with regard to full patents, two-thirds of the fee being returned in the case of either the rejection or withdrawal of a claim for a patent.

The forms are precisely the same as for obtaining a patent for an invention, etc. A petition, specification, and an affidavit or declaration on oath as to originality, must be filed, and a specimen or such essential portions of the pattern as may be needful to show the peculiarities of design, together with duplicate drawings, must be deposited at the Patent Office.

With regard to the kind of articles, the designs of which are usually patented, stoves and metal castings are by far the greatest number. In textile fabrics, very few are protected. Indeed, with the exception of the first-named articles, it is difficult to suppose that much originality in design could be legally claimed. In 1852 there were 126 applications for patents for designs, and of these 106 were granted and 20 were rejected.

In the majority of cases, copyright of design can only be valuable to the manufacturer so far as it enables him to secure to himself, by a rapid process, the exclusive immediate use of his invention. This is especially the case in textile fabrics; and the system of examination, with the delay in the decision, as practised in the United States, would be fatal to the value of one-half the designs registered in England.

17

Art Education

The question of art-education, as applied to manufactures, is beginning to claim serious attention in the United States, and though at present taking a totally different direction to that in which it has been the chief object of the more practical minds in England to direct it for the education of the artizan and the creation of a class of art-workmen, it is not the less significant that American manufacturers have reached that point at which it has become desirable that originality of thought should be infused into them by means of the designer, if not by the agency of well-instructed workmen. The latter, however, follows the former, as inevitably as effect follows cause; for the design once obtained, the very effort to realise it educates the workman in a greater or less degree in the process.

So far as institutions for the special purpose of teaching art in its application to industry have been at present carried, they are intended, with the exception of that attached to the Maryland Institution, Baltimore, for the instruction of females only. The first was established about four years ago at Philadelphia, chiefly through the efforts of Mrs Peter, the lady of the late British Consul to the State of Pennsylvania. Having visited several of the schools of design in England, and kindred institutions on the continent of Europe, Mrs Peter conceived that many well-educated females, the vicissitudes of whose families often compelled them to seek employment as a means of obtaining a livelihood, might, if properly instructed, render themselves useful in many departments of design as applied to manufactures, especially in the various kinds of textile fabrics, both printed, woven, and embroidered, as also in paper-hangings and hand-made lace. The result of her efforts was the establishment of a School of Design, partly supported by the contributions of herself and friends, and partly by the fees of the students, which were wisely graduated according to their means and the object for which they studied. Considerable difficulty, however, was experienced in carrying it forward for want of funds to fairly establish it at the outset, and as a permanent support, although it never appears to have lacked pupils; and after a period the committee of the Franklin Institute were induced to undertake the management of the school at Mrs Peter's request, as she removed after the death of her husband, the late William Peter, Esq., to her native city, Cincinnati, Ohio. The accidental death of Mrs Anne Hill, the principal of the school, whose talents and earnest devotion to her duties had given a position to the school which promised well for the future, again produced a change, and now the school is divided into two parts; those students who following one of the male teachers constitute a somewhat private establishment, and those who remain attached to the more public institution as originally founded. Each, however, is seeking assistance from the funds of the State of Pennsylvania, or to obtain a charter of incorporation and subscriptions to the formation of an endowment fund, as it has been found that current subscriptions and fees would not support the Institution. In consequence of the division above alluded to, and the embarrassed state of its funds, the connection of the original school with the Franklin Institute was dissolved, a committee of gentlemen taking the property and assuming the debts. It is now proposed to raise an endowment fund of $50,000 dollars (about £11,000 sterling), and thus provide means for carrying on the school by charter, as the fund properly invested would realize a satisfactory permanent income. Towards this object upwards of $17,000 had been subscribed in June last.

The business of the school is carried on in a large well-arranged house, situated at the side of one of the beautiful squares for which Philadelphia is so famous, which, laid out as walks and now filled with well-grown trees, afford in themselves lessons in art to the thoughtful student of design. This school formerly numbered seventy-four students, but there were only twenty-three on the books at the period of my visit in June. The classes, however, are usually indifferently filled during the summer months, and it is a rule for nearly all the scholastic institutions of the United States to take their long vacation during the hot months of the year. The rival establishment, conducted by a former male teacher, is known as the 'Institute for the instruction of Young Ladies in Design,' as contra-distinguished from the 'Philadelphia School of Design for Women,' and numbered 32 students. The fees to the latter school are $2 (not quite 10*s*. sterling) per quarter for those whose situation in life will not allow them to pay more, and $4 for those who can afford to pay that sum. The students who cannot afford the higher fee are certified by subscribers, in fact are nominees at *half-fee*. Ladies, whose means are ample, pay $10, $15, $20, or $24, that is from £2 5*s*. to £5 5*s*. per quarter, according to what they learn; and these graduated scales appear well adapted to an institution in which the student may enter any class, provided the fees of that class are paid, without reference, judging from the works, as to the amount of preliminary training in the elements of art. This, however, may be considered as the natural result of so large a dependence upon the fees of the students, as the teachers are rather in the position of private teachers of drawing, than the officers of a public institution; and therefore the choice of the student has more influence on the character of the instruction than sound principles, which, carried out quite independent of the likings or dislikings of the recipient of the instruction, must ever be the great value of a well-endowed public school, conducted on an exact plan of progressive education.

In connection with the Franklin Institute,* Philadelphia, is a drawing class for mechanical and architectural drawing, in which from 60 to 70 students receive instruction during the winter, on three evenings per week for 24 weeks, commencing on the first Tuesday in October in each year. The fee is $5 (about £1 3*s*. sterling) per quarter, or say £2 5*s*. for the session. No free hand-drawing is taught.

The Franklin Institute first commenced, in 1824, those annual exhibitions of manufactures which have since become so popular throughout the United States, and which have exercised so beneficial an influence in promoting and encouraging the early efforts of American manufacturers by premiums, and the public recognition of the benefits conferred upon the community at large by isolated industrial efforts.

The success of the School of Design for Women, at Philadelphia, under the late principal, Mrs Hill, and the management of Mrs Peter, appears to have suggested the foundation of similar institutions at New York and Boston. The New York School has only been established within the year, and the vacation prevented any inspection of its classes. The promised information as to sources of revenue, rate of fees, number of students, etc., has not been furnished; but it would appear that, from its recent formation upon the same plans as those of Philadelphia and Boston, the facts connected with its working would be of less importance than those of the last-named city, which, though more recently founded,

---

* The character of this Institute stands deservedly high, as under the title of the 'Franklin Institute of the State of Pennsylvania for the Promotion of the Mechanic Arts,' it has promoted secondary education in an efficient manner, by instruction in chemistry and mechanics, lectures, a permanent museum of models, minerals, and geological specimens, an admirable library, to which no light literature is admitted, even as a gift, and a reading room. Formerly, three courses of lectures were delivered every session by three permanently appointed professors, in chemistry, mechanics, and technicology; but now there are two courses of ten lectures each during twenty weeks, a different subject being taken each night of the week: thus, during the early part of the session 1852–53 the subjects were arranged as follows: Mondays, on Botany; Tuesdays, on Astronomy; Wednesdays, on Mining, Geology; Thursdays, on the influence of heat, air, and water on Natural Phenomena; Fridays, on Chemistry, as applied to the arts. Thus, a continuous course of oral and experimental instruction is kept up throughout the session, the pupils of the drawing class under twenty-one years of age being admitted free to the lectures.

The Institute also publishes a monthly journal, devoted to the promotion of the mechanic arts. This is the oldest periodical in the United States devoted to this subject, and was commenced in 1826, two years after the first annual exhibition of manufactures. Each number contains 72 pages octavo, at a yearly subscription price of $5. G. W.

seems to be upon a more permanent footing than that of Philadelphia.

'The New England School of Design for Women' was opened in October, 1851, under the superintendence of an Englishman, a former student of the Birmingham School of Design.* The rooms were adapted to the accommodation of 70 students, and this appears to have been the average number in attendance during the greater part of the time the school has been established. The objects of the school, as stated in the prospectus, are: '1. To educate a body of professed designers, capable of finishing original designs for manufactures, and other purposes where ornamental designs are required. 2. To teach the various processes of engraving, lithography, and other methods of transferring and multiplying designs. 3. To educate a class of teachers in drawing and design.'

As the school is intended for a standard or normal school, students are not admitted under fifteen years of age. The fees are $5 (about £1 3*s*. sterling) per quarter. The class hours are from 9 a.m. to 2 p.m. every week day except Saturday, and 'the school is open during all working hours for students in special departments, and for the industrial classes.'

The programme of instruction, without being very precise, is based upon the general terms of those of the English schools. All students are to go through an elementary course, after which they are to study their own speciality. On this elementary course hangs the whole question of absolute success, or that partial success only which carries on an institution like this for a time, but lays no solid foundation for future progress in itself or in its students. By the annual report for the year ending 30th September, 1852, it appears that the manufacturers of New England are alive to the importance of educating even a class of female designers (the question of art-workmen being as yet untouched); have supported the school with their subscriptions; and, better still, have given employment to those whose progress had been such as to enable them to be useful in designing or drafting for manufactures, as in the case of the two female students employed by the Lowell Manufacturing Company, quoted in Class XIX. Several pupils had learnt to draw on stone, and others to

---

* Mr W. J. Whittaker, an intelligent schoolmaster, who, feeling the importance of drawing as part of ordinary education, devoted a portion of his time to the acquirement of a knowledge of outline drawing. His artistic education, however, does not appear to have been carried beyond a fair power to copy 'Dyce's Outlines.' Prior to leaving England he had, in connection with Mr Hermann Krüsi, of the Home and Colonial Normal Schools, published a small work on 'Inventive Drawing on the principles of Pestalozzi,' the method indicated in which he had pursued in the school of which he was the teacher in Birmingham, and which he also followed in the United States when teaching drawing. He was no longer master of the above Institution at the date of my visit, but had commenced a private school of art in Boston, which gave evidence of care and attention to the wants of his pupils. This is now abandoned, and he is devoting his attention to the establishment of similar schools in other towns in the State of Massachusetts. Thus the New England School is the only school of the kind in Boston. G. W.

engrave and draw on wood. Several of the examples in the school were very creditable in manipulation, but entirely devoid of that sound artistic basis, always to be discovered, when existing, in the slightest work; and the want of which can never be concealed in the most elaborate production.

Lectures are delivered on botany, and a female teacher is specially engaged in giving instruction in this science to all the pupils.

The fees paid by the students have been already stated above, but a subscription of $20 per annum gives the subscriber the privilege of nominating one student *free*. Ladies subscribing $3 per annum in support of the institution are enabled to vote at the annual meeting in the election of the committee, etc., but have no privilege beyond that. The expenses have hitherto exceeded the annual income, being $3,500, the proceeds amounting only to $2,700, – $1,500 in subscription and $1,200 in students' fees. The House of Representatives of the State of Massachusetts has therefore recently sanctioned a grant of $1,500 (about £350 sterling) per annum, for three years, in support of the school, as an experiment, thus giving it the advantage of State sanction, and raising its income to the amount which its present wants demand.

The general character of the instruction given, both in the school at Philadelphia and Boston, has relation more to pattern-making for manufactures, and teaching drawing as an amusement or accomplishment, than inculcating and enforcing the steady pursuit of the arts of design in their highest forms, whether as applied to manufactures or otherwise. Still, enough has been done to show that, with teachers more efficiently trained to the precise purpose for which the schools are founded, with such examples around as would tend to elevate the minds of the students to the purer forms of classic, and the severer science of mediaeval and oriental art, alike in form, colour, and proportion, a progress of a very extraordinary character would be the result. The rapid progress made by the students of these schools is another evidence of the influence of the primary education which it is the good fortune of the children, male and female, of the United States to receive. As in the manufactories the youths enter with minds prepared by sound instruction to grapple with the technical difficulties of their future trades, so in these schools young girls commence their study of art with an amount of useful knowledge eminently calculated to open their minds to the influence of the instruction imparted.

The radical defect of the schools, however, is that which could be easily remedied – the want of a really useful and practical series of ornamental casts, and a severe course of elementary outline examples. The constant use of the former would train the eye both of teacher and pupil to the appreciation of a higher and better style of art than the comparatively mean casts of the ordinary builders' and decorators' ornaments hung about the class rooms; whilst study in a more precise elementary course would give the requisite power to appreciate fully the great principles of design, which a correct analysis of the examples could not fail to give. Pattern-making would become in reality pattern-designing, and artistic principles and practice prevail, where manipulation without power, and efforts without meaning, are now the only result.

The School of Design, forming part of the 'Maryland Institute, for the promotion of the Mechanic Arts,' at Baltimore, approaches much nearer in its character to the standard and objects of such schools in Europe, and is under the management of an Englishman, Mr Minifie.*

The school was opened in 1851, and then consisted of three departments–the Primary, in which elementary free-hand drawing, etc., was taught; the Architectural, in which the principles in construction and the styles of various periods were studied; and an Engineering Class, in which the drawing of machinery, mathematics, and the general principles and rules of calculation, as applied to mechanical engineering, formed the subjects of instruction. This plan appears to have been changed since Mr Minifie undertook the direction, and the school now forms one department divided into six classes:–1st. An Elementary Class; 2nd. A Geometrical Class; 3rd. A Mechanical Class; 4th. An Architectural Class; 5th. An Engineering Class; and 6th. An Artistic Class; the titles of which will sufficiently indicate the studies pursued in each.

On the plan of the classes being first made public, about 1,000 persons registered their names to attend; of these 800 applied for seats at the opening of the school; ultimately 400 students attended, and were divided into the six classes, each meeting one evening per week, from 50 to 70 students being the attendance in each class, instructed by a teacher and assistant teacher, and occasionally by the professor, whose attendance is not continuous. A staff of fourteen teachers was engaged in the work of the school last session. No charge is made to members and junior members of the Institute, but a small fee is about to be imposed.

The period of the year prevented the inspection of the classes in operation, as they are only carried on during the winter, as in the Franklin Institute at Philadelphia. The results of the instruction during the session of 1852–3, the second of the operations of the school, and the first under the arrangement last indicated, as shown in the drawings of the students displayed in the Institute, were very creditable.

The school, thus incorporated with the Maryland Institute, appears calculated to be of great service to the artizan class of the city of Baltimore, and if the system of instruction, as laid down, be carefully and steadily carried out, results of no ordinary character may be fairly predicted, more especially if better examples and models in ornamental art be obtained for the purposes of study, as the standard of art in this department is almost universally very low in the United States, because it is based upon modern adaptations of antique details, often so utterly transformed in their use as to be scarcely recognizable.

The study of drawing in the public schools of the various States is exceptional, and no settled system exists. In the Girard College at Phila-

---

* Mr Minifie is the author of 'A Text-Book of Geometrical Drawing for the use of Mechanics and Schools,' a work of repute both in the United States, where it was first published, and in England.

delphia, drawing is taught from models of geometrical solids; but the early progress of the students is retarded by the fact, that not being first taught to draw lines and simple forms projected upon a plane surface, and to be copied as an exercise for the hand and the eye in the apprehension of real form, they work in great uncertainty, and only attain proficiency, even in the delineation of lines, after a lengthened period. This is the experience resulting from all attempts to begin with models only; the power to draw the lines by which the forms of those models are to be represented, being still wanting.

The drawing class of the High School of the City of Philadelphia is conducted upon the system laid down by a former professor, Mr Rembrandt Peale, whose excellent little work on 'Graphics' is the text-book of the class.* Unfortunately, the operations of the class were suspended prior to the general examination of the pupils, therefore it could not be seen at work. Formerly the more elementary studies were pursued in the Primary and Grammar Schools of Pennsylvania, as well as the High Schools; but the instruction is now almost entirely confined to the latter,-most probably from a want of teachers sufficiently well trained in drawing to act as instructors.

---

* Three copies of this work, presented by the publishers, Messrs E. C. and J. Biddle, Philadelphia, are transmitted with this Report.

Conclusion

Having thus considered the respective departments of industry coming within the range of the various classes committed to my charge to examine and report upon, it may be desirable to call attention to several peculiarities requisite to be understood, as affecting the demand for certain productions, in order to judge fairly of the aims of those whose object it is to supply the wants of their countrymen in those particulars.

In textile fabrics the consideration of that sterling quality in make, which always forms so important an element in the judgment of the European consumer, is comparatively unattended to by the Americans. If the article looks well, and is of such a quality as to stand ordinary wear and use for a period, varying, of course, with the views of the buyer, even a slight advance in the price for a better quality of equal appearance only, will scarcely be obtained by the dealer. The material being expected to last for a single season, is purchased of a quality to do that, and no more. The next season the customer supplies himself again.

In furnishing fabrics, such as damasks, carpets, etc., the same principle prevails. A change is desired at the end of two or three years; carpets and furnishing draperies are purchased of such a quality as to look well for that period, and then others of a different style are brought in to supply their places. And this habit of almost constant change is said to run through every class of society, and has, of course, a great influence upon the character of goods generally in demand, which, although by no means devoid of those intrinsic qualities which are the essentials of good manufacture, yet are made more for appearance, and less for actual wear and use, than similar goods are in England.

The development in the department of ornamental manufactures, both in textile fabrics and in metals, as also in furniture and decoration generally, has been alluded to under their respective heads, as also under that of art-education. There is no appearance of any attempt to strike out a national style, although the many peculiar features of the country, the habits of the people, and undoubted originality in the mechanic arts, would lead to the inference that a gradual repudiation of European modes and forms will take place; and that in art, as applied to the utilities of life, true principles in the education of the people in this respect are alone needed to produce results of a very satisfactory character. At present the co-mingling of totally different styles of decoration in architecture, and the adoption of European designs for totally different purposes to those for which they were originally intended, are amongst the least of the errors committed in a vague seeking after novelty.

In reporting upon the state of Art-education, so far as it has been carried, the great object has been to call the attention of the Americans themselves to the defects of existing efforts, in a friendly but earnest spirit. The future success of the schools of design, as already established in the United States, is likely to be seriously retarded by the absence of good examples suited to the wants of the student, and calculated to carry the instruction in the right direction; inasmuch as the working out of a true system of art-education depends so largely on a proper attention to this point.

That the taste of the people is rapidly advancing with the increased facilities of communication with Europe, and the higher aims to which education is tending, must be evident to every observer. The works in gold and silver, the character of the higher class of furniture, and the silks imported from France and used in the houses of the more wealthy classes, the love of pictures and articles of *vertu*, all show that an almost Oriental love of splendour is rapidly taking the place of those simpler habits and tastes which characterised the severer tendencies of the earlier European inhabitants of the North American Continent and their immediate descendants. With building materials of an unequalled character, the granite and freestone of New England, the marbles of Pennsylvania, and the bricks of Baltimore, give facilities for architectural display, which is gradually developing itself, alike constructively and ornamentally.

The compulsory educational clauses adopted in the laws of most of the States, and especially those of New England, by which some three months of every year must be spent at school by the young factory operative under 14 or 15 years of age, secures every child from the cupidity of the parent, or the neglect of the manufacturer; since to profit by the child's labour during *three-fourths* of the year, he or she must be regularly in attendance in some public or private school conducted by some authorised teacher during the other fourth.*

---

* The following are the clauses in the laws of the Commonwealth of Massachusetts as affecting the education of children employed in manufactories.

'No child under the age of fifteen years shall be employed in any manufacturing establishment, unless such child shall have attended some public or private day-school, where instruction is given by a teacher qualified according to law to teach orthography, reading, writing, English grammar, geography, arithmetic, and good behaviour, at least one term of eleven weeks of the twelve months next preceding the time of such employment, and for the same period during any and every twelve months in which such child shall be so employed.' –St. 1836, ch. 245, §1; 1849, ch. 220, §1.

'The above prohibition does not apply to any child who shall have removed into this commonwealth from any other State or country, until such child shall have resided six months within this commonwealth.'–1849, ch. 220, §1.

'The owner, agent, or superintendent of any manufacturing establish-

This lays the foundation for that wide-spread intelligence which prevails amongst the factory operatives of the United States, and though at first sight the manufacturer may appear to be restricted in the free use of the labour offered to him, the system re-acts to the permanent advantage of both employer and employed.

The skill of hand which comes of experience is, notwithstanding present defects, rapidly following the perceptive power so keenly awakened by early intellectual training. Quickly learning from the skilful European artizans thrown amongst them by emigration, or imported as instructors, with minds, as already stated, prepared by sound practical education, the Americans have laid the foundation of a wide-spread system of manufacturing operations, the influence of which cannot be calculated upon, and are daily improving upon the lessons obtained from their older and more experienced compeers of Europe.

Commercially, advantages of no ordinary kind are presented to the manufacturing States of the American Union. The immense development of its resources in the West, the demands of a population increasing daily by emigration from Europe, as also by the results of a healthy natural process of inter-emigration, which tends to spread over an enlarged surface the population of the Atlantic States; the facilities of communication by lakes, rivers, and railways; and the cultivation of European tastes, and consequently of European wants; all tend to the encouragement of those arts and manufactures which it is the interest of the citizens of the older States to cultivate, and in which they have so far succeeded that their markets may be said to be secured to them as much as manufacturers, as they have hitherto been, and will doubtless continue to be, as merchants. For whether the supply is derived from the home or foreign manufacturer, the demand cannot fail to be greater than the industry of both can supply. This once fairly recognised, those jealousies which have ever tended to retard the progress of nations in the peaceful arts, will be no longer suffered to interfere by taking the form of restrictions on commerce and the free intercourse of peoples.

The extent to which the people of the United States have as yet succeeded in manufactures may be attributed to indomitable energy and an educated intelligence, as also to the ready welcome accorded to the skilled workmen of Europe, rather than to any peculiar native advantages; since these latter have only developed themselves as manufacturing skill and industry have progressed. Only one obstacle of any importance stands

---

ment, who shall employ any child in such establishment contrary to the above provision, shall forfeit a sum not exceeding fifty dollars for each offence, to be recovered by indictment, to the use of common schools in the town where such establishment may be situated.'–St. 1836, ch. 245, §2; St. 1842, ch. 60, §2; 1849, ch. 220, §3.

'It is the special duty of the school committees in the several towns and cities of the commonwealth to prosecute the owners, agents, or superintendents of manufacturing establishments, for employing children under fifteen years of age, who have not received the instruction above described.' –St. 1842, ch. 60, §1.

in the way of constant advance towards greater perfection, and that is the conviction that perfection is already attained. This opinion, which prevails to a larger extent than it would be worth noting here, is unworthy of that intelligence which has overcome so many difficulties, and which can only be prevented from achieving all it aspires to, by a vain-glorious conviction that it has nothing more to do.

In concluding this report I cannot do so without expressing my obligations, as an individual member of the Commission, for the courtesy, attention, and ever-ready kindness with which all my inquiries were responded to by all classes of Americans with whom my duties brought me in contact in the United States.

I have, etc.
signed GEORGE WALLIS
Society of Arts and Government School of Art,
Birmingham, December 31*st*, 1853.

Appendix No. 1

Table of Produce, etc., of the Cotton Mills at Lowell, Massachusetts, January, 1853

Manufacturing Company	*Number of Mills*	*Spindles*	*Looms*
Appleton Company	3	17,920	700
Boott Cotton Mills	5	51,866	1,432
Hamilton Manufacturing Company	4 *and Print Works*	47,168	1,340
Lawrence Manufacturing Company	5	44,800	1,382
Massachusetts Cotton Mills	6	45,720	1,571
Merrimack Manufacturing Company	6 *and Print Works*	71,072	2,114
Suffolk Manufacturing Company	3	17,528	590
Tremont Mills	2	16,608	620
Total, exclusively cotton	34	312,682	9,749
Lowell Manufacturing Company	1	8,050 *Cotton only*	205 *Cotton only*
	35	320,732	9,954

Females employed	*Males employed*	*Yards made per week*	*Cotton consumed per week in lbs.*	*Kind of goods made*
400	120	150,000	60,000	No. 14 shirtings and sheetings.
870	262 *including male tenders*	300,000	90,000	Drillings no. 14; Sheetings, shirtings, jeans, and printing cloth. no. 30
750	406	250,000	80,000	Prints, flannels, ticks, and sheetings, no. 14 to 30
1,200	200	260,000	95,000	Printing cloths, sheetings and shirtings, nos. 14 to 30
1,250	300	475,000	150,000	Sheetings no. 13; Sheeting and drillings, no. 14
1,650	650	377,000	84,000	Prints and sheetings, no. 21 to 40
400	100	120,000	50,000	Drillings, no. 14
400	100	155,000	46,000	Sheetings and shirtings, no. 14
6,920	2,138	2,087,000	655,000	
800 *In cotton, wool, and carpets*	500 *In cotton, wool, and carpets*	50,000 *Cotton only*	90,000 *Osnaburgs*	Osnaburgs
7,720	2,638	2,137,000	745,000	

Appendix No. 2

Cotton: Class XI

Mr Le Weeks to Mr Wallis

Sir, *Hannahatchei, Stewart County, Georgia, July 25, 1853*

A communication has been received at this Bureau from an intelligent planter in Georgia, accompanied by several bolls of cotton, asking information upon the following points relating to the harvesting, ginning, packing, and manufacture of cotton; viz.

'*1st.* Does the well-matured cotton boll yield its staple, or fibre of different lengths, in one or the same boll or lock; or does the fibre of an individual seed exhibit an approximate uniformity in its length before separation by the gin?

'*2nd.* What per-centage of loss would be considered a fair average in your establishment for a twelvemonth stock on the gross weight of cotton consumed, by reason of dust, grit, or sand?

'*3rd.* What is the per-centage of loss, as above, caused by vegetable substances, as leaves, grass, trash, motes, etc.?

'*4th.* What is the per-centage of loss arising from shortened or divided fibre, caused by ginning, and which flies off during the process of manufacturing? Also, how much, in your judgment, is the strength and durability of your fabrics diminished in consequence of such breakage or division?

'*5th.* Has the general condition of staple (as to length and strength) deteriorated for the last twenty years?'

The object of the communication above referred to appears to have been intended to disabuse and correct a sentiment prevailing in some of the cotton-growing States among the planters, that it is really non-essential, in *point of value*, to the manufacturer, whether the staple is broken by ginning, or whether it is foul from grass or other extraneous matter, except so much as it may happen to be increased thereby in weight. The question has also arisen whether there may not be an improvement made in the cotton gin, should it prove that the fibres attached to each seed are of approximate or uniform lengths.

Any information which you may be able to impart relative to the points herein considered will be acceptable to this Bureau, and will be duly appreciated by the cotton-planters in the South.

You will understand these inquiries to apply only to short staple or upland cotton.
I have, etc.
signed HENRY LE WEEKS

Mr Le Weeks to Mr Wallis

Sir, *Hannahatchei, Stewart County, Georgia, July 28, 1853*
I would respectfully submit the following question to your consideration, in relation to 'Sea Island' or 'Long Cotton,' desiring your attention and answer thereto at an early date after the receipt thereof.

Does this species of cotton present a large comparative difference in the length, strength, and firmness of its staple or fibre? And, if so,

Does this difference make it practically necessary that a process of assorting should be made a pre-requisite to its more successful manufacture?

The object in view is to forward the cotton in such condition as will best meet the wants and convenience of the manufacturers. It is highly probable that the necessary assorting could be accomplished during the process of gathering and ginning while in the planter's hands, with more facility than after it has been packed in bales.

Your answers are anticipated with much interest by
Yours, etc.
signed HENRY LE. WEEKS

Mr Wallis to Mr Le Weeks
Society of Arts and Government School of Art, Birmingham,
Sir, *October, 11, 1853*
I have now the honour to acknowledge the receipt of your communications of the 25th and 28th July, transmitted through His Excellency John F. Crampton, British Minister at Washington, to whose care I address this reply.

The queries contained in both your communications appeared to me of so much importance, that I considered it would be more satisfactory to you to have the opinion of one or more gentlemen, whose practical knowledge of the cotton trade would be a guarantee for the accuracy of their answers. I, therefore, communicated with Thomas Bazley, Esquire, President of the Manchester Chamber of Commerce, and Henry Houldsworth, Esquire, also of Manchester, two of our most eminent cotton-spinners, both of whom are producers of the finest yarns spun in Europe; and at the suggestion of the first-named gentleman, I also sent copies of your queries to Robert Hyde Greg, Esq., of Norcliffe, near Manchester, an extensive consumer of short staple cottons. I have now the satisfaction to quote their replies to my communications, so far as they affect the questions you have put to me. Mr Bazley says

Manchester, September 22, 1853
'In the same pod of Sea Island cotton, short as well as long fibres are found and associated. Evenness of length of fibre is most desirable, and

the consumers will always pay a higher price for long and equal stapled cotton; than for short and irregular cotton.

'If possible, cotton ought to be grown with its fibres of uniform length; but when they are irregular, any process which does not injure the cotton and sorts them, would be an advantage which would be appreciated.

'*Bowed* Georgia cotton, like any other, is found to consist of fibres of various lengths, but the most valuable is the most regular.
signed THOS. BAZLEY'

Mr Greg replies *seriatim* to your queries

September 26, 1853

'Query 1st. I believe, though I seldom have opportunities of examining, that the *matured boll*, of all sorts, contains fibre of unequal length, that is, some of each is *short*, but not a large per-centage of the boll.

'2nd. The loss of spinning on a twelvemonth's account will be perhaps,

1	oz. in the lb. of fine		
$1\frac{1}{4}$	"	"	fair
$1\frac{1}{2}$	"	"	good middling
$1\frac{3}{4}$	"	"	middling
2	"	"	ordinary
2 to 4	"	"	inferior

The latter loss depends chiefly on dust and sand, nothing else weighing heavy. From all these some fibres and some oily waste again find their way into coarser fabrics, selling from the mill at $\frac{1}{2}$*d.* per pound possibly. What I have given is the difference between the cotton weighed in, and yarn weighed out, and often includes some *damp*.

'3rd. Motes and vegetable matter amount to little in the better qualities, and perhaps one-half in middlings.

'4th. When staple is short, we put in 5, 10 or 20 turns per cent more in spinning, and of course, in that operation produce that per-centage less yarn.

'5th. *Deterioration*. About twenty years since, the *boweds* or Eastern cotton began to grow worse, shorter staple and much injured or broken, and "napped" in ginning, and for ten years we never bought a *bag*–that being the form it then came in–but confined ourselves wholly to Gulf cotton, Orleans, or Mobiles, which (some might be as short as the boweds or Uplands) gave plenty of choice, and, I know not why, was seldom injured by the gin. About five years since there seemed to be a change in the Eastern cotton, the staple improved,–little injury from the gin; and the *bag* was wholly abandoned, and I now buy Orleans and boweds (Uplands) indifferently. During many years the *name* of boweds excluded half the buyers, and $\frac{1}{2}$*d.* per pound less price was established solely from the name. This is now wearing away. I have heard that planters, finding the disrepute of their article, imported Orleans seed, and took other steps to improve. The cotton come this year is shamefully dirty. *Straws and broken stalks and grass* are quite unmanageable, independently of the objection of loss of weight; and *we* have cotton lying in Liver-

pool, and there are tens of thousands of bales no spinner will buy at any price. Leaves fall out in spinning, motes are struck out, sand and mud form only 1, 2, 3 or 4 ounces of waste, and may be estimated; but straws and grass bits clog the machinery, stop the frames, spoil work, and the operatives leave the mill and will not work the cotton at all. This defect reduces the value to a consumer 25 per cent., but this cotton is bought and sold by speculators, and now a great accumulation exists in Liverpool which will ultimately pass into consumption at a low figure. During many years past we have noted but very little mischief from the *gin*, though it may doubtless break the staple a little. In the Upland cotton *formerly* the damage was extreme, reducing the value 2*d.* per pound at 10*d.*
signed ROBERT HYDE GREG'

Mr Houldsworth, having mislaid the copy of your questions transmitted to him, says, 'My reply, referring as it does to all the points of importance that occur to me, will probably contain, in addition, much that was not specially asked for.' The reply is as follows

October 8, 1853

'In the spinner's estimate of long staple cotton, it may generally be considered that *strength* is the first requisite. A fair and uniform length of staple next (say from 1⅛ to 1⅜ inch for medium purposes, and 2 inches long for the very finest), then *fineness*, and, lastly, *getting up*, *cleanness*, and *colour*.

'The spinners of the lower qualities attach most importance to *strength* of fibre and *cleanliness*, and the spinners of the finest qualities much importance to the *getting up*.

'Within certain limits a very considerable and prejudicial difference in the length, strength, and fineness of the staple of the same sample of Sea Island cotton prevails. This difference appears as if it arose in part from different degrees of maturity or development of the fibres. It exists in the greatest degree in the finer qualities of Egyptian cotton.

'It not unfrequently happens that cotton containing much fibre, worth 3*s.* per pound, is deteriorated 6*d.* to 9*d.* per pound by the mixture of short, coarse fibres in the same locks.

'The writer has some reason to think, from an examination of the few pods he has seen, that great difference of quality of fibre exists in the same pod, and in the fibres attached to the same seed, – a difference that might possibly be lessened by giving the latest developed fibres a longer time to ripen, without any sensible damage to the earlier fibres.

'The difference in the quality of the fibres in the same lock of cotton is very different in different brands and in particular seasons.

'It is probable the inequality of fibre in respect to the three important particulars of *strength*, *length*, and *fineness*, arises from more causes than one, and is influenced by the selection of the seed, the cultivation and character of the soil, the point of maturity of the pods when gathered, and the seasons. Of these the last only is beyond the control of man.

'With regard to seed, the success of Mr Kinsey Burden, of South Carolina in improving the length of staple of the finest class of Sea Island cotton, about the year 1825, by a selection of seeds, affords a presumption that the *strength* of fibre may be improved by similar means.

'In respect to soil and cultivation, it is presumed these must influence the fibre in some degree.

'Care in plucking the pods at some given and uniform point of their development, when the largest number of the inclosed fibres are fully developed, and but few injured by over ripeness, appears a likely means of improving the uniformity of staple; the more particularly as the striking want of regularity of Egyptian fibre, in all the three important particulars of strength, length, and fineness, is supposed to arise from the *wholesale* mode of gathering the crop practised there, (viz.), by whole fields at a time, instead of by single pods as they ripen.

'The fibre should not be exposed to the air after being plucked, or as little as practicable, and it should undergo the least possible shaking or handling before and in the process of packing.

'The fibres of extra fine cotton are inclined to twist together and mat and form nips when handled, and are so matted together by some planters in their processes of cleaning and packing as to be deteriorated 1*s*. per pound.

'The cotton cannot be sent home too open, as far as respects the subsequent processes it has to undergo in being spun, – the best state being that in which it is when separated from the seed by *hand*, and in that state put into the bag.

'The best mode of learning the opinions of spinners on experimental samples of Sea Island cotton, appears to be that adopted by Mr K. Burden, 30 years ago (viz.), for the planter to inclose in one or more of the ordinary bags of his crop, packed at different periods during the season, small packets containing 4 to 10 lbs. weight of the cotton, the quantity of which (as estimated by the spinner) he wishes to learn, and to inclose in such pockets letters addressed 'To the spinner into whose hands this pocket falls,' *defining the points on which the grower seeks information, and asking for a communication on the subject by post, giving at the same time the grower's address.*

'The writer of the foregoing views was thus led to open a correspondence with Mr Kinsey Burden, which was continued for some years, and led to communication with the agricultural societies of South Carolina, and otherwise to the introduction of select seed-cotton now forming an important branch of the Sea Island trade.

signed 'H. HOULDSWORTH'

I can add nothing to these answers which would in the least enhance their practical value, coming as they do from gentlemen whose large experience and eminent practical knowledge cannot fail to render their views useful to the cotton-planters of the United States, and it has afforded me great pleasure to have had the opportunity of procuring their opinions upon the points raised.

I have, in conclusion, to apologise for the delay in making this communication; but, owing to the absence of Mr Houldsworth from home,

and other circumstances, he was unable to reply before the 8th of October, and his communication did not reach me until yesterday.
I have, etc.
signed GEORGE WALLIS

Memorandum of the substance of a communication from Mr Le Weeks to Mr Wallis, dated November 28, 1853

On raising 'Upland' Cotton Seed

For the last ten years there has been an annual importation of seed into south-western Georgia. The seed, however, is not unfrequently grown in that cotton-growing region, and is often produced by the planter himself; as occasionally in passing through a plantation when in full season of fruit-making, a stalk will be found remarkable for the sound healthy appearance of every part, for the thick-set fruit or bolls, for the number and closeness of its limbs, and also the large size of the bolls well filled with a good bodied fibre. This attracts the planter's attention, and he has these bolls carefully saved from time to time as they mature. In due season he plants the seed thus obtained entirely away from his other crop, and soon obtains seed enough for his own use, and that of others also.

Probably thirty varieties of short staple or 'upland' cotton can be enumerated in the county of Stewart (Georgia) alone, each at the outset of its career being celebrated for some peculiarity that ostensibly commended it to the notice of the planter. In this way varieties have been mixed and mingled until it is certain that there is scarcely a crop grown in the Chatahoochee river and its vicinity, of a pure species. Twenty years ago it was chiefly understood that all cotton was grown from varieties known as 'petit Gulf,' 'royal Gulf,' and 'green seed.'

It may be concluded, inferentially, that to this cause, viz., the mixing of varieties, may be attributed many of the perplexities of the spinner from irregular fibre. As to length, two kinds have had to be abandoned, the '*silk*' and '*Mastodon*,' which always promised finely as field prospects; but from the length of fibre were wholly impracticable in ginning. The fibre is too long for the kind of gins used in south-western Georgia.

Some planters contend, in order to justify the mixing of varieties, and that too with some plausibility, that the tendency of cotton in the climate of the Southern States of America is to deteriorate, and that in order to recover the loss, it is necessary to plant mixed varieties as a remedy.

Mr Le Weeks' own opinion on this point, although he confesses to having many respectable authorities against him, is that '*every crop planted should be a pure variety*,' and to preserve against the falling off in successful production, there should be a change of the seed, say from river lands to high uplands, from gray sandy soils to red clayey or stiff soils, and *vice versâ*.

On the preparation of Land and the cultivation of 'Upland Cotton'

The ordinary preparation of land is, first, running a furrow with a small plough, and repeating the same at such a distance apart as the land will allow, according to productiveness. This distance varies from 2½ feet to 5 or 6 feet, the controlling object being to have the space nearly 'locked' by the extension of the limbs or side branches when the plant is maturing, care being taken not to be too close, as the 'bottom,' or first matured bolls, need air and sunshine to open them. At that period they are under the growing foliage of the plant above them, and if they are too much in the shade they rot in the unopened state.

The land is then made into 'ridges' by turning on each side of this furrow one-half of the earth each way to the first or guide furrow. At planting, a small plough is run on the top or centre of this ridge, a field-labourer follows and strews the seed with a cast of his hand, say ten to twenty seed per foot. Another labourer follows with a horse drawing a plough-stock with a straight-edge plank or iron fixed, instead of the plough, and covers up the seed by striking the loose earth into the furrow and filling it up.

After the seed is up, the plough is run round the cotton, turning up the earth from the young stems. The hoes (about 10 inches wide) follow and chop through the drills the width of the hoe, leaving a few young plants, say at every 10 inches. Another plough follows the hoes and throws the earth back again upon the cotton to preserve the ridge.

The next time the cotton is gone over, it is put to a 'stand,' *i.e.*, all superfluous plants are taken out, leaving one only standing at the proper interval to be cultivated.

This is a time of much interest and anxiety to the cotton planter, for after he has brought his crop to a 'stand' every stalk or plant lost is a loss to his crop. The disasters at this stage, as also previously, are:

1st. '*Rust*,' '*sore shin*' or the sickly dwindling and death of the young plant, it being as tender as a cucumber, bean, or other vegetables which bring up their seed on their first development.

2nd. 'Lice,' an insect in all respects similar to those which infest cabbage, etc. These prey upon the sap of the leaves and exhaust it, producing death to the plant.

3rd. '*Cut worms*' or '*wire worms*,' so generally known to gardeners, burrowing by day, and appearing at night in search of food, cutting the stem or main stalk off. These often continue to infest the plant until it is 10 to 20 inches high, and then ascend near to the extremities and lop off the buds or ends of the branches.

After the crop begins to make its first 'set' or crop of bolls, say in August, the 'boll worm' is then the common enemy, its immediate progenitor being a *moth* or *miller*, somewhat smaller than the silkworm species. This insect too is nocturnal, lays one or more eggs in the extremity of the limbs. The young worm on appearing feeds first on the little tender bolls just 'set,' then advancing towards the centre, takes the next, as it is a few days older and the worm stronger, and so on until the fully matured boll is attacked, and here this insidious insect finds its farewell banquet. Thus this worm will in two months destroy one-half of the planter's prospective crop. It is, however, the least of the disasters to which the cotton-plant is subject. Showers of rain too are very dangerous when the blossom is open, as one drop of water falling within the blossom the first day it opens, ruins it.

Blossoms open white in colour the first day, and red the second, which is also the last. They open from 10 A.M. to 3 P.M., and on the third day fall off.

On gathering and housing 'Upland' Cotton

The time of commencing to gather cotton varies from 10 to 30 days in different years. This variation is brought about essentially by the difference in the seasons, owing to cold backward springs which always prevent the plant from growing off. Long droughts also suspend its growth when planted, as also wet weather after the first bolls are grown.

Generally some portion of the cotton is 'picked' out during the last half of August. An acre of land which will produce 1000 lbs. weight of cotton *unginned* will yield from 50 to 150 lbs. the first picking. If the weather be pure, dry and sunny, it will in three or four weeks be 300 to 500 lbs. for the next 'picking.' The balance will be opened soon after the next killing frost, say in November. Thus may be understood the planter's technicalities of 'bottom,' 'middle,' and 'top' cotton. 'Bottom' and 'top' cotton are inferior to the middle; as the 'bottom,' is near the ground it does not 'flush out' when opened, and is likely to be dirty from earth. The 'top' is grown when night dews, etc., are cooler, and is more tardy in maturing. Besides, the vigour of the plant is somewhat exhausted, hence its inferiority is a want of perfect development. The 'middle' cotton is that which constitutes the most desirable sample, provided rains and winds do not injure its appearance.

The process of picking is as follows: Each field-hand or labourer has a canvas sack suspended around his neck, and hanging down nearly to his knee, into which he may put from 10 to 20 lbs. Having a large basket standing half way, when his sack is full he places its contents therein, and again proceeds with his 'picking.' At noon and in the evening a waggon and team is sent round to convey the baskets when filled to the gin-house. This proceeds until the whole crop is 'saved' or housed.

A fruitful source of the admixture of irregular fibre of unequal length, strength, and fineness, may occur here in continuing to gather the '*bottom*,' '*middle*,' and '*top*' cotton, and housing these various grades indiscriminately in the same bulk.

A bulk sometimes equal to the whole capacity of the gin-house, as to room and power to support weight, will be got together; and as the gin stands on one side, and its feeding is done from the bulk without selection, a portion of the *first* and the *last* 'picking' may be packed together in the same bale.

G. W.

Appendix No. 3

Woollen and Worsted: Class XII

Description of the Bay State Mills, Lawrence, Massachusetts
(With Illustrative Plans)
Extract from the Report of the Sanitary Commission of the State of Massachusetts, 1850

The manufacturing establishments at Lawrence have been erected under favourable circumstances. They were planned and constructed under the scientific skill and practical experience which had been acquired by wise and successful men, in a series of years, in other places, aided by excellent water-power, ample capital, and under reasonable national and state legislation. The results have appeared to us so admirable, and so highly worthy of imitation, that we have supposed we could not perform a more useful public service than to give a more particular description of one of them. It will show, in a favourable light, the intelligence, the enterprise, and the liberality that has generally presided over all the affairs of this town.

The establishment belonging to the 'Bay State Mills' is devoted to the manufacture of cassimeres, shawls, and other fancy woollen goods, and was planned and erected under the general superintendence of Samuel Lawrence, Esq. It is the largest mill of the kind in the world, and will consume, when in full operation, more than 2,000,000 lbs. of wool annually. The mills occupy a parallelogram of 1,000 feet in length by 400 feet in breadth, between the canal and the Merrimack river. Buildings are erected on the outer borders of this site, affording a spacious central area. That on the river side is 1,000 feet in length and 40 feet in breadth, with two wings, at right angles, at the ends, 240 feet by 40 feet, and outer porches for ingress and egress. The whole is three stories in height, excepting the centre, 52 feet by 42 feet, which is five stories high. On the side next the canal is another line of buildings, 800 feet in length, 38 feet in width, and two stories in height, designed for counting-rooms, store-houses, watch-houses, and other purposes. Within the interior the three principal mills are erected, each 200 feet by 48 feet 8 inches, containing, including the basement and the attic, 9 working floors. All these mills are substantially built of brick, and covered with slate. The rooms are 11 to

PLANS OF THE BOARDING HOUSES FOR OPERATIVES

BELONGING TO

THE BAY STATE MILLS, LAWRENCE, MASSACHUSETTS.

Enlarged from the Sanatory Report of the State of Massachusetts.

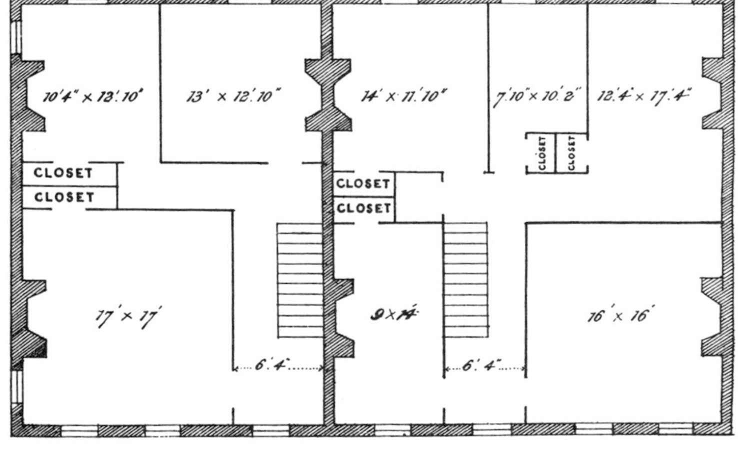

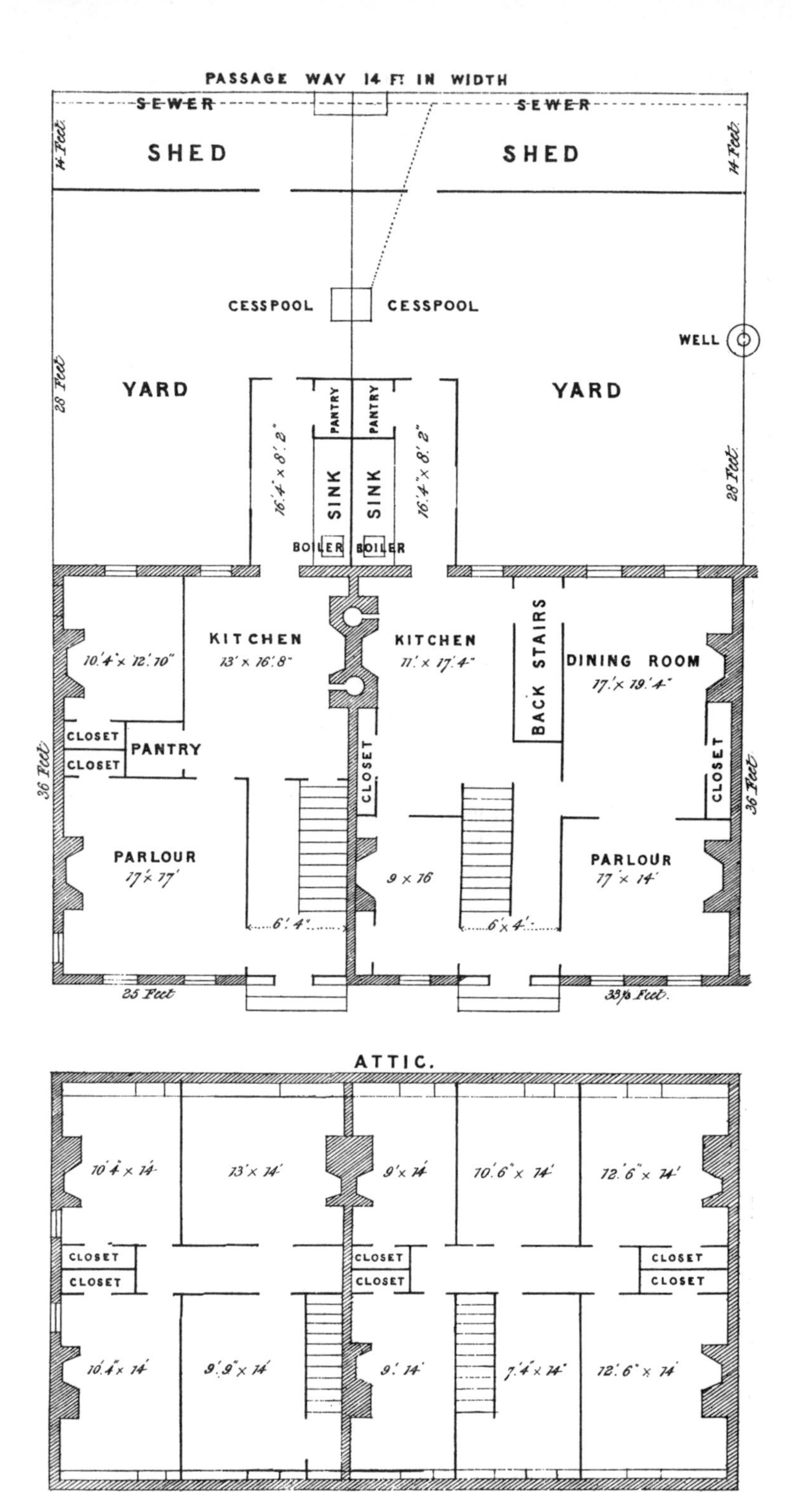
PASSAGE WAY 14 FT. IN WIDTH
SEWER
SEWER
SHED
SHED
14 Feet
14 Feet
CESSPOOL
CESSPOOL
WELL
28 Feet
YARD
YARD
28 Feet
PANTRY
PANTRY
16.4" × 8.2"
SINK
SINK
16.4" × 8.2"
BOILER
BOILER
KITCHEN
13' × 16.8"
KITCHEN
11' × 17.4"
BACK STAIRS
DINING ROOM
17' × 19.4"
10.4" × 12.10"
CLOSET
CLOSET
PANTRY
CLOSET
CLOSET
36 Feet
36 Feet
PARLOUR
17' × 17'
9 × 16
PARLOUR
17' × 14'
6' 4"
6' × 4'
25 Feet
33½ Feet
ATTIC.
10.4" × 14'
13' × 14'
9' × 14'
10.6" × 14'
12.6" × 14'
CLOSET
CLOSET
CLOSET
CLOSET
CLOSET
CLOSET
10.4" × 14'
9.9" × 14'
9' 14'
7.4" × 14"
12.6" × 14'

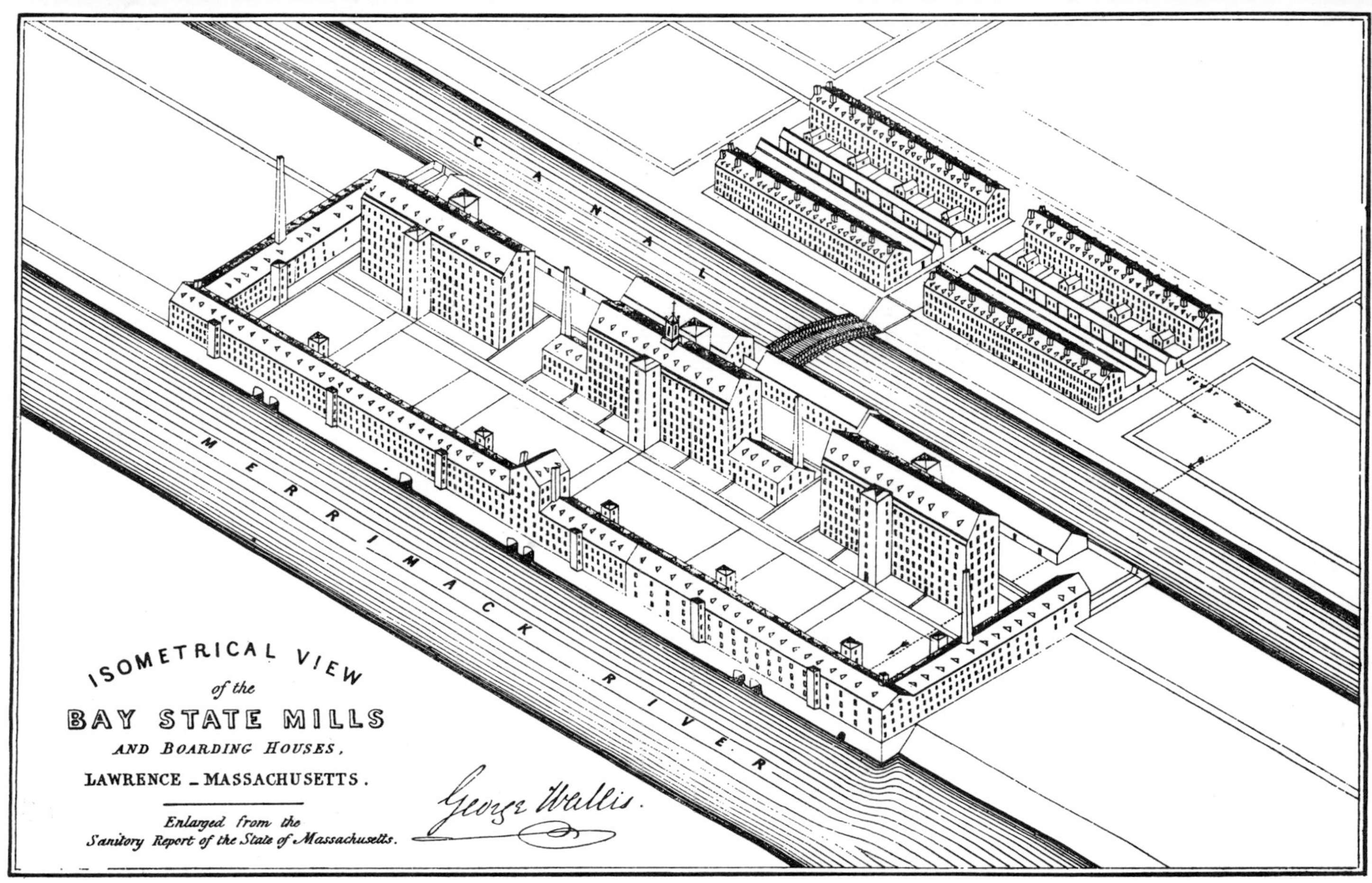
CANAL
MERRIMACK RIVER
ISOMETRICAL VIEW
of the
BAY STATE MILLS
AND BOARDING HOUSES,
LAWRENCE _ MASSACHUSETTS.
Enlarged from the
Sanitary Report of the State of Massachusetts.
George Wallis

13 feet in height, and are warmed with steam, and lighted with gas. The apparatus for warming consists of wrought-iron pipes, 1, 3, or 4 inches in diameter, placed in three, four, or five parallel lines around the interior of the building, immediately under the windows in each story, maintaining in all the rooms, at all times, a uniform temperature of about 68°. A structure to supply the steam is situated in each wing of the river building, and has 12 boilers, and a chimney 135 feet in height. All the rooms are provided with hydrants, to which force-pumps and hose are attached, that may be used in case of fire, for washing, and other purposes. A person is employed in each room to keep every part of it and the stairways clean. Each mill is also provided with extra porches, one in front and the other in the rear, and with four iron ladders reaching from the bottom to the top, for ascent or descent in case of fire, or for any other purpose. The motive power for these mills is obtained from seven breast-wheels of the first class, 23 feet 4 inches in length, 26 feet in diameter, and of 125 horse-power each, two of which are placed in each principal mill, and one in the river mill. Eight mill-powers were purchased by this company.

The boarding-houses are on the opposite side of the canal, and consist of four blocks, substantially built of brick, and covered with slate, each 250 feet in length, 36 feet in breadth, 3 stories high, of 10, 9, and 8 feet respectively, with 4 L's in the rear, 1 story high, to each block. Each block contains 8 tenements, and each tenement, except the end one, $33\frac{1}{3}$ feet in width and 36 in depth, exclusive of the L, and contains 20 rooms, including the attic, and is designed to accommodate 36 boarders. The location and size of the rooms will appear from the accompanying plan and illustration. The end houses are 25 feet in width, a little smaller than the others. The houses in each block, except the end ones, are like that on the right of the plan here presented. As you enter this tenement on the left there is a small room appropriated exclusively to the mistress of the house. At the right are two dining-rooms, connected by folding doors, each forming pleasant sitting-rooms at other than meal times. Passing through the entry you enter the kitchen, which is furnished with all necessary conveniences. Beyond this is the back kitchen, containing a large boiler and conveniences for other household purposes. In the rear of this is the wash-room, from which you pass into a large yard, inclosed by a high tight fence, having at the end the wood shed, 14 feet wide, and the privies, the whole bordering on a common passage-way 14 feet wide. Under each alternate fence is a double cesspool, serving for two houses, and having an underground passage leading to the common sewer under the sheds. A well of pure water is connected with every four tenements, and all are supplied with soft water, for washing and other purposes, by cast-iron pipes leading from cisterns in the mills to the sinks in the several houses. On the second floor is the parlour and also the sick-room, a small chamber with a fireplace, designed for an invalid who may need seclusion and extra warmth. Besides these, are sleeping apartments for the boarders in the second and third stories and in the attic, designed to accommodate two, four, or six persons each, according to the size of the room.

Each tenement cost about $4,000, exclusive of the land, and will compare to advantage with respectable dwelling-houses in Boston, and are much better than the average in country villages.

To protect the health of the inmates, underground sewers are constructed under the sheds in the rear of each block, through which a current of water, supplied by iron pipes connected with the canal on the left or above the block, is constantly running, carrying off all the contents of the privies, cesspools, and other filth, and passing at right angles under the canal, discharging them into the river, preserving the houses perfectly free from offensive smells. A plan of these sewers may be seen in the accompanying illustration. *Thirty thousand dollars* was expended by this company in their construction alone, for the benefit of the health of the operatives!

Labour begins, or the gate closes, at five o'clock a.m. from May 1 to September 1, and at ten minutes before sunrise the remainder of the year. A first bell is rung about forty minutes before, to allow time to prepare for work. *Labour ends* at half-past seven p.m. from September 20 to March 20; at seven from May 1 to September 1; and at fifteen minutes after sunset for the remainder of the year. Dinner during the whole year is at half-past twelve p.m. Forty-five minutes are allowed for each meal.

The number of operatives at present employed in these mills is 1,867, of whom 956 are males, and 911 are females. When entirely completed and in full operation, they will employ about 2,500, and require a town population of 7,500. The principal part of the operatives work by the job, the males earning on the average about $5 80 cents per week, and the females about $2 75 cents per week, besides board, which is $1 50 cents to $2 per week for males, and $1 25 cents for females. The females are principally inmates of the boarding houses. Most of the males, however, have houses of their own, or board elsewhere.

The boarding houses for the accommodation of the operatives in these mills, as in other manufacturing establishments, are owned by the corporation. They have been erected, not for an investment of capital on which a profitable income is to be anticipated, but as a means of preserving a proper supervision over the operatives employed and for their benefit. Boarding houses of this kind generally afford less than 4 per cent interest on the capital invested. Some afford no income at all, and even become an annual expense to the owners. They are kept in repair and rented to the tenants, subject to such regulations and restrictions as the company see fit to establish. The rent and price of board are fixed by the company, but both are subject to such alteration as the circumstances of the times, and of all the parties interested, shall render just and proper.

The tenants of the Bay State boarding-houses now pay $150 each annually, which is about *three per cent* on the cost. The furniture of the houses is obtained and owned by the tenants themselves, and they furnish provisions and other articles of consumption for the inmates. They now receive $1 25 cents per week for the board of females, and $1 75 cents to $2 for males. The fare provided is of a plain, substantial, and wholesome kind, well prepared, neatly served, and in sufficient quantities. Operatives are under no compulsion to board in one tenement rather than in another; it is for the interest of the boarding-house keepers, therefore, that the bill of fare should be attractive and satisfactory. The keepers are sometimes men with wives and families, but they are generally widows, or females who have been accustomed to perform the principal part of the business of providing for their families, and who desire a remunerating means of subsistence. Applications for these situations are generally numerous, but they can be obtained by none but persons of known capacity and respectability; and whenever indications of a different character are manifested, the obnoxious keeper is immediately ejected. Males and females are not allowed to occupy the same house, not even a man with his wife, as boarders.

Several classes of regulations to be observed by the inmates of these houses are printed and placed conspicuously in each house. One code is as follows:

'1. The tenants must not underlet any part of their tenement, nor board any persons not employed by the company, unless by special permission; and in no case are males and females to board in the same house.

'2. The tenants must, when required by the agent, give a correct account in writing of the number, names, character, habits, and employment of their boarders, and whether they are habitual attendants on public worship. They must also, on the first Monday of every month, send to the counting-room a list of all the boarders they have taken, and of all who have left their houses during the preceding month. They must also, at the same time, render a list of the names of all such boarders as have required the services of a physician, on account of sickness, during the same period.

'3. The doors must be closed at 10 o'clock in the evening, and no one admitted after that time, unless some reasonable excuse can be given.

'4. The boarders must not be permitted to have company at unseasonable hours.

'5. All improper conduct among the boarders, and all rude and disorderly deportment, must be prevented by the tenants, if possible, and if persisted in must be reported to the agent.

'6. It is confidently expected that all children over 12, and under 14 years of age, living in the houses, be kept constantly at school.

'7. It is indispensable that all who live in the houses should be vaccinated, and this will be done at the expense of the company, by a physician, at the counting-room, for all those employed by the company and for the families of the tenants.

'8. The health of the inhabitants requires that particular attention should be paid to the cleanliness and daily ventilation of the rooms.

'9. No water, nor filth of any kind, must be thrown out in front of the houses, nor be allowed to remain in the cellars, backyards, or sheds.

'10. Ashes must not be kept in wooden vessels, nor will any carelessness be allowed in the use of fire or lights. Neither camphine nor any other explosive compound used for lights will be allowed on the premises.

'11. The rooms must not be mutilated nor defaced, and in all cases where the plaster of the walls is broken, either by driving in nails, screws or pins, or by rubbing with furniture, or by any carelessness, or by any other means beyond ordinary use and wear, the injury will be repaired, and the cost thereof charged to the person leasing the house.

'12. A suitable chamber for the sick must be reserved in each house, so that they may not be annoyed by others occupying the same room.

'13. Window glass must not be allowed to remain broken longer than one day.

'14. Wood and coal will not be permitted to be taken into the cellars, nor from them, through the front windows.

'15. The closest supervision will be exercised to enforce these rules, and the tenants themselves are particularly required to pay close attention to them, and to insist upon their observance on the part of their boarders.

'16. No tenement will be leased to persons of immoral or intemperate habits, and any tenant who, after occupancy, shall be found to be of such habits, or to receive boarders of such habits, will be notified to vacate the premises.

☛ *The tenants are particularly desired to lend their aid in the preservation of the trees in front of the houses, and to give immediate information to the agent if any injury be done them.*'

Similar regulations are issued by the Atlantic Cotton Mills, besides an additional code, one section of which is the following:

'A proper observance of the Sabbath being necessary for the maintenance of good order, all persons in the employ of this Company *are expected to be constant in attendance at public worship*, and those who habitually neglect this regulation, or who are known to attend improper places of amusement, will be discharged.'

The execution of these and other police regulations of the whole establishment is entrusted to the general agent, who, by his known capacity, his experience, and his character, is fitted for the station. Under his wise and systematic supervision, the boarding-houses and all the departments of these extensive mills are managed with the same care as a small well-regulated family.

The influence of the system by which the boarding-houses are regulated is immensely beneficial, whether we consider it in a social, moral, or sanatory point of view. It is an influence which is felt by all the operatives, at all times, while they are out of the mills as well as in them. In the boarding-houses too, a care, attention, and oversight is frequently exerted by the landlady over her boarders, which is nearly allied to that which a kind parent exerts over her children, and which produces almost as strong a mutual attachment in the one case as in the other.

NEW YORK INDUSTRIAL EXHIBITION.

SPECIAL REPORT OF MR. JOSEPH WHITWORTH. PRESENTED TO THE HOUSE OF COMMONS BY COMMAND OF HER MAJESTY, IN PURSUANCE OF THEIR ADDRESS OF FEBRUARY 6, 1854. LONDON: PRINTED BY HARRISON AND SON,

Introduction

Having been unavoidably prevented, as explained in the General Report of the Commission, from making a report upon the Machinery exhibited in the New York Industrial Exhibition, I have drawn up in a concise form the results of observations made while visiting the principal seats of those manufactures which came within my department.

The statement thus prepared embraces a variety of subjects, somewhat miscellaneous in their character, and which do not conveniently admit of a regular classification. I have endeavoured, however, as far as possible to overcome this difficulty, and to adopt an arrangement by which the substance of the information which I collected may be distinctly and fully appreciated. This Report does not affect to embrace the whole scope of American manufactures, nor even to exhaust the interest of those particular departments upon which it touches; it is merely intended to direct attention to such facts connected with the Machinery of the United States as came within my observation, and which it appeared desirable should be known to those engaged in mechanical and industrial pursuits in this country.

To the general reader many of the descriptions of manufacturing processes will, I am aware, be uninteresting, and in some cases perhaps unintelligible; for, looking to the persons for whose information I have more immediately written, I have thought it proper to adhere to those technical terms which are in use among men who are more or less acquainted with the application of mechanical science. For instance, when in describing a cotton mill in America I have said, 'One man can attend to a mule containing 1,088 spindles, each spinning 3 hanks, or 3,264 hanks on the average per day,' I am aware that I am using technical language incomprehensible to the ordinary reader, but these few words convey all the requisite information to those practically acquainted with the subject. I have not, therefore, attempted to impart a popular tone to the Report; and the very few general reflections which I have ventured to offer, are given at its close.

Some works have been noticed for their novelty and interest, others for their practical utility as affecting our native industry, and in some instances the information which I have collected has been directed to convey a general idea of the extent to which particular branches of manufactures are developed, of the conditions, as to management, under which they are carried on, and of the causes to which their flourishing condition is chiefly due.

The accidental absence of principals at the period of my visits, the limited time at my disposal, and my anxiety to embrace as wide a field of observation as possible, have severally constituted impediments where more minute information might have seemed desirable; but whatever may be the defects of this necessarily imperfect Report, I desire to record my sincere acknowledgments for the great courtesy and the kind attention which I received on all hands during my visit to the United States; I am the more bound to do so as the plan which I have pursued has precluded me from mentioning the names of many gentlemen who so greatly facilitated the objects of my visit, by showing me over their own establishments, accompanying me to those of others, and affording me all the information in their power.

1

Steam Engines and Machinery

New York, Philadelphia, Baltimore, Pittsburgh, Buffalo, Boston, Lowell, Lawrence, Holyoke, Worcester, Hartford, and Springfield

1. *Number of Establishments visited.* The vast resources of the United States are now being developed with a success that promises results whose importance it is impossible to estimate. This development, instead of being, as in former cases, gradual and protracted through ages, is by the universal application of machinery effected with a rapidity that is altogether unprecedented. Upwards of thirty establishments visited in different parts of the States, and employing in the aggregate from 6,000 to 7,000 men, afforded direct evidence that the greatest energy and attention are brought to bear upon the manufacture of machinery.

2. *Marine Engine Works.* The principal marine steam engine works are in New York, but there are large establishments of a mixed character in almost every town of importance; to particularise these and to give a full description of each is unnecessary. They are similar in character, and it frequently happens that it is only in some of the details that there is anything to remark upon. The practice which prevails of combining various branches of manufacture in the same establishment, would also render separate descriptions of each somewhat complicated. For instance, in some cases the manufacture of locomotives is combined with that of mill gearing, engine tools, spinning and other machinery. In others, marine engines, hydraulic presses, forge hammers, and large cannon were all being made in the same establishments. The policy of thus mixing together the various branches is objectionable, but the practice doubtless arises, in addition to other causes, from the fact that the demand is not always sufficient to occupy large works in a single manufacture.

Wherever facilities of carriage or other causes render it practicable to confine establishments to a special branch, the most advantageous results ensue. This was very evident in numerous and striking instances which will be hereafter noticed.

It is to the introduction of railroads that the advantageous subdivision of manufactures is to be chiefly ascribed. The operations of large establishments are no longer confined to particular localities: the facilities of transport being so great, they are enabled to supply their special products not only to a district or a kingdom, but to the world at large.

It is evident that isolated notices of details made in various places must have a disconnected character, and it is therefore convenient to mention in separate paragraphs what appeared worthy of remark.

3. *Locality of Works. Beam Engines.* Nearly all the marine engine works in New York are conveniently situated on the water side, and have slips attached to their yards, where vessels may be moored. The rise and fall of the tide is generally so small as to obviate the necessity of docks. Steam engines having their beams above the deck, are commonly used for the river and ferry boats, and they have generally a very long stroke. The following are some of the dimensions of a beam engine of one of the New York river steamers.

Length of stroke, 12 feet; length of beam, 24 feet; depth of beam in the middle, 10 feet. The form of the beam is that of an elongated parallelogram, the outer frame or skeleton is made in one piece of wrought iron, inside of which is fitted another frame of cast iron, carrying the axis.

4. *River Steamers for Shallow Waters.* A steam-boat running on the Ohio from Pittsburgh to Cincinnati, had a pair of direct acting engines with 32 inch cylinders and 8 feet stroke. There was no main crank shaft connecting the two paddle wheels, but each engine worked its own wheel independently of the other. This arrangement enables the boat to be steered with greater facility round the sharp turns encountered in the tortuous course of the river. The framework and outer bearings of the paddle-wheels are supported by suspension rods, which are, as it were, slung over beams, and framework strongly constructed and fixed in the centre of the vessel. The main deck is 280 feet long, and 58 feet wide. The paddles are 38 feet in diameter, having twenty-four floats, 12 feet wide by 28 inches in depth. For shallow rivers, flat-bottomed steamers propelled by a paddle-wheel at the stern are commonly used. Two were being built of iron in New York, drawing only 2½ feet of water, which are intended for the passage across the isthmus of Panama by the Nicaragua route.

5. *Lake Steamers. Propellers.* A marine engine establishment at Buffalo was principally occupied in making engines for screw steamers intended for lake and river navigation. The propellers of those intended to run in shallow waters are made with four, and sometimes six blades, each, and revolve with rather less than half their diameter immersed in the water. The blades are made of wrought iron, and bolted on to a cast-iron boss, fitted on the propeller shaft, so that a blade broken or damaged by coming in contact with 'snags,' or other obstructions, may be easily replaced. Some of the dimensions of the machinery in a lake boat, used for carrying cargo, are as follows:–The propeller is 16 feet in diameter, with

a pitch of 17 feet 3 inches, making 60 revolutions per minute; the cylinders are 36 inches in diameter, with a 10 feet stroke, and the speed attained averages about eleven miles per hour. The vessel has an upper deck for the accommodation of about sixty passengers. Small high-pressure steam engines for flour mills, agricultural implements, and other machinery, are made in the same establishment.

6. *Caloric Engines.* Ericsson's caloric engines were undergoing repair and alteration in New York, with the view of making the heated air act alternately upon each side of the piston, similarly to steam in an ordinary engine. The bottom of the cylinder is made of wrought iron, and arched. The upper part, in which the piston works, is of cast iron, and is connected to the lower part by bolts: it was this lower portion which proved defective in strength.

It was expected that the vessel, with her caloric engines, would be again ready for sea in the course of two or three months.

2

Process of Casting, Cooling, etc., Railway Wheels, and Annealing–Railroad Spike Making, Nail and Rivet Making–Cast Steel Works–Engine Tools

Pittsburgh, Philadelphia, Lawrence, Worcester

7. *Iron Castings.* The iron castings in some of the establishments were very good, and cylinders from 8 to 14 feet in diameter were well bored, with a finishing feed of cut of about three-eighths of an inch per revolution, which is at a width of cut at least three times as great as that ordinarily given in English works.

At Pittsburgh a large casting for a hydraulic press was cooled by the following method:–Water is introduced into the interior of the core by a pipe, which extends to the bottom, and fills it previous to casting. Provision is made for the escape of the air by making the core fluted.

When the metal is poured into the mould it immediately heats the water, which is then drawn off by an escape pipe at the top of the core, and a supply of cold water is continually running in at the bottom. Heat is thus gradually taken from the mass, and the whole cools uniformly. The casting was 10 inches thick, and weighed 7 tons. It took from three to four days in cooling.

The best charcoal pig iron was selling in Pittsburgh at $45 per ton, having risen within a short period from $30 per ton.

Major Wade, of Pittsburgh, has made many experiments on the tensile strength of this iron. He found that it required a force of 45,000 pounds to tear asunder a bar an inch square. Some of the results of his experiments have been published by the United States Government in the 'Ordnance Manual.'

8. *Pickling Castings.* The process of 'pickling castings,' as it is called, is performed in the following manner:

The castings are placed on two wooden stages, covered with lead, each being 20 feet by 12 feet wide, and supported by two rollers, about 18 inches from the floor. The trough containing the pickle (which consists of

$2\frac{1}{2}$ parts of water to 1 of acid) is of the same length as the stages, which are inclined towards it, to enable it to receive the drainings. The diluted acid is poured over the castings by hand from a long ladle, and when they are dry, the operation is repeated as often as necessary. The stages are then inclined in the opposite direction, and cleansed from the coating of acid and sand by a powerful stream of water directed upon them from a hose pipe.

In England when the process of pickling is adopted for light castings, it is usual to immerse them in the liquid. The American method was probably adopted in consequence of the high price commanded by manual labour.

9. *Annealing Railway Wheels.* The leading and trailing wheels of locomotives, and railway carriage-wheels, are commonly hollow cast-iron disc wheels.

The process of annealing adopted in a large manufactory in Philadelphia is as follows:

The wheels are taken from the moulds, as soon after they are cast as they can bear moving without changing their form, and before they have become strained while cooling. In this state they are put into a circular furnace or chamber, which has been previously heated to a temperature about as high as that of the wheels when taken from the mould; as soon as they are deposited in this furnace or chamber, the opening through which they are passed is covered, and the temperature of the furnace and its contents is gradually raised to a point a little below that at which fusion commences. All the avenues to and from the interior of the furnace are then closed, and the whole mass is left to cool gradually as the heat permeates through the exterior wall, which is composed of firebrick $4\frac{1}{2}$ inches thick, inclosed in a circular case of sheet iron $\frac{1}{8}$ inch thick.

By this process the wheel is raised to one temperature throughout before it begins to cool in the furnace, and, as the heat can only pass off through the medium of the wall, all parts of each wheel cool and contract simultaneously. The time required to cool a furnace full of wheels in this manner is about four days. By this process wheels of any form, and of almost any proportions, can be made with a solid nave.

The manufacture of these wheels was commenced in 1847, and in 1850 15 tons weight were cast per day. The foundry and works as now completed are calculated to turn out 40 tons per day.

In another establishment the wheels while hot are lifted from the mould, and the centre part is placed in a hole communicating by means of a flue with a high chimney, and the edge is packed round with sand. A draft is thus created which cools the mass of iron near the centre of the wheel, and in some measure prevents it from contracting unequally during the operation.

At a foundry in Worcester, the wheels when cast were taken hot from the moulds, and immersed in a pit of white sand, where they are left to cool gradually.

In order to obtain the best chill, it is considered necessary to use cold blast iron made with charcoal.

10. *Railroad Spikes.* There is a large demand for railroad spikes in the United States.

On nearly all the railroads the rails are laid on transverse wooden sleepers, and are simply fastened down by large iron spikes with projecting heads, except at the junction of two rails, where the ordinary chair is employed.

In a manufactory at Pittsburgh, a machine was at work which made these spikes, each weighing ½lb., at the rate of 50 per minute. They are packed in kegs, each containing 300. Seven men only are employed on the works, and they manufacture 5 tons of spikes per day.

11. *Nails and Rivets.* In another establishment at Pittsburgh, 250 men are employed in manufacturing bar iron, rods, sheets, and nails. The iron is manufactured both with anthracite coal and charcoal.

They have 51 machines for making cut nails, many of them are self-acting in the feeding for the smaller sizes, the strip of iron is inserted in a tube, which is made to revolve alternately half round each way. They make 2000 kegs of such nails per week, each weighing 100 lbs., and containing sizes from fourpenny up to tenpenny nails.

A rivet-making machine was at work which made rivets weighing 7 to the pound, at the rate of 80 per minute. Its main shaft carried two cams, one a side cam which gave the motion for cutting off and holding the iron between the dies, the other a direct cam for forming the head of the rivet.

The cams for the nail machines are made of chilled cast-iron, and that part of the lever which acts against the cam is faced with a plate of bell-metal. Several large grinding stones were used, having mouldings on their peripheries for restoring the dies when worn.

It is usual in England to soften the dies by annealing previous to restoring their shape, and again hardening them subsequently. The shape may be thus more perfectly restored, but at a greater cost, and the operation of softening and re-hardening deteriorates the quality of the steel.

12. *Cast Steel.* The manufacture of cast steel is not carried on to any great extent. Some works have been started in Pittsburgh, which have hitherto met with great difficulties, but they are now more successful. Workmen were obtained from Sheffield, but they were intractable, and failed to give satisfaction to their employers.

There were 2 converting and 9 melting furnaces, producing upwards of 2000 lbs. per day. The steel sells at 17½ cents per pound (8¾*d.*)

13. *Engine Tools.* The engine tools employed in the different works are generally similar to those which were used in England some years ago, being much lighter, and less accurate in their construction, than those now in use, and turning out less work in consequence. The proportion of slide to hand lathes is greater than in the generality of English workshops.

Planing and drilling machines are commonly used; but there are comparatively very few horizontal or vertical shaping machines, and a considerable amount of hand labour is therefore expended on work which could be performed by machines much more economically.

The foundries are, for the most part, large and well arranged, and are furnished with good powerful cranes.

Great anxiety is now manifested by many manufacturers to have engine tools of a better description than those in use; and before long there will, no doubt, be great improvement in this respect.

3

Buttons – Daguerreotype Frames – Pins – Hooks-and-Eyes – Cutlery, etc.

Waterbury

14. *New Manufacturing Towns.* The energetic character of the American people is nowhere more strikingly displayed than in the young manufacturing settlements that are so rapidly springing up in the Northern States.

A retired valley and its stream of water become in a few months the seat of manufactures; and the dam and water-wheel are the means of giving employment to busy thousands, where before nothing more than a solitary farmhouse was to be found. Such, in a few words, is the history of Waterbury and all the Naugatuck settlements of Holyoke, Chicopee, Lowell, and Lawrence. Many others might be mentioned, but allusion is now only made to those visited.

15. *Waterbury.* Waterbury is situated in the Naugatuck valley, about 24 miles north of New Haven. It contains many manufacturing establishments, carried on principally by joint stock companies. Besides other firms, there are 28 companies of which the greater number are employed in the manufacture of rolled and sheet brass, copper, wire buttons, German silver, pins, cutlery, hooks-and-eyes. The others are employed in manufacturing hosiery, felt, cloth, webbing, covered buttons, umbrella trimmings, leather, etc.

The official statements of these companies show that their respective capitals vary in amount to a very great extent, and that they are in some cases remarkably small:

There are 2 with a capital of $6000 (about £1200).
5 between $10,000 (£2000), and $20,000 (£4000).
13 – – $20,000 – – $100,000 (£20,000).
And the rest – – $100,000 – – $250,000 (£50,000).

Great facilities are afforded in many of the States for the formation of manufacturing companies. The liabilities of partners not actively engaged in the management are limited to the proportion of the capital subscribed by each, and its amount is published in the official statements of the Company. In the case of the introduction of a new invention, or a new manufacture, the principle of limited liability produces most beneficial results. Persons who from their connections or occupations are likely to be interested in, or profited by, the new invention or manufacture, readily associate together and subscribe capital to give the new proposal a fair trial, when they are assured that their risk will not extend beyond the amount of which they may choose to contribute.

The cost of obtaining an act of incorporation is very trifling; in one case where the capital of the company amounted to $600,000 (£120,000), the total cost of obtaining the act of incorporation was 50 cents (2*s*. 1*d*).

16. *Button Machinery, Buttons, etc.* Upwards of 200 men are employed by one of the companies in the manufacture of buttons, hinges, daguerreotype plates and frames, etc.

The round-shaped button is formed by two punches, one working inside the other, each being driven by a separate eccentric, and the inside punch having the longer stroke. By this arrangement the disc is forced through the die, and drops into a box, thus saving the labour of picking out, which is necessary where a single punch and solid die are used.

The spindle of the polishing lathe in which the button is fixed whilst being burnished, makes 10,000 revolutions per minute.

17. *Daguerreotype Frames.* The lathe in which the oval frames used as settings, for daguerreotypes, are turned, has an oval chuck, and a stationary cutting tool fixed to the slide rest, for 'trueing out' the previously punched oval. Two milling tools are used, one for forming the bevilled edge, the other for ornamenting the face of the oval frame.

The milling tool, as it revolves, is allowed to swivel so as to accommodate itself to the oval. When the bevilled edge has been formed, the first milling tool is removed and another substituted while the work revolves.

One workman is able [to] turn, and ornament by milling, two gross of frames per day.

18. *Pin and Hook-and-Eye Manufactory.* No description of the machinery used in pin making can be given, as the process of 'papering' is all that is permitted to be seen.

The pins are all papered by machinery; they are placed in a shallow feeding dish in an inclined position, so as to allow them to descend gradually as they are shaken by a quick vibratory motion. They fall from the spout of the feeding dish upon the centre of an inclined shallow trough, about 18 inches long, through which runs lengthwise a slit sufficiently wide to admit the shank of a pin, and yet suspend it by its head. It being a matter of chance, when a pin falls from the spout, whether it will drop into the slit or slide down the trough, a sufficient number are allowed to descend to insure the filling of the slit by those which happen to fall favourably. The superfluous pins slide down into a box, from which they are again lifted from time to time to the upper feeding dish. The descending line of suspended pins is conducted by the slit (which is curved at its lower end) to a sliding frame which is worked by the woman who attends the machine.

The frame carries a dozen grooves, and in each of these a pin is deposited as it passes under the slit; the pins are thus arranged in a row, with their points all turned the same way. The sheet of paper for receiving them is placed by the attendant on a grooved table, and deep folds are pressed into it at equal distances, and into the cross ridges thus formed, a row of pins is pushed by the carrying frame at every thrust forward.

Under no circumstances whatever are strangers allowed to enter the rooms in which the pin-making machines are at work. The workmen employed are obliged to enter into a bond, and find two sureties, that they will not disclose anything relative to the machinery. The company preferred keeping their mode of operation a secret in this way to taking out a patent.

19. *Hook-and-Eye Making.* Three different descriptions of machines are employed in making hooks-and-eyes,–the wire being let in on one side of the machine, and a completed hook or eye dropped out on the other. The machines appeared to make them at the rate of about 100 per minute.

About 80 hands are employed, who are said to make 1,200 packs of pins, each containing 3,360 pins, and 2,500 gross of hooks-and-eyes, per day.

20. *Cutlery.* The cutlery and file works were conducted on a limited scale. Many beautifully finished knives were exhibited, and were said to command a higher price than those of a similar class imported from England. The artizans are employed principally on piece-work.

In the cutlery department a workman was pointed out who earned $70 (about £14) per month, while the earnings of others occupied on precisely the same kind of work only amounted to $30 (£6) per month. Thus it will be seen that each workman does the best he can for himself, irrespective of others, and reaps the reward due to his superior skill and industry.

4

Locks – Porcelain-handles – Clocks – Pistols and Guns

Pittsburgh, New Haven, Worcester, Hartford

21. *Lock making*. The manufacture of locks appears to be rapidly extending. In an establishment at Pittsburgh employing 350 men in making locks, coffee mills, copying presses, etc., good work was being turned out. Another at New Haven, Connecticut, employs about 200 men in making locks and lock-handles. The latter are made of coloured clays, so mixed as to present a grained appearance. They are first moulded by hand, then turned in a self-acting lathe with great rapidity, and are afterwards baked in a furnace.

Padlocks are made here of a superior quality to those of the same class ordinarily imported from England, and are not more expensive.

22. *Clock making*. The celebrity attained by New England in the manufacture of clocks gave a peculiar interest to a visit to one of the oldest manufactories of Connecticut; 250 men are employed, and the clocks are made at the rate of 600 per day, and at a price varying from $1 to $10, the average price being $3.

The frames of the clocks are stamped out of sheet brass, and all the holes are punched simultaneously by a series of punches fixed at the required distances. The wheels also are stamped out of sheet brass, and a round beading is raised by a press round their rims for the purpose of giving them lateral strength. They are cut by a machine having 3 horizontal axes, carrying each a cutter placed about 4 inches apart. The first cutter is simply a saw, and the second rounds off the teeth. In cutting an escapement wheel, the first cutter is made to cut each tooth entirely round, and then either the second or third axis with its cutter is used for finishing. The pullies on the three axes are driven by one driving pulley with three straps working over and in contact with each other.

The plates forming the clock faces, and other discs, are cut out by circular shears. The beaded rims intended to go round the clock faces, varying in size from 15 inches downwards, are stamped in concentric rings out of a disc, and then made of the required form by means of dies and a stamping press.

The ogee form given to the wooden framing of the common clock is formed by a revolving cutter of the required shape, making 7000 revolutions per minute, over which the piece of wood is passed by hand,–the requisite pressure downwards being given at the same time.

A circular cutter fixed on a horizontal axis is also used for roughly planing the back parts of the wooden clock. Its diameter is about 18 inches, and it has 4 lateral projections, carrying 4 cutters, 2 gouges, and 2 chisels. These revolve round a fixed circular centre plate, of about a foot in diameter, against which the work is pressed as it is passed along. Each clock passes through about 60 different hands: more than half of the clocks manufactured are exported to England, and of these a large portion are re-exported to other markets.

And it is worthy of remark, that the superiority obtained in this particular manufacture is not owing to any local advantages; on the contrary, labour and material are more expensive than in the countries to which the exportations are made; it is to be ascribed solely to the enterprise and energy of the manufacturer, and his judicious employment of machinery.

23. *Guns and Pistols.* In a large manufactory at Hartford, from 400 to 500 men were employed in making revolving pistols at the rate of from 1200 to 1500 per week.

Self-acting machinery and revolving cutters are used for making all the separate parts, and the tools are made and repaired in a machine shop which is attached to the works.

In another establishment at Worcester, Connecticut, 175 men were at work, manufacturing guns, rifles, and pistols. Revolvers were made in large numbers with barrels on the old principle, and were all proved by hydraulic pressure.

Further particulars will be given respecting this manufacture when the Government establishments are noticed.

5

Wood Working

Lowell, Buffalo, Philadelphia, Baltimore, Worcester, New Haven

24. *Labour-saving Machines.* In no branch of manufacture does the application of labour-saving machinery produce by simple means more important results than in the working of wood. Wood being obtained in America in any quantity, it is there applied to every possible purpose, and its manufacture has received that attention which its importance deserves.

It would be difficult to point in any country to a more successful application of machinery to the working of wood than was made in England long ago in the manufacture of ship blocks, by the late Sir Isambard Brunell, aided by the late Mr Maudslay: other instances of mechanism most ingeniously adapted to similar purposes might also be cited. It cannot therefore be said that in England nothing has been done in this branch of manufacture; but it must be confessed that the improvements which have been made have not been extended, as they might have been, to ordinary purposes, though in this respect a desire for progress is now evidently manifested.

A house in Liverpool is importing the best machines of the kind in use in America, and is making great efforts to introduce them generally in England.

25. *Saw Mills, Lowell.* The trees sawn up in the Lowell saw mills are floated down from the interior of the country by river; they are docked in a basin in the timber yard, and are dragged up an inclined plane into the interior of the mill as they are wanted.

In an upper story are placed two large saw frames, and between them travels an endless chain, running along the shop floor over pullies, and extending down the inclined plane nearly to the edge of the basin. To any part of this endless chain may be hooked another chain, which, being passed round one or more trees as they lie in the basin, drags them up into the mill and deposits them alongside the saw frames.

Shingles, used for covering the roofs and sides of houses, are made in vast quantities. A circular saw cuts them 16 inches long, from 3 to 9 inches wide, and of a thickness tapering from $\frac{3}{8}$ to $\frac{1}{8}$ of an inch, at the rate of from 7,000 to 10,000 per day, according to the nature of the wood.

Timber is also cut up into laths 4 feet long, at the rate of from 60 to 100 per minute, by a circular saw attended by two men.

26. *Saw Mills, Buffalo.* In these saw mills boards were sawn into 'sidings,' that is, long wedged-shaped boards for the sides or roofs of houses, by a circular saw at the rate of 17 feet per minute. The board is introduced at the back of the saw, and moves in the direction in which it revolves. It thus cuts with the grain, and the strength of the cut assists the forward feeding motion of the board.

Subdivision of manufacture is advantageously adopted as a system.

Many works in various towns are occupied exclusively in making doors, window frames, or staircases by means of self-acting machinery, such as planing, tenoning, morticing, and jointing machines. They are able to supply builders with the various parts of the woodwork required in buildings at a much cheaper rate than they can produce them in their own workshops without the aid [of] such machinery. In one of these manufactories twenty men were making panelled doors at the rate of 100 per day.

Portable sawing machines, driven by horse-power, are commonly used for sawing up logs of wood for fuel, particularly at the various stations on the railroads, where the wood intended for the consumption of the locomotives is stored in piles.

The 'horse-power machine' consists of a stout frame supporting a railway about 7 feet long, on which run the rollers of an endless travelling platform. The axles of the rollers are of iron, $\frac{5}{8}$ in diameter, stretching across the rails, and are connected together by a series of links, each about twelve inches long, so as to form an endless chain, which passes over a fixed segment at one end and the chain wheels at the other. The travelling platform is made by planks of wood about 12 inches broad, $1\frac{1}{4}$ inches thick, fastened transversely to the endless chain. It is inclined at an angle of about 7° to the horizontal line, and the horse being placed on the platform pushes it backward from under him, which causes the chain wheels at the end of the frame to revolve, and the motion thus obtained is conveyed to the circular saw or other machine required to be driven. Some horse-power machines are made to admit two horses abreast. They are found very useful to farmers; when requisite they are mounted on wheels, and may be easily taken from place to place to saw up trees which could not conveniently be moved entire.

27. *Planing Machines.* A numerous variety of planing machines are in common use. For flooring boards, Woodworth's machine is found to answer very satisfactorily. In planing mills at Philadelphia, four of them were working in one room side by side; they have three cutters on each horizontal axis, having a radius of 6 inches, and making 4,000 revolutions per minute. The cutters are said to be capable of planing from 2,000 to 3,000 feet of work without being sharpened with the oil-stone, and from 20,000 to 30,000 feet without being ground.

They plane boards 18 feet long, varying in width from 3 to 9 inches, at the rate of 50 feet per minute. At the same time that the face of the board is planed, it is tongued and grooved by cutters revolving with a radius of about 3 inches, on vertical axes on each side of the board.

The chips made by the four planing machines are driven through large pipes, and fall into a trough about 20 inches wide, running across the room immediately under the cutters. In this trough works an endless chain, on which are fixed wooden scrapers that carry along the chips as they fall, to a recess at the side of the room. Here they are carried off by the scrapers of another endless chain running up an inclined plane. The pulleys on which this side chain works are larger in dimensionst han those of the transverse chain which works inside. The transverse chain thus deposits its chips in the trough of the inclined plane, and they are carried up to a hole in the ceiling of the fuel room, adjoining the boiler house; through this hole they fall into the fuel room, and fill it up, if necessary, to the top.

28. *Daniel's Planing Machine.* Where an accurately smooth surface is required, Daniel's planing machine is employed. It consists of an upright frame, in which a vertical shaft revolves, having horizontal arms, at the ends of which are fixed the cutters. The work is carried along on a travelling bed under the cutters, which are driven at a very high speed.

29. *Box Making.* In a box manufactory at Worcester a machine, made on Woodworth's principle, planed boards 10 inches wide simultaneously on both sides. Boxes are made in great numbers, from boards which are tongued and grooved by what is called a 'matching machine,' and then put together as that operation is finished.

The tonguing and grooving cutters are fixed on horizontal axes, and the workman passes the boards alternately over one or the other, as the sides require to be tongued or grooved.

Other varieties of planing machines are also in use, known by the names of their different inventors; some of them have fixed vertical or horizontal cutters, others vertical or horizontal revolving cutters, and various others combinations, according to the purposes for which they are intended.

30. *Spill or Match-making Machine.* This machine makes 900 round spills, 15 inches long, $\frac{1}{8}$ inch diameter, per minute; so that if each spill were cut into 5 matches, each 3 inches long, 4,500 would be produced every minute.

The spills are cut from pieces of straight-grained timber, made of such a length as to pass between two grooved feeding rollers, which hold the timber so that its under surface is level with the lowest parts of a row of tubular cutting tools, or long sharp-edged punches.

The cutting tools are thus arranged: 5 pieces of steel are fixed side by side in a horizontal bar. Each piece of steel is perforated with 3 long holes, lying close together, and having their ends sharpened like the cutting edges of a hollow punch. A line of 15 tubular cutters is thus formed, and motion is given to the horizontal bar, in which they are fixed by a crank which impels them against the timber. This is depressed at each stroke sufficiently to allow each cutter to cut out its spill, which passes through and falls out behind.

The cost of this machine would not exceed £20,and when the number of matches, all nicely rounded, which it is capable of producing, is contrasted with the number which could be produced by a hand-instrument in the same time, it will serve as a simple and striking illustration of the advantageous employment of matter in the form of machinery to do the work of man.

31. *Last and Boot-tree Manufactory*. A machine, constructed on Blanchard's principle, is used for making lasts.

A pattern last, and the block of wood from which another last is to be cut, are fixed upon, and revolve round, a common axis, being connected with the centres of a headstock fixed on a frame made to oscillate from below. As the pattern revolves it is kept continually pressed against a knob of iron by a spring, and as the block of wood revolves it is shaped by a circular cutter, revolving on a fixed axis, with its cutting edges in a line with the face of the knob. The pattern and the copy revolving simultaneously on a common axis, as the surface of the pattern is pressed against the projecting knob, the oscillating frame is made to move, so that the revolving cutters shape from the block a surface exactly corresponding to the surface of the pattern, and the copy occupies relatively to the cutters the same position which the pattern does relatively to the knob of iron. About 18 men are employed, who make 100 pair of lasts per day, exclusive of boot-trees.

32. *Furniture Making*. Labour-saving machines of all kinds, sawing, planing, boring, shaping, and jointing machines, are very advantageously employed in the manufacture of furniture. An apparatus of a very simple character is used for shaping the arms and legs of chairs. Two vertical cutters are made to revolve in opposite directions, at the rate of about 1,700 revolutions per minute, on axes projecting above a bench.

The cutters are about 4 inches in diameter, and between them and the bench are loose washers or rollers, against which the pattern to which the work is fastened is pressed so as to guide the cut. The cutters revolve in opposite directions, and the work may be pressed against either one or the other, so as to suit the cut to the direction of the grain of the wood, without the workman having the trouble of reversing its position.

33. *Agricultural Implements. Ploughs*. Labour-saving machines are most successfully employed in the manufacture of agricultural implements. In a plough manufactory at Baltimore eight machines are employed on the various parts of the woodwork. With these machines seven men are able to make the wooden parts of 30 ploughs per day.

The handle pieces are shaped by a circular cutter, having four blades, similar to those of smoothing planes, fixed on a horizontal axis, with about 2 inches radius, and making nearly 4,000 revolutions per minute. The work to be shaped is fastened to a pattern, which is pressed against a loose roller on the axis of the cutter as the workman passes it along, and it is thus cut of exactly the same shape as the pattern.

All the ploughs of a given size are made to the same model, and their parts, undergoing similar operations, are made all alike. Some of the sharp edges of the wood are taken off or chamfered by a cutter revolving between two cones; these guide and support the work as it is pressed down edgewise on the cutters, and passed along by the workman.

34. *Ploughs* (*continued*). The other machines in use consisted of a circular and vertical saw, and machines for jointing, tenoning, drilling, and for making round stave rods, and giving them conical ends, the whole being of a simple and inexpensive character.

The curved handle pieces of the ploughs, which require to be steamed and bent, are obtained already shaped from the forests where they are cut, and are advantageously supplied to the large manufacturers. The prices of the ploughs vary from \$2½ to \$7.

The price of Pig-iron in Baltimore was \$40 per ton.
Pine timber – – \$20 per 1,000 feet.
Ash – – – – \$24 per 1,000 feet.
Oak – – – – \$25 per 1,000 feet.

35. *Mowing Machines*. In a manufactory at Buffalo, mowing machines were being made in large numbers, 1,500 having been supplied this summer. Two were in operation in a field a short distance from the town; each was drawn by 2 horses, and could mow on an average 6 acres of grass per day.

The machine is similar in its construction to the common reaping machine, but it has only one wheel, furnished with projections to prevent it from slipping. This wheel gives motion to the cutters, and supports one side, the other rests on a runner like that of a sledge. It has a pole to which two horses are attached in the ordinary way, and the driver sits on a seat fixed behind the cutters.

36. *Churn Making, etc.* In an establishment at Worcester, 250 hands are employed principally in making ploughs, hay-cutters, churns, etc. Templates and labour-saving tools are used in the manufacture of these implements, which are sold in very large numbers.

The churns consist of a double case, the inner one being of zinc, which receives the milk or cream, and in which the arms revolve, the outer one being of wood. It is found by experience that butter is formed most rapidly when the milk or cream is churned at a certain temperature, and in order to obtain this temperature, which is indicated by a thermometer inserted in the churn, warm or cold water is introduced between the inner zinc and outer wooden casing, as may be required.

37. *Carriage Making*. Many of the carriages, especially those technically called 'waggons,' are made of an exceedingly light construction, and are intended generally to carry two and sometimes four persons.

Their wheels are frequently made with only two felloes, which are bent round by the operation of steaming, and are strengthened at the joining with iron clamps. The wheel of a carriage constructed to carry four persons had felloes only 1½ inches square. They are generally made of white oak, and the spokes are obtained ready shaped from shops where their manufacture forms a special trade.

It would seem as if the elasticity of these carriages peculiarly fitted them for the very bad roads on which they in general have to run, and it is evidently a principle with the Americans to use up their light carriages and save their horses.

Every man in America who is able to keep his waggon is free to do so, unfettered and unquestioned, consequently their use is so general that it may be said to be almost universal. Their manufacture is one of great importance, and supports a vast number of wheelwrights and artizans of that class, who from the nature of their employment attain great skill and aptitude, enabling them to turn their hands to almost any variety of work, and rendering them a most useful and important class.

6

Stone-planing works – Brick making from dry clay

Staten Island, New York, Washington

38. *Stone-planing Machines.* In an establishment at Washington, which has but recently commenced operations, there were two planing machines and a grinding or polishing machine. Considerable difficulties have hitherto attended the employment of machinery for planing stone such as granite, and stone of similar formation. These difficulties have, however, been surmounted most successfully by the construction of planing machines such as are used in stone works in New York and Washington, in the former of which upwards of 400 men and 10 machines are employed.

The planing machine consists of an upright frame, in which revolves a vertical shaft, carrying 3 horizontal arms. At the extremities of these arms, are fixed circular cutters inclined outwards about 45° from the perpendicular, or about the angle at which the workman would hold his chisel. They are about 10 inches in diameter, and $\frac{3}{4}$ inch thick, made of steel, and bevilled on both sides, leaving a sharp edge. They are fitted upon axes, and are at liberty to revolve loosely in their bearings as their edges strike the stone. The cutters are carried round by the shaft at the rate of about 80 revolutions per minute when planing freestone, and 60 when planing granite.

The stone is moved forward on a bed to which it is keyed; the cutters strike its surface obliquely as they are carried round on the revolving arms, turning at the same time on their own axes, and chipping and breaking off the projecting portions of the stone at every cut.

The machine planed the face of a stone slab, 4 feet long and 2 feet wide, in seven minutes.

Another modification of this machine, which is not so economical, is employed when it is necessary that the face of the stone be left in lines as it came from the tool. The stone is keyed on a travelling bed, and passed under a frame, in which works a sliding carriage driven by a crank; in this carriage is fixed the circular cutter at the required angle, and as the stone is carried along, the cutter is driven backwards and forwards across its face at right angles to the direction in which it moves, and chips off parallel breadths of stone at every cut.

The cutters can be used for planing from 300 to 400 square feet of free-stone surfaces, and about 150 square feet of granite, without being ground.

39. *Stone Polishing Machine.* The stone is polished by a flat circular disc of soft iron which is made to revolve horizontally. The axis of a disc is fixed at the end of a heavy frame, which moves round a strong centre shaft in a radius of about 12 feet. The polishing disc revolves at the rate of 180 revolutions per minute. It is driven by a strap to which motion is given by a driving pulley fixed on the centre shaft. The disc is guided, and its pressure regulated by hand. It will polish about 400 square feet of surface in a day of 10 hours.

40. *Brick Making with Dry Clay.* A machine for making bricks from dry clay was in operation on Staten Island, about nine miles from New York. The works are carried on under extensive sheds, near to the water side, and are connected with a wharf by a railway, which also extends to the bed from which the clay is dug. A large moveable shed is erected on the bed of clay at the terminus of the railway.

In dry weather the clay is collected by slicing it from the surface with a kind of shovel having a sharp edge, which is drawn by two horses, and will hold about two barrowfuls. In wet weather the surface of the clay is harrowed to the depth of 2 or 3 inches by a triangular wooden frame carrying 9 teeth, a process which in the powerful rays of an American sun soon causes the moisture to evaporate. It is then taken off by the scoop or shovel above described, and conveyed to the shed, whence it is carried by rail to the machine shed. It is deposited close to a cylindrical screen, revolving on a fixed axis, which has projecting loose bars. The screen is about 8 feet long and 3 feet in diameter, and consists of bars $\frac{5}{8}$ inch square, rivetted on two cast-iron wheels, which form the ends; the bars are about $\frac{1}{16}$ inch apart, and the clay is riddled through them. The screen is inclined, and the clay is fed in between the arms of the wheels, and as it revolves the small pulverized particles fall through the bars, while the large stones pass out at the lowest end.

The clay is next raised by elevators, which are fixed to an endless leather belt about a foot wide, to the height of about 12 feet, and conveyed to rollers. It is ground and shovelled into hoppers which feed the moulds; these are 8 inches long, 4 inches wide, and 4 inches deep. The clay is dropped into the moulds, which are placed six in a row to the depth of 3 to $3\frac{1}{4}$ inches, according to the quantity of moisture it contains, and is afterwards compressed to the thickness of 2 inches in the following way.

Six presses or rams, fixed in a heavy frame, are raised by a cam, and being allowed to fall, exert very great pressure by their impact on the clay. The blow is repeated, and then the bricks are powerfully compressed above and below by revolving cams: 36 bricks are made per minute. They are at once conveyed to the kiln, which is under the shed at a short distance from the machine.

After being burnt, they are separated into three shades of colour, of light and deeper reds. The best burnt bricks are equal in quality to the best English stocks, and were selling at $12 per 1000.

41. *Brickmaking from Dry Clay* (*continued*). In a brick yard at Washington, Sawyer's machine, which had been in use for 16 years, makes about 1800 bricks per hour from dry clay, by compression only.

The clay is obtained from a pit close by. As it is dug out it is carted up an inclined plane to the floor over the room where the machine is at work.

A roller weighing 1600 lbs., and making 60 revolutions per minute, grinds it upon a grating through which the pulverized particles fall into the room below. There it is shovelled into a hopper which supplies the brick moulds by feed-pipes. Three bricks are made at one time, being compressed by top and bottom pistons or pressers, which are connected together by long iron rods, and from the top part are suspended levers, with toggle joints worked by cranks. The bricks were sold at the rate of $6½ per 1000, and were of a medium quality between English seconds and stock bricks.

7

India Rubber Manufactory – Fishing-net Making Machine – Flour Mills – Elevators

New Haven, Baltimore, Pittsburgh, Buffalo

42. *India Rubber Manufactures.* India rubber is applied to a great variety of purposes, and its manufacture here is attended with very great success.

By a new process of hardening, the substance becomes of the consistency of horn. In that state it is manufactured into combs, walking sticks, and other articles.

43. *India Rubber Overshoes.* The India rubber in its rough state is first cut up by shears into small pieces. It is then put through a machine similar to that used for tearing and cleaning rags intended to be made into paper. The water used in the operation is drawn off from time to time through a wire grating.

The material thus chopped up, and cleaned, is passed through rollers, where it is sufficiently ground. It is then put through other rollers, where it is kneaded, and worked up with the necessary composition. The India rubber, so mixed, is passed in the form of an endless web through 4 rollers placed vertically one above the other, and comes out a broad web fit for use.

The 'gumming process' is performed by three rollers, Nos. 1, 2, 3. Nos. 1 and 2, the two lower ones, revolving side by side, and No. 3 revolving above and in contact with No. 2. The India rubber is fed between Nos. 1 and 2, and the cloth to be gummed or covered is passed between Nos. 2 and 3, taking up from No. 2 a thin and equally spread coating of rubber.

The India rubber cloth is cut out from the sheet by workmen, in the shape required to form shoes. The parts so shaped are put together by women, who form them on lasts, closing the joints by cohesion after touching them with camphine. Each woman finishes an entire shoe, and about 1400 pairs are made daily. The shoes are then covered with a coat of varnish, and taken to the stove drying room, where they are subjected to a heat of from 250° to 280°, and allowed to remain a night.

To provide for an equal distribution of heat in the drying room, two large heating stoves are placed underneath, each in a separate compartment. These are fed with fuel from the outside, and the heat is admitted into the drying room above, through several apertures pierced in the floor.

Thermometers are placed at the side of the room, and can be inspected through glass from the outside.

44. *Fishing Net Machines–Baltimore.* These machines combine the general features of the power loom and the lace machine.

They are made from 6 to 7 feet wide, according to the size of the mesh. One machine nets a ¾inch mesh, and can be used for netting meshes of 1½ inch and 2¼ inches. It works at the speed of 12 picks per minute, and a complete course of 100 knots is made in the width of this machine, at each pick of the shuttle.

One woman can do the work of upwards of 100 hand netters. The meshes are made rectangular, in the direction of the length of the net, and not diagonally, as in hand-made nets. The cost of the machine is $800 (about £160).

The manufacture of sailcloth is carried on in the mill where these machines are at work.

The throstles for spinning yarn for the sailcloth spin 6 hanks to the pound. The carding engine sliver is carried by the railroad system along a trough to the drawing frame. The main cylinder of the carding engine is 36 inches in diameter, and the doffing cylinder 13 inches, the former making 135 revolutions and the latter 7 revolutions per minute. In the fly-frame the front roller makes 200 revolutions per minute, and the flyer from 1,900 to 2,000.

By some shipowners sailcloth made of cotton is preferred to that made from hemp. Fishing nets made by hand are here also manufactured of cotton.

45. *Corn and Flour Mills–Pittsburgh.* These mills employ 40 persons, including clerks and all others engaged in the various departments, and are capable of producing 590 barrels of flour per day, each containing 196 lbs.

The grain is brought in bulk in boats alongside the building, and is raised by an elevator consisting of an endless band, to which are fixed a series of metal cans revolving in a long wooden trough, which is lowered through the respective hatchways into the boat, and is connected at its upper end with the building where its belt is driven. The lower end of the trough is open, and as the endless band revolves, six or eight men shovel the grain into the ascending cans, which raise it so rapidly that 4,000 bushels can be lifted and deposited in the mill in an hour.

The grain is next allowed to descend by a shoot or trough (the descent being regulated by traps) into a large hopper, resting on the platform of a weighing-machine; its weight is then registered, and afterwards, by drawing a trap in the bottom of the hopper, the grain is allowed to descend by another shoot to a lower story.

It is next raised by an elevator to the highest story of the mill, where it is cleaned by passing through three different machines. The greatest care and attention is bestowed on this process, in order to insure the perfect cleansing of the grain preparatory to being ground. The grain is then conducted to the stock-hoppers, which feed eight pairs of grinding-stones.

A short length of the feeding-pipe of each pair is made of glass, through which the grain, as it descends, can be seen. The stones are 4 feet in diameter, and make 232 revolutions per minute.

The meal, when ground, is conveyed by means of a spiral conveyor to the cooling chamber, where a rake, revolving horizontally, is substituted for the old 'hopper boys.' The meal is raked from the circumference to the centre, where it falls through a hole and is taken to the bolting machine; it is there sifted, and separated into different qualities of flour. It is then conveyed to hoppers, from which it descends by spouts into the barrels in which it is packed.

46. *Elevators.* The business of unloading vessels is followed as a special trade.

On the wharves in Buffalo may be seen in many places large signs announcing that elevators are kept for hire. They are used for raising grain from vessels, storing it in warehouses, and transferring cargoes of wheat from one vessel to another, the grain in the last case being raised by the elevator cans and then allowed to descend by a trough or shoot which guides it in any required direction.

8

Manufacturing Companies – Civil Engineering – Cotton Mills – Carpet Manufactory – Woollen and Felt Cloth Making – Sewing Machinery – Cotton Gin

Lowell, Lawrence, Holyoke, New Haven, Waterbury

47. *Textile Fabrics.* The manufacture of textile fabrics is extending, particularly in the New England States. Many new towns, founded for the purpose of carrying on this branch of manufacture, have in a short time attained considerable importance.
48. Lowell, Lawrence, and Holyoke, in Massachusetts, may be cited as instances well worthy of notice.

Lowell is situated on the banks of the Merrimack, about 25 miles north of Boston. It contains twelve large manufacturing establishments, belonging to different companies; of these, eight manufacture cotton goods, possessing in the aggregate about 350,000 spindles and 10,000 looms, and employing about 7,000 women and 2,000 men. Two of them manufacture woollen goods, carpets, rugs, and broadcloths (one also combines the manufacture of cotton goods to some extent), possessing about 20,000 spindles and 600 looms, and employing about 1,500 women and 1,000 men. One is a bleaching concern, employing 250 men. One a machine shop, employing 700 men.
49. The capital stock of the companies varies in amount from $300,000 (£60,000), to $2,500,000 (£500,000); the total for the whole being $14,000,000 (£2,800,000).

The interiors of the mills are kept in a state of great cleanliness. The rooms are lofty and properly ventilated: their white ceilings and walls, combined with the blue hangers and columns, have a pleasing appearance. The courtyards of many of the mills are laid out with flower-beds, interspersed with shrubberies, or shaded by lofty trees, and great care seems to be taken to keep them in good order.
50. Water power is used for driving the machinery in all the mills. It is

obtained by means of a large and deep canal which is cut from the river at some distance above the town. At the head of the canal, where it joins the river, are floodgates, worked by large screws, all of which are driven by a small turbine. The masonry of the canal is constructed of blocks of granite, some of which are of very large dimensions. The water thus conducted by canal is employed by the various companies in driving powerful turbine water-wheels. The mill gearing conveying the power from the turbine to the different parts of the mill is exceedingly well constructed.

The civil and mechanical engineering works employed in adapting water power to driving the machinery of the Lowell mills have been most ably executed.

51. The first mills commenced operation in Lowell in 1823. Eight of the companies have been founded since 1830. In 1828 the town contained 3532 inhabitants: in 1850 its population was 33,385! It has four banks and two institutions for savings, and a hospital established by the several companies for their sick operatives.

52. *Lawrence*. Lawrence is situated about 26 miles north of Boston, on the Merrimack river, 9 miles below Lowell. The first dwelling-house was erected in September 1845, and in 1850 the town contained 8500 inhabitants, and upwards of 1000 dwellings. There were 15 schools, attended by 1000 scholars, and conducted by 16 teachers. The town now contains upwards of 13,000 inhabitants.

53. *Water Power*. The water power was obtained by building a dam which has a 25 feet fall of water, 900 feet broad. The dam is constructed in the form of a curve of solid masonry, imbedded in and bolted to the rock. It is 35 feet thick at the base, and averages 32 feet in height. Its cost was $250,000 (about £50,000).

The water is taken from the river above the dam by a canal just a mile in length, 100 feet broad at its upper end, and 60 feet broad at the lower, 12 feet deep in the middle, and 4 feet at the sides. Its total cost, including locks and other structures connected with it, was $200,000 (about £40,000).

54. *Cotton Mills*. Seven large incorporated manufacturing companies have already commenced operations, and others are about to be established. The largest cotton mills employ about 1200 hands. They have a frontage of 600 feet, and consist of a centre block of six stories 106 feet wide, and two wings of five stories, 64 feet wide. The machinery is driven by three turbine water-wheels, 8 feet in diameter.

55. *New Cotton Mill*. Another establishment, lately erected, is now being fitted with machinery for the purpose of manufacturing mousselines de laines, barèges, and other light fabrics. It is six stories high, each averaging 13 feet, 750 feet long, and 72 feet wide. There is also another building in the course of erection which will be 1200 feet long, two stories high, with two wings, each 200 feet long and three stories high, and is intended to be used for printing, dyeing, etc. These mills are built of good bricks, measuring each 8 inches by 4 inches and 2 inches thick.

The contract price for laying 1000 bricks, including cost of material, is

£1 16*s*. The entire erection was found to cost 50 cents (or 2*s*. 1*d*.) per square foot of flooring laid down.

The wool-combing machinery will be obtained from England. One woman working one of these machines, will be able to comb 1,000 lbs. of wool per day, while a skilled hand wool-comber is only able to comb from 8 to 10 lbs. in the same time.

The machinery of these mills will be driven by turbine water-wheels, of between 500 and 600 horse power in the aggregate.

56. *Woollen Factory*. The mills of this establishment are built in the form of a parallelogram, round three sides of which run buildings from three to five stories high. The front is formed by three detached mills, each 200 feet by 48 feet, and 9 stories high, including attic and basement.

2,300 hands are employed in the manufacture of cassimeres, shawls, felt cloth, and other woollen goods.

57. *Felt Making*. The whole waste from the mills mentioned in the preceding paragraph is worked up in the manufacture of felted cloth. The felt-making machines occupy but a small space.

A sliver of wool is taken from the carding engine and passed between two endless cloths; these carry it over a narrow steam box, where it is steamed, and it is then passed under a vibrating pressing-plate, which operations cause the fibres to curl and interlace with each other, and so form a cloth.

The machinery of these mills is driven by seven breast wheels, each 26 feet in diameter.

58. *Machine Shop*. There is a large machine shop employing 500 hands in the manufacture of spinning and other machinery. It is 4000 feet long, 64 feet broad, and contains four stories of from 16 to 13 feet high. The forge shop is 230 feet long, 53 feet wide, and 17 feet high, and contains 32 forges. The foundry is 150 feet long, 90 feet wide, and 22 feet high.

59. *Holyoke–Cotton Mills*. Holyoke is a manufacturing town situated on the banks of the Connecticut river. A short notice of its history will serve to explain the way in which manufacturing companies are established in the United States. In 1847, a company was formed for the purpose of turning to account the water power supplied by the river Connecticut, buying up the water privileges, and purchasing land to form the site of a manufacturing town. The company subscribed a capital of $4,000,000, and was incorporated by the State of Massachusetts in 1847.

It succeeded in obtaining the water privileges, and upwards of 11,000 acres of land, besides other tracts in the vicinity. A dam, more than 1,000 feet long, was constructed across the river, in the summer of 1849.

The site of a town has been laid out with streets from 60 to 80 feet wide, calculated for a population of 200,000 inhabitants. It contains already upwards of 5000 inhabitants, and it is officially stated, that the average sum appropriated for the education of each child was in 1852, $3 72.

There is a 60 feet fall of water, which can be used by two sets of mills on different levels, affording power sufficient to drive the machinery of 100 large mills.

60. *Cotton Mills.* Two cotton mills employing 1,100 hands, a machine shop employing 365 hands, and a paper mill are already at work; others are in the course of erection. One of the mills was spinning yarn, Nos. 70 and 90, and making it into cloth of excellent quality.

Self-acting mules were used, and twelve piecers were minding 13,056 spindles: three hanks per spindle were spun in a day of 11½ hours.

One girl is able to weave of this yarn on four looms, 100 yards per day. Upwards of 70 girls were brought from Scotland a short time ago. The machinery is driven by turbine wheels.

In some mills, gearing is employed for driving the heavy shafting, but generally belts are much preferred; of these, some had a breadth of 20 inches, and were driven at the speed of nearly 1800 yards per minute. In some cases, in order to obtain sufficient adhesion, without having recourse to too tight a belt, the pulley is covered with leather, which is put on with white lead, and fastened with copper rivets.

61. *Hosiery.* A large establishment at Waterbury is occupied exclusively in the manufacture of under-vests and drawers. The cloth waistbands of the latter are stitched by sewing machines, working at the rate of 430 stitches per minute. These machines have been worked with entire success for the last eighteen months.

The manufactured goods and the sewing machines are all that are shown to visitors. No stranger is ever permitted to see the hosiery looms; workmen, directors, and president all enter into a bond not to disclose anything connected with the machinery of the company.

62. *Shirt Making by Machinery, New Haven.* In a shirt manufactory at New Haven, entire shirts, excepting only the gussets, are sewn by sewing machines. By the aid of these machines one woman can do as much work as from twelve to twenty hand sewers. The workwomen work by the piece, and are frequently able to finish their estimated day's work by two o'clock, and when busy, work overtime.

63. *New Cotton Gin.* This gin has, instead of saws, a card cylinder 8 or 9 inches in diameter, covered with coarse wire teeth, with considerably more bend or hook than the ordinary card tooth. The cylinder revolves against a spirally fluted cast-iron roller, the tooth being about $\frac{1}{10}$ inch, and the space between the teeth $\frac{3}{10}$ inch broad. To save the expense of turning and fluting the roller, it is cast in lengths of about six inches, which are bored and turned at the ends, and then put together, the tooth and space being left as they are cast.

In contact with the card cylinder, a cylindrical brush, 28 inches diameter, is made to revolve. The card cylinder makes 200 revolutions, the fluted stripper 400 in a contrary direction, and the cylindrical brush 800 revolutions per minute.

When the raw cotton is introduced with its seeds between the card cylinder and the stripper (which are placed so far apart as to stop the seeds from passing), the hooked teeth of the card take hold of the fibres and pull them from the seed, which is held up against the roller as long as any fibres cling to it for the card teeth to hold by: the seeds are then released, and fall to the ground. The spirally fluted roller causes the position of the seed and cotton to be continually changing. The cotton fibres, as they are taken round by the teeth of the card cylinder, are brushed off by the rapid revolution of the cylindrical brush, and carried to the bin.

The machine is about 60 inches wide, and can gin 1,500 lbs. of cotton per day. Its cost is $350 (£70).

9

Railways – Railway Carriages – Large Four-masted Ship – Fire Companies – Fire Engines

64. *Railways.* In the construction of railways, economy and speedy completion are the points which have been specially considered. It is the general opinion that it is better to extend the system of railways as far as possible at once, and be satisfied in the first instance with that quality of construction which present circumstances admit of, rather than to postpone the execution of work so immediately beneficial to the country; to the future is left further progress and improvements.

A single line of rails nailed down to transverse logs, and a train at rare intervals, are deemed to be sufficient as a commencement, and as traffic increases, additional improvements can be made.

65. *Railway Crossings.* Bridges are seldom thought necessary to carry the common highways across the railroads where they intersect, gates are even in many cases dispensed with, and a notice of 'Look out for the locomotive when the bell rings' is considered a sufficient warning, and wayfarers are left to take care of themselves. Sharp curves and steep inclines are frequently submitted to for the sake of economy.

66. *Railroad Inclines.* The railroad that connects the eastern and western parts of Pennsylvania, bringing the towns on Lake Erie, and the great western rivers into direct communication with Philadelphia and the Atlantic, consists at present of a single line of rails carried over the lofty ridges of the Alleghany Mountains by a series of inclined planes. These are five in number, and the summit of the highest is 2,600 feet above the level of the sea. The trains are dragged up each incline by a rope attached to a drum worked by a stationary engine. They are drawn across the plateaux which intervene between the inclines, in some cases by horses, in others by small locomotives. A new road is, however, being constructed which will cross the mountains by one long winding incline. The ascent will be so gradual in its circuitous course, that a locomotive will be able to ascend and descend with its train of carriages. It is calculated that four hours will be saved by the substitution of this new route, and the dispensing with the stationary engines.

It is doubtful whether the delay would not have been very considerable, had the construction of the railroad been postponed until means had been found for executing these great works in the first instance.

67. *Street Railroads.* It is a common practice to detach the carriages from the engine at the outskirts of towns, and draw them by horses along rails laid down in the streets. Many objections may be made to this system, and it seems on the whole disadvantageous; a circuit of rails carried round the town would be more preferable.

68. *Railroad Cars.* The construction of the railroad cars or carriages commonly used in the United States has been frequently described. They are very long, and are supported at each end on four-wheeled trucks, on which they swivel when turning the sharp curves, which are of ordinary occurrence.

A car constructed for 60 passengers measured 40 feet long, 8½ feet wide, and 6½ feet high, inside measure; small benches with reversible backs, having each two seats, are ranged parallel to each other down both sides of the carriage, leaving a passage clear from the door at one end to that at the other. The car afforded upwards of 2,200 cubic feet of space, or 37 feet per passenger. Its weight was 11 tons, giving a dead weight of about 3½ cwt. per passenger. The cost of a 60 passenger car is about $2,000 (£400).

A contrivance has been lately tried for excluding the dust by connecting the different carriages together by india-rubber curtains at the ends, the air being admitted through the roof of the first carriage.

The object sought to be obtained is, a current of air running through the entire train, and always setting outwards from the interior of the carriages. The results did not appear to answer fully the expectation which had been formed.

69. *A Four-masted Clipper Ship.* A large clipper ship of 4,000 tons was being built at Boston; the length of keel was 287 feet, length on deck 320 feet, extreme breadth of beam 52 feet, and depth of hold 30 feet. Her keel is of rock maple in two thicknesses, the frame is of seasoned white oak, dowelled, and bolted together through the dowelling with 1¼ inch iron. The frame inside is diagonally cross-braced with iron, the braces being 4 inches wide and ¾ inch thick, bolted through every timber: these braces extend from the floor-heads to the top timbers, and form a perfect network of iron over all her frames fore and aft. She has five depths of midship keelsons, each 16 inches square, three tier of sister keelsons, 15 inches square, bolted vertically and horizontally. There are four tiers of bilge keelsons on each side, 15 inches square. Ceiling from bilge to lower deck 15 inches, scarped and bolted edgewise.

She has three full decks, securely fastened with fore and aft knees; the hanging knees are extra fastened, having in the lower hanging knees 18 bolts, 1¼-inch iron; middle deck, 20 bolts; also upper-deck hanging knees, 20 bolts, and all of oak. Beams in lower deck 14 by 16, in middle deck 15 by 17, in upper deck 12 by 16, and some 12 by 20 inches. Lower deck main hatch is 14 by 20, middle deck 14 by 16, and main deck 14 by 11. She has a hurricane-deck over all, merely for working the ship, thereby obviating the difficulty in obstructions from houses, spare spars, water casks, etc.

Her main mast is 126 feet long, 98 above deck, diameter 44 inches; masts built of hard pine, to carry two stationary yards with trusses, the same as used on lower yards; fore and main deck alike, and those on the mizen-mast the same as those on the other masts above the lower yards, so that except the courses all her sails will have duplicates on every yard fore and aft. Her main yard is 110 feet long, the others in proportion; she will have a fourth mast, principally to lead the mizen-braces, to prevent the difficulty arising from mizen-braces leading forward, and hauling the mast out of place.

The model of the ship was said to promise a combination of swiftness, buoyancy, and beauty that has never been excelled. Notwithstanding her vast size, such is her length and buoyancy that when loaded ready for sea her draught of water will not exceed 23 feet, a common draught for ships half her size.

70. *Fire Companies.* The fire companies are formed in many towns of volunteers, who do not receive pay, but enjoy certain immunities from taxes and militia service.

The parade-day of the fire companies of a town is considered as a fête, the companies of other towns are invited to attend, and test the qualities of their respective engines in a trial as to which can throw the highest stream of water. At a meeting of firemen, held in New Haven, 36 companies attended, each dressed in a distinctive uniform, and averaging about 50 strong. A prize was given to the company whose engine succeeded in throwing the highest stream of water.

The engines played against a pole 150 feet high, through hose 450 feet long. Two engines, one having a 10-inch, the other an 8-inch cylinder, threw a stream 143 feet high, and carried off each a prize. These engines, however, were surpassed the following day by another engine with a 10-inch cylinder, which threw a stream over the pole.

10

Government Works

New York, Boston, Washington, Springfield

71. *Navy Yard, New York.* This navy yard is situated on Long Island, opposite the city. It covers a considerable extent of ground, and has many large storehouses and workshops, and gives employment to between 400 and 500 men. It contains the most capacious dry dock in the United States, constructed to admit vessels of the largest size.

They may be completely docked, and the water pumped out, in four hours and a quarter. The quantity of water to be removed is about 610,000 cubic feet. It is pumped out by a condensing engine, with a 50-inch cylinder 12 feet stroke, and 32 feet beam.

The cut-off motion is self-adjusting, so that more steam is admitted into the cylinder as the height to which it has to pump the water increases. The framing of the engine is in the form of Gothic columns, supporting arches, all painted and bronzed. All the work not painted is highly polished. The whole is surrounded by a bronzed rail, and a cast-iron flooring, ornamented with stars in relief, covers the floor.

The engine-house is about 60 feet square, and 50 feet high. The boilers are placed in a fire-proof room adjoining; they are three in number, 26 feet long, 7 feet in diameter, and are ordinarily used at a pressure of 50 lbs. The engine works two draining pumps, each 63 inches in diameter, having 8 feet stroke, one being connected to each arm of the beam.

The whole cost of the dock and its appendages is estimated at $2,000,000 (about £400,000).

72. *Navy Yard, Washington.* In this yard there are from 500 to 600 men in the various departments, employed in the manufacture of ordnance, marine engines, chain cables, anchors, etc.

Experiments were being made with a large gun carrying 240 lbs. shot, and also with a pendulum mortar.

The quality of the metal of which each gun is composed is tested and registered, and a sample piece preserved.

73. *Boston Navy Yard.* The Boston Navy Yard is of great extent, and contains three large sheds for ship-building, one of which is now used as a store for timber; another is occupied by the ship 'Virginia,' which has been on the stocks for more than 20 years; the third is empty. The manufacture of rope is carried on on a very extensive scale. A building 1,360 feet long contains a rope walk where a length of upwards of 1,200 feet of rope may be made. They have also the means of making 24-inch cables.

A set of machinery is used for making sheaves for ship blocks. An ingenious machine was employed for boring the sheaf, and recessing it on both sides for receiving the bush. Two lathe headstocks are mounted on a frame, and carry the small revolving cutters for making the recesses. An universal concentric chuck with three 'jaws,' having a large hole in its centre, is mounted between the headstocks. This carries the work, and has a vertical adjustment.

74. *Springfield Armoury.* Springfield Armoury is beautifully situated on an eminence overlooking the town. The various buildings together form a quadrangle; the grounds, which are tastefully laid out, occupy an extent of about 40 acres. I was conducted over the establishment by the Commanding Officer, Colonel Ripley, and the master machinist, Mr Buckland, who is the inventor of the principal machines employed in the manufacture of fire arms.

The front building, which has a handsome centre tower, is used as an arsenal for muskets. It contains 100,000 muskets, stacked with beautiful uniformity.

The barrels are made in mills, situated on the banks of a small river at some distance. The lighter parts of the musket, as the stock, the lock, guards, etc., are manufactured in the workshops attached to the Armoury.

The machines employed in the manufacture of the musket stocks are worthy of particular notice. By the kind courtesy of Colonel Ripley, facilities were afforded me for observing the time occupied in each operation.

The stocks are purchased rough from the saw for 28 cents (or 1*s.* 2*d.*) each.

		Time occupied	
		min.	sec.
1	They are roughly turned in Blanchard's machine, which has been in operation nearly 30 years	4	11
2	While one stock is being turned, the attendant is able to face and slab another by a circular saw	3	30
3	The stock is next taken to what is called a 'spotting' machine, where the sides are cut flat in different parts, to serve as bearings, or points to work from in future processes; this is done by 2 pairs of horizontal cutters, one pair at each end, and 3 single cutters in the centre	0	7
4	Next to a 'barrel-bedding' machine, where a groove is cut for the barrel; this is done by 4 bits set with their guides in a row, in a sliding frame combined with a horizontal cutter, with	1	7½

	Operation	min.	sec.
	a vibratory motion given by hand for shaping the groove conically, and a vertical bit for recessing		
	The next operation, that of finishing the groove by chisel, is performed by hand in 1 min. 42 sec. 1 min. 42 sec.		
5	The stock is then sawn to the required length	0	11½
6	A 'bed' is next recessed for the side plate, the sides of the stock are flattened by 2 vertical cutters, and the bed is recessed by a horizontal bit	1	11¾
7	The edges of the stock are then faced by a horizontal cutter	0	14
8	The stock is next taken to a butt-plate machine, where a bed is recessed and screw-holes made for the butt-plate by a horizontal bit and screw, and also a vertical bit and screw	0	21½
9	Next to a 'band-fitting' machine, where 3 horizontal cutters cut 3 straight bands, and a fourth bevils the upper or bayonet band	0	22¾
10	Next to a 'band-finishing' machine, where 4 horizontal cutters round off the parts intervening between the bands	0	28½
11	The stock is then turned a second time, in order to smooth its surface, 1st from the butt to the breech tang	8	35
12	2nd, from the breech tang to the end of the stock	5	28
13	It is next taken to the 'lock-bedding' machine, where the bed for the lock is recessed and shaped by 5 vertical bits set with their guides in a circular frame, at equal distances from each other. The driving-strap is made to run on a loose pulley fixed above the circular frame, and as each bit is brought into operation the band drops from the loose pulley on the driving pulley of the bit which is brought underneath it, and is raised, when the operation is finished, to its former position, ready to descend on the pulley of the next bit. The cuttings are blown away by 2 fan-pipes	0	46¾
14	Next to the 'guard-bedding' machine, it was similar in its construction to the former, but it carried 4 bits instead of 5, and recessed and shaped the bed for the guard	0	51
15	The holes for the side screws were then bored	0	15
	then for the tang screw	0	8¾
16	The stock was lastly taken to the 'band, spring, and ramrod-fitting' machine. A vertical revolving cutter grooved recesses for the band-springs, a horizontal cutter recessed the groove for the barrel	0	55¼
	A hand operation then finished off the whole. Time, 35 seconds 35 sec.		
	Total time of machine operations	28	45¼
	Total time of hand operations	2	17
		31	2¼
	Allowance for double simultaneous operations during turning	8	58
	Man's time given to the whole operations of making a complete musket-stock	22	4¼

The complete musket is made (by putting together the separate parts) in 3 minutes. All these parts are so exactly alike that any single part will, in its place, fit any musket.

The general principle adopted in the construction of these machines is that of guiding the cutter in its course, by a shaper or 'former,' that is, a pattern made exactly of the form in which it is required that the work should be shaped.

The number of muskets made in the year 1852 amounted to 19,800.

75. *Coast Survey Office, Washington.* Workshops are attached to the Coast Survey Office, where copies of the standard weights and measures of the United States are made. The office supplies the capital of every State in the Union, in addition to the standard weights and measures, with three very accurate balances.

No. 1 is constructed to weigh from 50 lbs. down to 10 lbs.

No. 2 to weigh from 10 lbs. to 1 lb.

No. 3 to weigh from 1 lb. to $\frac{1}{10,000}$ oz.

The estimated cost of the three is $4,600, about £900. The latter balance was tested, and deflected by the $\frac{1}{10,000}$ oz. In weighing 1 lb., the effect of the addition of $\frac{1}{200}$ grain was instantly visible.

When the balance is not in use, the beam is made to descend by means of a screw, so that two external cones, placed on its under side, rest in two internal cones fixed on the supporting frame. Two steel discs attached to the chains of the scales are, by the same screw, made to descend, and rest upon the frame. Thus the whole balance is supported, and there is no continued strain on the knife edges.

A lever worked by hand acts upon the two short vertical rods placed under the scales, and adjusts them evenly before commencing the operation of weighing.

The balance stands on four feet, each adjustable by screws.

The full set of standard weights and measures supplied by the office consist of–

1. A set of standard weights from 1 lb. to 50 lbs. avoirdupois; and 1 lb. troy.
2. From 1 oz. down to $\frac{1}{10,000}$ oz. troy.
3. A yard measure.
4. Liquid measures. The gallon and its parts down to half-pint inclusive.
5. A half-bushel measure.

Twenty-one States have been supplied, and other sets are being prepared for the remaining States. There are 13 workmen employed.

The United States standard yard has been obtained from a 7 feet standard procured from England. It is made of gun-metal, about 2 inches broad and $\frac{3}{8}$ inch thick, and has a thin strip of silver, $\frac{1}{5}$ inch broad, let into it through its entire length. It is divided into small divisions, each being an aliquot part of an inch. The standard was obtained by taking the mean of a great number of measurements made from different points in the 7 feet scale.

A set of standards has been presented to France, and a set of French standards was presented to the United States in return.

It is a matter of surprise that while the people of the United States have long felt and appreciated the benefits of their decimal monetary system, the old English system of weights and measures has not yet been abolished by the Legislature. Its inconveniences are much complained of, and custom has tried to remedy its evident defects to a great extent by adopting the plan of reckoning by 100 lbs. (instead of the cwt. or 112 lbs.) and by 1000 lbs.

Monetary accounts are kept, and calculations are made with the greatest facility in dollars and cents, the dollar (4*s.* 2*d.*) being divided into 100 cents (a cent ½*d.*). Convenient coins called 'dimes' are in circulation, 10 cents being equal to one dime, and 10 dimes making a dollar. Quarter dollar and half dollar pieces are also commonly used; there appears to be no reason why a decimal system would not afford equal advantages, if applied, as it doubtless will be eventually, to the scales of weights and measures.

The Coast Survey Office has custody of two instruments used in measuring by means of end measurements. A base line 7 miles long was measured on an island near Charleston, in about six weeks. The measuring instruments were supported on two adjustable stands.

They were made on the compensating principle, and inclosed in a double case of tin, to prevent, as much as possible, their being affected by changes of temperature. The ends of the instruments were of agate, one flat, the other having a knife edge. The latter was made to slide, and was connected by a bell-crank lever to a spirit level, which indicated when the end measures were in contact, according to the method employed by Bessel, in making standard measures in Prussia.

11

Electric Telegraphs

76. The advantages to be derived from the adoption of the Electric Telegraph, have in no country been more promptly appreciated than in the United States. A system of communication that annihilates distance was felt to be of vital importance, both politically and commercially, in a country so vast, and having a population so widely scattered. It met accordingly with great encouragement, both from the federal Government at Washington and from the local Governments of the various States.

Distances are now to be measured by intervals, not of space, but of time: to bring Boston, New York, and Philadelphia into instantaneous communication with New Orleans and St Louis–to centralize in Washington, at any given moment, information gathered simultaneously from the far corners of the thirty-one provinces of the Union, is to extend throughout the confederacy bonds of the most intimate connexion.

In the operations of commerce, the great capitals of the North, South, and West are moved, as it were, by a common intelligence; information respecting the state of the various markets is readily obtained, the results of consignments may be calculated almost with certainty, and sudden fluctuations in price in a great measure provided against.

If, on the arrival of an European mail at one of the northern ports, the news from Europe report that the supply of cotton or of corn is inadequate to meet the existing demand, almost before the vessel can be moored intelligence is spread by the Electric Telegraph, and the merchants and shippers of New Orleans are busied in the preparation of freights, or the corn-factors of St Louis and Chicago, in the far west, are emptying their granaries and forwarding their contents by rail or by canal to the Atlantic ports.

There may be, no doubt, similar general advantages derived everywhere from the introduction of the Electric Telegraph, but they are such as affect with peculiar benefit a country like the United States, consisting of confederated provinces, differing one from another in climate, in productions, in laws, and institutions, and in some cases in the character of their inhabitants.

77. The introduction of so important a system was not left to the unaided exertions of private enterprise. In 1844, Congress made a liberal grant in order to put in operation the first telegraph line that was erected in the States – that between Washington and Baltimore – and before seven years had elapsed the Committee on Post Offices and Post Roads presented to the Senate, in 1851, their report on the route which they had selected for a gigantic telegraph line, nearly 2,500 miles in length, connecting San Francisco with Natchez on the Mississippi, and thence with the vast network of lines that already covered the Atlantic States. Such was the rapid development of this system of communication, supported by the federal Government, and fostered by that of individual States, which passed general laws authorising the immediate construction of telegraph lines, wherever they would be conducive to the interest of the public, and affording every facility to the formation of companies for that purpose.

If a company, or even a private individual, should propose to construct a telegraph line, and can show that it would be beneficial to the public (and as to this proof there is generally but very little difficulty), he may obtain an Act authorising him to proceed, as a matter of course; no private interests can oppose the passage of the line through any property; there are no committees, no counsel, no long array of witnesses and expensive hearings; compensation is made simply for damage done, the amount being assessed by a jury, and generally on a most moderate estimate. With a celerity that is surprising a company is incorporated, the line is built, and operations are commenced. Similar facilities are also afforded in many of the States by general laws authorising the construction of railroads.

78. There are in the United States between 20 and 30 joint-stock electric telegraph companies, and an endeavour is now being made to amalgamate them into one company. The systems generally in use are those of Morse, Bain, and House. The conflicting claims of these three systems have formed the subject of protracted litigation, and, without going into minute details, it may be sufficient to mention the distinctive features of each.

Professor Morse employs receiving magnets which close local circuits attached to each telegraph office, and so charge another magnet called the register magnet; this acts upon an armature attached to a lever which presses down a metallic point upon a cylindrical roller, a strip of paper is passed at an uniform rate between the metal point and the roller, and by the indentation of lines and dots, representing letters of the alphabet, the message transmitted is registered at the rate of about 20 words per minute. It has been also found practicable to receive and understand messages by merely listening to the click of the armature, and the attendant writes down the words of the message, trusting solely to his ear for interpreting them correctly. The cost of a Morse Register is about $40, £8.

The system patented by Mr Bain is called the Electro-chemical Telegraph. He employs a metal disc, carrying a prepared sheet of paper, on which lines and dots are marked by the decomposition of a metallic point, acted upon by an electric current. The metal disc revolves at an uniform rate by the agency of clock-work, and the point or pen is made to move over the paper laid on the disc in the direction of a spiral. No receiving magnet is necessary, and a comparatively weak current of electricity, traversing long wires, leaves instantaneously a mark upon the prepared paper.

The system is capable of being so modified as to transmit messages with great rapidity. Spaces representing alphabetic lines and dots are punched out from a long strip of paper, called the message-strip; this is passed rapidly between a cylinder and a toothed lever connected respectively with the electrodes of a battery; when the tooth of the lever falls in the spaces punched in the strip, it will be in connection with the cylinder, and close the circuit; when the paper (which, being dry, is a non-conductor) intervenes, the circuit will be broken. When the circuit is closed, the pen fixed at the other end of the wire will mark on the paper lines and dots exactly corresponding to those punched out of the message-strip.

In Mr House's system the message is printed by the telegraph instrument itself. The electric current is made to act by rapid pulsations on an 'axial magnet' that opens and closes a valve connected with a pneumatic printing-machine. The machine, which is ably contrived, is worked by manual power, and prints messages at the rate of about 20 words per minute; its cost is about $250 (£50).

Grove's batteries are generally used in all the systems.

79. The aggregate length of the telegraph lines in the United States exceeded, in 1852, 15,000 miles, of which–

12,124 were worked on the system of Morse.
1,199 . . . Bain.
1,358 . . . House.

These numbers have been considerably increased this year, and many amalgamations of the different companies working under Morse's and Bain's patents have been made.

In November, 1852, eleven different companies had offices in New York.

Extracts from the annual reports of two companies, one whose line extends from New York to Washington, in competition with a rival line; the other, a western company, whose line connects Pittsburgh with Louisville, will serve to give an idea of the amount of business carried on.

80. *New York and Washington Line*. Messages are charged at the rate of 50 cents (2*s*. 1*d*.) for 10 words, and 5 cents (2½*d*.) for every additional word, for transmission from New York to Washington, a distance of 270 miles.

	Messages	*Receipts*
In the month of July, 1851, there were sent	13,463, producing	$4,991 62, about £1,000
In the month of January, 1852,	23,962	$11,352 97, about £2,230
In the month of June, 1852,	25,298,	$11,832 03, about £2,360
Total for year ending June, 1852	253,857,	$103,232 37, about £20,000

The capital of the company is $370,000 (about £70,000); it has 7 wires from New York to Philadelphia, 6 from Philadelphia to Washington, connecting together 13 stations. The number of persons employed on this line, including the staff of clerks and out-door surveyors and messengers, amounts to 125.

It should, however, be borne in mind that the distances through which messages have to be transmitted in America are so great that a slight delay in time is not of that immediate importance which it is in our own country; especially when it is considered that in England the great proportion of business is crowded into the 'office hours' of the day, while in America it is spread over the entire 24 hours.

The average cost of constructing the line is estimated at $185 per mile.

Pittsburgh and Louisville Telegraph Line:

Length of line, 450 miles.
Words transmitted in 1850, 3,602,760.
Total receipts $73,278,
about £14,500.

81. The most distant points connected by electric telegraph in North America are Quebec and New Orleans, which are 3,000 miles apart, and the network of lines extends to the west as far as Missouri, about 500 towns and villages being provided with stations.

There are two separate lines connecting New York with New Orleans, one running along the sea-board, the other by way of the Mississippi, each about 2,000 miles long. Messages have been transmitted from New York to New Orleans, and answers received, in the space of three hours, though they had necessarily to be written several times in the course of transmission.

When the contemplated lines connecting California with the Atlantic, and Newfoundland with the main continent, are completed, San Francisco will be in communication with St John's, Newfoundland, which is distant from Galway but five days' passage. It is, therefore, estimated that intelligence may be conveyed from the Pacific to Europe, and *vice versa*, in about six days.

82. The cost of erecting telegraph lines varies according to localities, but the expenses upon the whole are estimated to average about $180 (about £35) per mile throughout the States; the moderate amount of this estimate is, in a great measure, to be attributed to the facilities afforded by

the general telegraph laws for the formation of companies and the construction of lines.

The electric telegraph is used by all classes of society as an ordinary method of transmitting intelligence.

Government despatches and messages, involving the life or death of any persons, are entitled to precedence, next come important press communications, but the latter, if not of extraordinary interest, await their regular turn.

The leading newspapers of New York contribute jointly towards the expenses of daily telegraphic communications. The annual sum paid by the 'Associated Press' averages $30,000 per annum.

The following is the tariff for the press despatches:

	Under		200 miles,	1	cent	per	word.
Between	200	and	500	,,	2	,,	,,
Between	500	,,	700	,,	3	,,	,,
Between	700	,,	1,000	,,	4	,,	,,
Between	1,000	,,	1,500	,,	5	,,	,,
	1,500	,,	over	,,	6	,,	,,

Assuming 3 cents as the mean average, the total amount of matter received by telegraph for the 'New York Associated Press' amounts to a million words per annum, or about 600 columns of a London newspaper of the largest size, averaging about two columns per day. Supposing six papers to be associated together, the share of each would annually amount to about $5,000, or £1,000, for two columns of telegraphic intelligence daily.

Commercial men use the electric telegraph in their transactions to a very great extent. In 1852 there were transmitted by one of the three telegraph lines that connect New York and Boston between 500 and 600 messages daily. The sums paid on this line by some of the principal commercial houses who used it averaged in 1852 for each from $60 (£12) to $80 (about £16) per month. On other lines the leading commercial houses were estimated to pay from $500 to $1,000 (£100 to £200) per annum for telegraphic despatches.

83. Interruptions occur most frequently from the interference of atmospheric electricity; in summer they are estimated to take place on an average twice a week, but many contrivances have been adopted for obviating this inconvenience, such as lightning arrestors, etc., which are generally known; the number of interruptions have been thereby reduced about 30 per cent. Other accidental causes of interruption occur irregularly from the falling of the poles, the breaking of the wires by falling trees, and, particularly in winter, from the accumulated weight of snow or ice.

The electric current is made to act through long distances, by using local and branch circuits, and relay magnets, in those systems where it would be otherwise too weak to operate effectually.

In Mr Bain's system, a weak current is found sufficient for very long distances; between New York and Boston, a distance of 270 miles, no branch or local circuit is required. In some cases, where both Morse's and Bain's telegraphs are used by an amalgamated company in the same office, it is found convenient, in certain conditions of the atmosphere, to remove the wires from Morse's instruments, and connect them with Bain's, on which it is practicable to operate when communication by Morse's system is interrupted.

It is generally believed that by laying insulated wires underground the interruptions will be reduced so as to be altogether inconsiderable. The expense of the process, however, is regarded as a great impediment in the United States, where cheapness of construction is an object of the highest consideration.

84. The application of the electric telegraph is not confined to the transmission of messages from one part of the States to another: in the form of a local or municipal telegraph, it is employed as an important instrument of regulation and intelligence in the internal administration of towns.

No adaptation of the system can be more interesting and useful than that which is made for the purpose of conveying signals of alarm and intelligence in the case of fire.

This system has been very completely developed in Boston.

The city is divided into seven districts, each provided with a powerful alarm bell. Every district contains several stations, varying in number according to its size and population. There are altogether in the seven districts 42 stations. All these stations are connected with a chief central office, to which intelligence of fire is conveyed, and from which the alarm is given; two telegraph wires are employed, a return wire being used to complete the circuit, and provide as completely as possible against accidental interruption or confusion.

At each of the 42 stations, which are placed at intervals of 100 rods throughout the city, there is erected in some conspicuous position a cast-iron box containing the apparatus for conveying intelligence to the central office. The box is kept locked, but the key is always to be found in the custody of some person in the neighbourhood, whose address is painted on the box door.

On opening this door, access is gained to a handle which is directed, by a notice painted above it, to be turned slowly several times. The handle turns a wheel that carries a certain number of teeth, arranged in two groups, the number of teeth in one representing the district, in the other, the station; these teeth act upon a signal key, closing and breaking the circuit connected with the central office as many times as there are teeth in the wheel. Signals are thus conveyed to the central office, and, by striking the signal bell a certain number of times, the district and station from which the signal is made is indicated.

An attendant is always on the watch at the central office, and on his attention being called to the signals by the striking of a large call bell, he immediately sets in motion his alarm apparatus, and by depressing his telegraph-key, causes all the alarm bells of the seven districts to toll as many times in quick succession as will indicate the district where the fire has occurred, the alarm being repeated at short intervals for as long a time as may be necessary.

The signal-boxes erected at the stations contain, in addition to the signal-handle, a small electro-magnet, an armature, and a signal-key, so that full and particular communications can be made between each box and the central station, the clicks of the armature forming audible signals. They have also an apparatus called a 'Discharger of Atmospheric Electricity,' for preventing the occurrence of injuries during thunder-storms.

By this system certain information is given to the central office at the earliest possible moment of the exact locality in which a fire may have broken out, and the alarm is immediately spread over the entire city.

Every one who is aroused by the alarm is enabled to tell at once whether interest or duty calls him to the scene of action, and the exact point to which assistance is summoned. Should the alarm be given in the night, those whose attention is awakened may ascertain from the tolling of the bell the precise quarter in which danger threatens, and should they have been needlessly disturbed, may rest in peace, and find in the knowledge that they and theirs at least are in safety, a consolation for broken slumbers.

85. Telegraph wires in towns are almost universally carried along the tops of houses, or on poles erected in the streets, instead of being conveyed in pipes underground. So little difficulty is met with on the part of proprietors of houses, that telegraph lines are in some cases erected by private individuals for their own particular use. As an instance, may be mentioned the case of a large manufacturer in New York, who has an office in one part of the city, while his works lie in a contrary direction. In order to keep up a direct communication between both, he has erected a telegraphic wire at his own expense, and carried it over the tops of the houses intervening between his office and his works, having obtained without any trouble the permission of their various owners.

86. It has been already stated that the number of miles of telegraph lines erected in the United States exceeded, in 1852, 15,000; and it may be useful, in conclusion, to give a list of them, and the number of wires they had each in operation.

	MORSE LINES	No. of Wires	Length in Miles
1	New York to Boston, *viâ* Springfield	3	250
2	New York to Buffalo, *viâ* Troy	3	513
3	New York to Dundirk	1	440
4	New York and Erie Railroad Telegraph	1	460
5	New York to Washington, *viâ* Baltimore	7	260
6	New York to New Orleans (sea-board)	1	1,966
7	Boston to Portland, *viâ* Dover	2	100
8	Philadelphia to Lewiston, Del.	1	12
9	Philadelphia to Pittsburgh, *viâ* Harrisburg	1	309

	MORSE LINES, *contd.*		
10	Philadelphia to Pottsville, *viâ* Reading	1	98
11	Baltimore, *viâ* Wheeling, to Cumberland	1	324
12	Baltimore to Harrisburg, *viâ* York, Pa.	1	72
13	Washington to New Orleans, *viâ* Richmond, Pa.	1	1,716
14	Harper's Ferry to Winchester, Pa.	1	32
15	York to Lancaster, *viâ* Columbia, Pa.	1	22
16	Reading to Harrisburg	1	51
17	Pittsburg to Cincinnati, *viâ* Columbia	1	310
18	Cincinnati to St Louis, *viâ* Vincennes	1	410
19	Cincinnati to Maysville, Ky., *viâ* Ripley	1	60
20	Columbia to New Orleans, *viâ* Natchez	1	638
21	Cincinnati to St Louis, *viâ* Indianapolis	1	400
22	Cincinnati to Sandusky	1	218
23	Cincinnati to Dayton and Chicago (Ohio wire)	1	100
24	St Louis to Chicago, *viâ* Alton, Ill.	1	330
25	St Louis to Independence, Mobile	1	25
26	Alton to Galena, *viâ* Quincey	1	380
27	New Orleans to Balize	1	90
28	Cleveland to Cincinnati	2	125
29	Cleveland to Pittsburgh	1	150
30	Cleveland to Zanesville	1	150
31	Zanesville to Newark	1	40
32	Toledo to Terre Haute, *viâ* Fort Wayne	1	300
33	Sandusky to Mansfield	1	40
34	Columbus to Portsmouth, O.	1	90
35	Columbus to Lancaster, O.	1	25
36	Lancaster to Logans Port	1	15
37	Milwaukie to Green Bay	1	200
38	Milwaukie to Galena, *viâ* Madesa	1	250
39	Banesville to Marietta	1	66
40	Buffalo to Milwaukie, *viâ* Cleveland and Chicago	2	400
41	Buffalo to Queenston, Canada, *viâ* Lockport	1	48
42	Buffalo to Detroit, *viâ* Cleveland	1	400
43	Syracuse to Oswego, N. Y.	1	40
44	Troy to Montreal, *viâ* Burlington	1	278
45	Portland to Calais (Maine)	1	350
46	Calais to Halifax, *viâ* St John's	1	400
47	Worcester to New Bedford, *viâ* Providence	1	97
48	Worcester to New London, *viâ* Norwich	1	74
	Total	61	13,124

	BAIN LINES (some of these are now amalgamated with the Morse Lines).		
1	New York to Boston (Merchants)	2	250
2	New York to Buffalo	2	513
3	Boston to Portland	1	100
4	Boston, *viâ* Burlington, to Ogdensburgh	1	350
5	Troy to Saratoga	1	36
	Total	7	1,249

	HOUSE LINES, belonging to one Company.	No. of Wires	Length in Miles
1	New York to Washington	2	260
2	New York to Boston	2	250
3	New York to Buffalo	3	513
4	Buffalo to Cincinnati	1	325
	Total	7	1,348

These lines are being rapidly extended. It is intended to connect Cincinnati with Louisville and St Louis.

	CANADIAN LINES	
1	Quebec to Suspension Bridge, Niagara	155
2	Quebec to New Brunswick Frontier	220
3	Montreal to No. 7 State Line, Highgate	47
4	Montreal to Bytown	115
5	Hamilton to Port Sarina, on L. Huron	143
6	Niagara to Chippewa	14
7	Brantford to Simcoe	33
8	Kingston to Hamilton	256
	Canadian Lines	983
	Morse Lines	13,124
	Bain Lines	1,249
	House Lines	1,348
	Total Lines in North America	16,704

12

The Patent System

87. The patent system of the United States is based upon the first Article of the Constitution, which says:

'The Congress shall have power to promote the progress of science and the useful arts by securing, for limited times, to authors and inventors the exclusive right to their respective writings and discoveries.'

Letters patent, securing this exclusive right, and extending to all the States of the Union, are granted by the federal Government. The Patent Office, from which they issue, is attached to the department of State, and its chief officer, 'called the Commissioner of Patents, is appointed by the President, by and with the advice of the Senate.'

All patents are issued in the name of the United States, signed by the Secretary of State, and countersigned by the Commissioner of Patents.

'All actions or suits, arising under any law granting or confirming to inventors the exclusive right to their inventions or discoveries, are originally cognizable by the Circuit Courts, and an appeal lies to the Supreme Court of the United States.'

The Patent Office is conspicuous among the public buildings of Washington for the beauty of its architecture; it is built of white marble, and modelled after the Parthenon at Athens. When completed, it will be quadrangular, having a frontage of 413 feet, with a depth of 280 feet, and inclosing an open court, having an area of 270 feet by 112 feet. The principal entrance leads into a hall, round which are ranged glazed cases, containing models of inventions. From the right of the hall branches off a passage, on each side of which lie the various offices of the commissioner, examiners, and clerks, and the library. On the left are three large rooms, appropriated to the reception of models of all the inventions patented in the United States. At the extreme end of the hall is a double staircase leading to the superior apartments, one of which is used as a museum.

The various departments have each their respective offices; in some the applications for letters patent are received, in others the alleged inventions are examined (each examiner taking some particular subject), in others the drawings and specifications are preserved in portfolios, which slide on rollers in deep cases. There is also an extensive library, to which large additions are being continually made, for official use, and the convenience of applicants for patents.

The model-rooms contain upwards of 23,000 models, arranged in large glass cases, but there is no catalogue to assist or guide those who search for any particular model that may be required for inspection.

It not unfrequently happens that the subject of a patent undergoes considerable modification in the case of 'additional improvements' being added, or when the letters-patent are 're-issued,' so that the model in such a case does not eventually represent the invention for which the patent is virtually granted, and is therefore useless. If, however, the models were to be so classified that those which are valuable were separated from those which are worthless, and arranged in appropriate and distinctive series, they would then exhibit, step by step, the progress which had been made in the particular art or manufacture to which they belong, and with the aid of a well-digested catalogue they would be of great national utility. But an assemblage which comprehends indiscriminately models of every patented invention must, sooner or later, become so large that confusion only can result, and all practical purposes of utility must be defeated.

The models of the inventions for which patents have been refused are kept in the basement vaults, and it is doubtful if they serve any purpose sufficiently useful to repay the trouble of preserving them.

88. The Official Staff of the Patent Office is at present composed of

The Commissioner,	receiving	$3,000 per annum.
The Chief Clerk	receiving	2,000 per annum.
Six Examiners	receiving	2,500 per annum.
One Assistant, Class IV	receiving	1,800 per annum.
Twelve Assistants, Class III	receiving	1,500 per annum.
Seven Assistants, Class II	receiving	1,200 per annum.
Librarian	receiving	1,200 per annum.
Machinist	receiving	1,250 per annum.
Agricultural Clerk	receiving	1,500 per annum.

Twenty temporary Clerks at a variable salary.

The total amount of official salaries was, in 1852, $56,064 (about £11,200).

The amount of business transacted in the office in 1852 will appear from the following statement:

Applications for patents pending January 1852	155
Applications received during 1852	2,639
Patents issued	1,020
Rejections and suspensions	1,293
Applications undecided	481

Among the patents issued are reckoned

Re-issues	18	Extensions	3
Additional improvements	4	Designs	107

There were also filed

Caveats	996
Disclaimers	2
Patents expired	525

In order to convey a general idea of the different classes of invention for which the patents were granted, and the number in each class during the year 1852, they will be given in the classified form adopted in the Patent Office.

Class		
Class 1	Agricultural implements	114
2	Metallurgy, manufacture of metals, and instruments therefor	125
3	Textile manufactures and machines for preparing fibrous substances	69
4	Chemical process manufactures and compounds, including medicines, dyeing, colour-making, distilling, soap and candle-making, mortars, cements, etc.	50
5	Calorifics, comprising lamps, fireplaces, furnaces, stoves, preparations of fuel, etc.	50
6	Steam and gas-engines, boilers and furnaces	36
7	Navigation and maritime implements, vessels, diving apparatus, life-preservers, etc.	23
8	Mathematical, philosophical, and optical instruments, clocks, etc.	27
9	Civil engineering and architecture, and apparatus employed on railroads, bridges, waterworks, etc.	32
10	Land conveyances, roads, vehicles, wheels, etc.	72
11	Hydraulics and pneumatics, water-wheels, wind-mills, apparatus for raising or delivering fluids	16
12	Lever screw, and mechanical power applied to pressing, weighing, raising and moving weights	12
13	Grinding-mills, gearing, grain-mills, and mechanical movements and horse powers	41
14	Lumber-machines for dressing wood	44
15	Stone and clay manufactures, pottery, glass, bricks, stone-dressing, cements, etc.	29
16	Leather-dressing and manufactures, boots, harness, etc.	26
17	Household furniture, domestic implements, feather-dressing mattresses	43
18	Fine arts, music, painting, sculpture, engraving, books, paper, printing, binding, jewellery	48
19	Fire-arms, implements, and munitions of war	7
20	Surgical and medical instruments, trusses, bathing apparatus, etc.	13
21	Wearing apparel, toilet articles, and instruments for manufacturing them	16
22	Miscellaneous	20
		913
	Designs, etc.	107
	Patents issued	1,020

89. *Persons entitled to Apply for Letters Patent.* An original inventor only is entitled to apply for a patent; the mere introducer of an invention has no claim whatever. Every applicant is required to make oath 'that he verily believes himself to be the true and first inventor,' and whenever it appears that the inventor did so believe at the time of his application, the subsequent discovery that the invention had been used abroad will not invalidate his patent, provided it does not appear that the invention had been patented, or described in a printed publication.

If the inventor shall have assigned his entire interest previous to the application, and the assignment has been recorded in the Patent Office according to law, the patent will be issued to the assignee, but the inventor must still sign the application.

Upon the decease of an inventor, his legal representatives may (on the same conditions on which the inventor, if living, would stand) obtain a patent for the benefit of his heirs or devisees, even though the inventor had not during his lifetime taken any steps towards securing a grant.

Foreigners who have resided for one year previous to their application, and have made oath of their intention to become naturalised, have all the privileges of citizens.

90. *Subject-matter of Letters Patent.* It has been said that a patent is only granted for a *bonâ fide* invention, and not for a mere introduction. In other respects there is not much difference between the law of England and that of the United States.

It should be remarked, however, that all applications for grants are rejected in which the examiners of the Patent Office believe that they can discover any 'double use,' that is, an application of an old contrivance to a new purpose: so rigidly is this rule followed out, that in some instances patents have been refused for inventions confessedly valuable, and producing new manufactures, because in totally different branches of manufacture inventions somewhat similar had been already patented.

The law of the United States holds as good subject-matter for a grant:

'Any new or useful art, machine, manufacture or composition of matter, or any new and useful improvement on any art, machine, manufacture, or composition of matter.'

With regard to prior use, an inventor has very extensive privileges in the United States. He may, during the period of two years, publicly use, and even sell his invention, without invalidating a patent obtained at the end of that time; but a public use for a longer time is considered as an abandonment of his discovery to the public. A prior discovery will not invalidate a subsequent patent if it can be shown that it was laid aside without being perfected or reduced to practice. A discovery that has once been abandoned to the public, by being used, unprotected, for more than two years, or otherwise, cannot be subsequently reclaimed.

A patent will be granted for an invention patented in a foreign country at any time within six months of the date of the foreign patent, or afterwards if it can be shown that the foreign invention has not been brought into public use in the United States.

91. *Caveats.* A citizen or alien of one year's residence, intending to become naturalised, who wishes for time to perfect his invention, may file a caveat in the Patent Office, which, for the space of one year, will prevent any applicant from obtaining a patent for an invention of a similar character, without notice being given to the caveator.

This notice requires him to complete his application within three months, and if the two applicants appear to make similar claims, the commissioner will appoint a day for hearing both parties, that is, will 'grant an interference,' and will direct a patent to issue to him who shall prove to be the original inventor.

A caveat may be renewed from year to year during the time that an applicant is perfecting his invention. A foreigner cannot file a caveat. The fee for a caveat is $15 (£3).

92. On making applications for letters patent, there are six requisites to which attention must be paid:

1st. The petition addressed to the commissioner of patents.
2nd. The specification.
3rd. The oath of originality.
4th. Drawings.
5th. Model, or specimens of the invention.
6th. Proper fees.

The petition for a grant must be addressed to the commissioner of patents, and state that the requirements of the Act have been complied with. It must be signed by the inventor, or if he be dead, by his legal representatives.

93. *The Specification.* The specification differs but little in its character from an English specification.

It must be 'a written description of the invention, and the manner of using it, expressed in such full and clear terms as to enable a skilled person to use the same; and, in the case of a machine, must particularly specify and point out the part, combination, or improvement, claimed as an invention.'

It is not, however, construed with the same rigid severity as an English specification. For the Act, March 3, 1837, s. 9, enacts that 'Whenever by mistake, accident, or inadvertence, and without any wilful default, or intent to defraud or mislead the public, any patentee shall have in his specification claimed to be the original and first inventor or discoverer of any material, or substantial part of the thing patented, of which he was not the first and original inventor, and shall have no legal and just right to claim the same, in every such case the patent shall be deemed good and valid for so much of the invention and discovery as shall be truly and *bonâ fide* his own.' But it must in such a case be shown that the part really invented is a substantial part.

Power is also given to the patentee in such a case to bring an action for infringement; but if he obtain a verdict on the question of infringement, he is not entitled to costs unless he have filed a disclaimer 'prior to the commencement of the suit,' which may happen to be subsequent to the act of infringement.

There is a further provision 'that no person be entitled to the benefits of this enactment who shall have unreasonably neglected or delayed to enter at the Patent Office a disclaimer.'

The specification is drawn up and filed *before* the grant of the patent to which a copy is annexed–the two being construed together as one document.

The patent may be dated from the time of filing the specification and drawing, if it be more than six months previous to the actual issue.

94. The oath or affirmation is a statement annexed to the specification, signed by a person lawfully qualified to administer an oath, certifying that the applicant had made oath in proper form that he believed himself to be the original inventor of what he claimed, and that he did not know or believe that the same was ever before known or used. If he be a resident alien, intending to become a citizen, that circumstance also must be stated on oath.

95. Drawings, models, and specimens are always absolutely requisite whenever the case admits of their being supplied.

96. The fees payable are as follows:

If the application be made by a citizen, or foreigner who has resided for one year previous to his application, and made oath of his intention to become a citizen, the fee is $30 (£6)

If a subject of Great Britain $500 (£105)

If any other foreigner $300 (£63)

The principle on which this scale appears to have been drawn up will be alluded to hereafter.

When these six requisites have been furnished, the application is taken up in its turn, unless the commissioner should otherwise decide for convenience of classification, or unless the applicant has already taken out a foreign patent. For, since an American patent can only run for fourteen years from the date of a foreign patent, the applicant would suffer by delay.

97. The examination of the documents furnished to the Patent Office with the application is very strict, and the claims in respect of novelty undergo long consideration; all claims savouring of a 'double use' being, as was before stated, rejected.

If the drawings or specification are found defective, they are returned to the applicant with instructions how to alter them. If it be deemed advisable to prepare new papers, the originals must be returned, and if the character of the invention is materially changed after the application is filed, it must be withdrawn altogether, and a new one filed, two-thirds of the former fees being allowed. If, on examination, the claims are found to embrace what, in the judgment of the examiners, is considered to be old, the application is at once rejected, and notice is given to the applicant, with a full statement of the grounds of the rejection, and reference to prior inventions which he may examine in the office. He has also the opportunity of explaining his claims to the examiner in a private interview. If, in the end, he is dissatisfied with the decision of the Commissioner of Patents, he may appeal to the District Judge of Columbia, on paying

$25 (£5), and a still further appeal may be made by bill to a court of equity.

When, on an examination of an application, it is found that a caveat has been entered for a similar invention within the previous year, notice is given to the caveator to complete his application within three months; and if, on its coming in, the claims are found to be conflicting, an 'interference' is granted.

When an applicant claims anything embraced in an unexpired patent, and persists in his claims to priority, the commissioner will, at his request, 'grant an interference' between him and the patentee.

The Office has, however, no power over the patent, even though it decide in favour of the other party, and grant him also a patent for his claim.

The unsuccessful party is allowed an appeal to the District Judge of Columbia.

98. If additional improvements be made to an invention subsequent to the date of a patent, they may be added to the specification, and incorporated with the original patent: fee $15 (£3).

The object of this provision appears to be to save an inventor the extra expense and trouble of taking out a new patent, comprising his recent inventions, and founded on his original claims.

The incorporation of new claims in a patent bearing an antecedent date, seems open to great objection.

99. *Re-issue*. The specification and claims of a patent undergo considerable modification in the case of a re-issue, and the provision is one liable to entail great abuses.

Section 13 of the Act of 4th July, 1836, enacts that when a patent is inoperative, or invalid by reason of a defective or insufficient description or specification, if the error arise from inadvertency, accident, or mistake, and without any fraudulent or deceptive intention, the commissioner may, on the surrender of the old defective patent, 'cause a new patent to be issued for the same invention, and for the residue of the period then unexpired for which the original patent was granted, in accordance with the patentee's corrected description and specification; and the patent so re-issued, together with the corrected descriptions and specifications, shall have the same effect and operation in law, on the trial of all actions hereafter commenced, for causes subsequently accruing, as though the same had been originally filed in such corrected form, before the issuing of the original patent.' The fee for a re-issue is $15 (£3). But if it be found that the specification comprises several distinct substantive inventions, a patent is issued for each, and the fee for each is $30 (about £6).

It is obvious that there is scarcely any limit to the permutations and combinations which may be made out of the claims of a long, complicated, and ambiguous specification.

Modern improvements are seldom or never so new in character and form as to be entirely dissimilar to every previously patented invention. It frequently happens that there exist a variety of patents whose claims are similar in character, and which are all worthless from the lack of something which is hit upon by a subsequent discoverer, who supplies the one thing needful to success.

It may be conceived with what eagerness antedated patents would in such a case be 'surrendered'–with what ingenuity their claims would be strained and distorted so as to embrace the identical one thing, and embody it in a 're-issued' patent, bearing the same date as the old one, and thus rob the really meritorious inventor of the fruits of his labours.

100. *Extension of the term of a Patent.* The commissioner has power to grant an extension of a patent for seven years, if it shall be shown that the invention is in itself 'novel, useful, and important to the public, and that the inventor has not been adequately remunerated, though he used due diligence in introducing his invention into general use.'

The extension cannot be granted after the patent expires, so the application must be made a sufficient time before its expiration, to allow of a hearing and a decision, and yet not so long as to render the general statement of profits and expenses materially inaccurate.

101. *Assignment.* An invention may be assigned by an instrument in writing, either partially or entirely, before or after a patent is obtained.

If the assignment be of the entire interest, and precede the grant of the patent, it must be recorded in the Patent Office, and then the patent issues to the assignee. If the assignment be of the entire interest, and be made after the grant, it must be recorded within three months at the Patent Office to hold good against a subsequent purchaser without notice.

102. *Refusal of Letters Patent.* When an application for a patent is refused or withdrawn, two-thirds of the fees paid are returned, unless a caveat fee had been reckoned in the amount.

The money is paid in specie at the Patent Office, or is forwarded by mail to the applicant, or paid to his order.

103. *Designs.* A patent for a design is granted only to citizens, or aliens who have resided for one year in the United States, and made oath of their intention to become citizens.

The fees payable are only one-half what are paid for a patent for an art or manufacture, and the term of the grant extends only to seven years. All other regulations and provisions are the same as in the case of ordinary patents.

104. All applicants for letters patent have direct access to the examiners, and are frequently heard again and again while they explain the principles of their discoveries in personal interviews. Their applications, when defective, are returned, the defects pointed out, and also the best method of correcting them; sometimes an application is returned for correction as many as six times.

The principle of affording an applicant direct access to the Patent Office has many advantages; but it is doubtful if the long and laborious investigation which an application receives is ultimately beneficial. It seems impossible that such investigations can ever be complete and satisfactory, however great may be the skill of the examiners, however wide and extended their knowledge of arts, sciences, and manufactures.

The accumulations of business are necessarily very great, in spite of the most assiduous attention of the examiners to their duties, and the system which is adopted of assigning to each a separate department; a period of five or six months usually, and in some cases a much longer time, elapses before an application can be decided upon, a delay which is often attended with great inconvenience.

It may be doubted whether too much is not attempted by the examiners in undertaking to decide the important questions of novelty and utility, either in the case of applications allowed or rejected; and whether any such preliminary inquiries ought not to be limited to warning an applicant of what has been done or known before, and referring him to authentic sources of information, but allowing him upon such warning to take out letters patent at his own risk. In such a case the patent might safely be left to find its proper position and value when brought before the public, and there would be ample opportunity of testing its validity and utility both by public opinion and, if requisite, ultimately in a court of law.

The policy of requiring models of inventions in every case where they can be supplied, and forming them into one vast indiscriminate collection, has been already discussed.

The abuses arising from the system of extending and altering the claims of a specification in the case of re-issues, have also been noticed.

105. *Repeal of Letters Patent.* It appears to be a subject of much complaint, that there exist no means of repealing or quashing an invalid patent by legal proceedings, similar to those provided for by an English writ of *scire facias*. It may be true that an invalid patent is naturally disregarded by the public, but it is still a weapon of legal offence and annoyance which it would be as well to put aside altogether.

106. It remains to say a few words, in conclusion, on the scale of fees. These are extremely moderate in the case of citizens, and the privileges and protection afforded by letters patent are most properly placed within the reach of all citizen inventors. But foreigners labour under great disadvantages, being required, if British subjects, to pay nearly seventeen times as much as a citizen ($500), if any other foreigner, ten times as much as a citizen ($300).

The principle of retaliation, which alone can be alleged as a reason for drawing up a scale of fees so disadvantageous to foreigners, seems wanting in fairness, as least in the case to which it was most pointedly directed. For a citizen of the United States, on applying for letters patent in England, stands, and always has stood, on terms of perfect equality with British subjects.

The following is the scale of fees payable in connection with the grant of letters patent in the United States:

On application for a design	$15 (about £3)
On application for a caveat	20 (about 4)
On application for a patent in the case of a citizen or resident foreigner intending to become naturalised	30 (about 6)
On application for a patent in the case of a British subject	500 (about 105)

On application for a patent in the case of any other foreigner	300	(about 63)
On application for a Disclaimer	10	(about 2)
Addition of improvements	15	(about 3)
Re-issue	15	(about 3)
Extension of term	40	(about 8)
Appeal	25	(about 5)

Copies of letters patent are furnished by the Office on payment of ten cents (5*d.*) for every 100 words, and the cost of copying the drawings.

The charge for recording an assignment is for–

300 words and under	$1	0
Over 300 and not over 1,000 words	2	0
Over 1,000 words	3	0

Conclusions

The parts of the United States which I visited form, geographically, a small portion of their extended territory, but they are the principal seats of manufactures, and afford ample opportunities for arriving at general conclusions. I could not fail to be impressed, from all that I saw there, with the extraordinary energy of the people, and their peculiar aptitude in availing themselves to the utmost of the immense natural resources of the country.

The details which I have collected in this report show, by numerous examples, that they leave no means untried to effect what they think it is possible to accomplish, and they have been signally successful in combining large practical results with great economy in the methods by which these results are secured.

The labouring classes are comparatively few in number, but this is counter-balanced by, and indeed may be regarded as one of the chief causes of, the eagerness with which they call in the aid of machinery in almost every department of industry. Wherever it can be introduced as a substitute for manual labour, it is universally and willingly resorted to; of this the facts stated in my report contain many conclusive proofs, but I may here specially refer, as examples, to plough making, where eight men are able to finish thirty per day; to door making, where twenty men make 100 pannelled doors per day; to last making, the process of which is completed in 1½ minutes; to sewing by machinery, where one woman does the work of 20; to net making, where one woman does the work of 100. It is this condition of the labour market, and this eager resort to machinery wherever it can be applied, to which, under the guidance of superior education and intelligence, the remarkable prosperity of the United States is mainly due. That prosperity is frequently attributed to the possession of a soil of great natural fertility, and it is doubtless true that in certain districts the alluvial deposits are rich and the land fruitful to an extraordinary degree; but while traversing many hundred miles of country in the Northern States, I was impressed with the conviction that the general character of the soil there was the reverse of fertile.

It is not for a moment denied that the natural resources of the United States are immense, that the products of the soil seem capable of being multiplied and varied to almost any extent, and that the supplies of minerals appear to be nearly unlimited.

The material welfare of the country, however, is largely dependent upon the means adopted for turning its natural resources to the best account, at the same time that the calls made upon human labour are reduced as far as practicable.

The attention paid to the working of wood, some details connected with which I have included in the report, is a striking illustration of this. The early settlers found in the forests which they had to clear an unlimited supply of material, which necessity compelled them to employ in every possible way, in the construction of their houses, their furniture, and domestic utensils, in their implements of labour, and in their log-paved roads.

Wood thus became with them a universal material, and work-people being scarce, machinery was introduced as far as possible to supply the want of hands. The character thus given to one branch of manufactures has gradually extended to others. Applied to stone-dressing, for example, one man is enabled, as I have shown, to perform as much work as twenty masons by hand. So great again are the improvements effected in spinning machinery, that one man can attend to a mule containing 1,088 spindles, each spinning 3 hanks, or 3,264 hanks in the aggregate per day. In Hindoostan, where they still spin by hand, it would be extravagant to expect a spinner to accomplish one hank per day; so that in the United States we find the same amount of manual labour, by improved machinery, doing more than 3,000 times the work. But a still more striking comparison between hand and machine labour may be made in the case of lace making in England. Lace of an ordinary figured pattern used to be made 'on the cushion' by hand, at the rate of about three meshes per minute. At Nottingham, a machine attended by one person will now produce lace of a similar kind at the rate of about 24,000 meshes per minute; so that one person can, by the employment of a machine, produce 8,000 times as much work as one lace maker by hand.

The results which have been obtained in the United States, by the application of machinery wherever it has been practicable to manufactures, are rendered still more remarkable by the fact, that combinations to resist its introduction there are unheard of. The workmen hail with satisfaction all mechanical improvements, the importance and value of which, as releasing them from the drudgery of unskilled labour, they are enabled by education to understand and appreciate. With the comparatively superabundant supply of hands in this country, and therefore a proportionate difficulty in obtaining remunerative employment, the working classes have less sympathy with the progress of invention. Their condition is a less favourable one than that of their American brethren for forming a just and unprejudiced estimate of the influence which the introduction of machinery is calculated to exercise on their state and prospects. I cannot resist the conclusion, however, that the different views taken by our operatives and those of the United States upon this subject are determined by other and powerful causes, besides those dependent on the supply of labour in the two countries. The principles which ought to regulate the relations between the employer and the employed seem to be thoroughly understood and appreciated in the United States, and while the law of limited liability affords the most ample facilities for the investment of capital in business, the intelligent and educated artizan is left equally free to earn all that he can, by making the best use of his hands, without let or hindrance by his fellows.

It may be that the working classes exhibit an unusual independence of manner, but the same feeling insures the due performance of what they consider to be their duty with less supervision than is required where dependence is to be placed upon uneducated hands.

It rarely happens that a workman who possesses peculiar skill in his craft is disqualified to take the responsible position of superintendent, by the want of education and general knowledge, as is frequently the case in this country. In every State in the Union, and particularly in the north, education is, by means of the common schools, placed within the reach of each individual, and all classes avail themselves of the opportunities afforded. The desire of knowledge so early implanted is greatly increased, while the facilities for diffusing it are amply provided through the instrumentality of an almost universal press. No taxation of any kind has been suffered to interfere with the free development of this powerful agent for promoting the intelligence of the people, and the consequence is, that where the humblest labourer can indulge in the luxury of his daily paper, everybody reads, and thought and intelligence penetrate through the lowest grades of society. The benefits which thus result from a liberal system of education and a cheap press to the working classes of the United States can hardly be over-estimated in a national point of view; but it is to the co-operation of both that they must undoubtedly be ascribed. For if, selecting a proof from among the European States, the condition of Prussia be considered, it will be found that the people of that country, as a body, have not made that progress which, from the great attention paid to the education of all classes, might have been anticipated; and this must certainly be ascribed to the restrictions laid upon the press, which have so materially impeded the general advancement of the people. Wherever education and an unrestricted press are allowed full scope to exercise their united influence, progress and improvement are the certain results, and among the many benefits which arise from their joint co-operation may be ranked most prominently the value which they teach men to place upon intelligent contrivance; the readiness with which they cause new improvements to be received, and the impulse which they thus unavoidably give to that inventive spirit which is gradually emancipating man from the rude forms of labour, and making what were regarded as the luxuries of one age to be looked upon in the next as the ordinary and necessary conditions of human existence.

signed JOSEPH WHITWORTH

Index

Index

Index

Index

Index

Index

Index

Index